Beam's Eye View Imaging in Radiation Oncology

IMAGING IN MEDICAL DIAGNOSIS AND THERAPY

Series Editors: Andrew Karellas and Bruce R. Thomadsen

Published titles

Quality and Safety in Radiotherapy
Todd Pawlicki, Peter B. Dunscombe, Arno J. Mundt, and Pierre Scalliet, Editors
ISBN: 978-1-4398-0436-0

Adaptive Radiation Therapy
X. Allen Li, Editor
ISBN: 978-1-4398-1634-9

Quantitative MRI in Cancer
Thomas E. Yankeelov, David R. Pickens, and Ronald R. Price, Editors
ISBN: 978-1-4398-2057-5

Informatics in Medical Imaging
George C. Kagadis and Steve G. Langer, Editors
ISBN: 978-1-4398-3124-3

Adaptive Motion Compensation in Radiotherapy
Martin J. Murphy, Editor
ISBN: 978-1-4398-2193-0

Image-Guided Radiation Therapy
Daniel J. Bourland, Editor
ISBN: 978-1-4398-0273-1

Targeted Molecular Imaging
Michael J. Welch and William C. Eckelman, Editors
ISBN: 978-1-4398-4195-0

Proton and Carbon Ion Therapy
C.-M. Charlie Ma and Tony Lomax, Editors
ISBN: 978-1-4398-1607-3

Physics of Mammographic Imaging
Mia K. Markey, Editor
ISBN: 978-1-4398-7544-5

Physics of Thermal Therapy: Fundamentals and Clinical Applications
Eduardo Moros, Editor
ISBN: 978-1-4398-4890-6

Emerging Imaging Technologies in Medicine
Mark A. Anastasio and Patrick La Riviere, Editors
ISBN: 978-1-4398-8041-8

Cancer Nanotechnology: Principles and Applications in Radiation Oncology
Sang Hyun Cho and Sunil Krishnan, Editors
ISBN: 978-1-4398-7875-0

Image Processing in Radiation Therapy
Kristy Kay Brock, Editor
ISBN: 978-1-4398-3017-8

Informatics in Radiation Oncology
George Starkschall and R. Alfredo C. Siochi, Editors
ISBN: 978-1-4398-2582-2

Cone Beam Computed Tomography
Chris C. Shaw, Editor
ISBN: 978-1-4398-4626-1

Computer-Aided Detection and Diagnosis in Medical Imaging
Qiang Li and Robert M. Nishikawa, Editors
ISBN: 978-1-4398-7176-8

Cardiovascular and Neurovascular Imaging: Physics and Technology
Carlo Cavedon and Stephen Rudin, Editors
ISBN: 978-1-4398-9056-1

Scintillation Dosimetry
Sam Beddar and Luc Beaulieu, Editors
ISBN: 978-1-4822-0899-3

Handbook of Small Animal Imaging: Preclinical Imaging, Therapy, and Applications
George Kagadis, Nancy L. Ford, Dimitrios N. Karnabatidis, and George K. Loudos Editors
ISBN: 978-1-4665-5568-6

IMAGING IN MEDICAL DIAGNOSIS AND THERAPY

Series Editors: Andrew Karellas and Bruce R. Thomadsen

Published titles

Comprehensive Brachytherapy: Physical and Clinical Aspects
Jack Venselaar, Dimos Baltas, Peter J. Hoskin, and Ali Soleimani-Meigooni, Editors
ISBN: 978-1-4398-4498-4

Handbook of Radioembolization: Physics, Biology, Nuclear Medicine, and Imaging
Alexander S. Pasciak, PhD., Yong Bradley, MD., J. Mark McKinney, MD., Editors
ISBN: 978-1-4987-4201-6

Monte Carlo Techniques in Radiation Therapy
Joao Seco and Frank Verhaegen, Editors
ISBN: 978-1-4665-0792-0

Stereotactic Radiosurgery and Stereotactic Body Radiation Therapy
Stanley H. Benedict, David J. Schlesinger, Steven J. Goetsch, and Brian D. Kavanagh, Editors
ISBN: 978-1-4398-4197-6

Physics of PET and SPECT Imaging
Magnus Dahlbom, Editor
ISBN: 978-1-4665-6013-0

Tomosynthesis Imaging
Ingrid Reiser and Stephen Glick, Editors
ISBN: 978-1-138-19965-1

Ultrasound Imaging and Therapy
Aaron Fenster and James C. Lacefield, Editors
ISBN: 978-1-4398-6628-3

Beam's Eye View Imaging in Radiation Oncology
Ross I. Berbeco, Ph.D., Editor
ISBN: 978-1-4987-3634-3

Principles and Practice of Image-Guided Radiation Therapy of Lung Cancer
Jing Cai, Joe Y. Chang, and Fang-Fang Yin, Editors
ISBN: 978-1-4987-3673-2

Beam's Eye View Imaging in Radiation Oncology

Edited by
Ross I. Berbeco
Brigham and Women's Hospital
Dana-Farber Cancer Institute and Harvard Medical School

CRC Press
Taylor & Francis Group
Boca Raton London New York

CRC Press is an imprint of the
Taylor & Francis Group, an **informa** business

CRC Press
Taylor & Francis Group
6000 Broken Sound Parkway NW, Suite 300
Boca Raton, FL 33487-2742

First issued in paperback 2020

ISBN-13: 978-1-4987-3634-3 (hbk)
ISBN-13: 978-0-367-78193-4 (pbk)

Library of Congress Cataloging-in-Publication Data

Names: Berbeco, Ross I., editor.
Title: Beam's eye view imaging in radiation oncology / edited by Ross I. Berbeco.
Other titles: Imaging in medical diagnosis and therapy ; 25.
Description: Boca Raton, FL : CRC Press, Taylor & Francis Group, [2017] |
Series: Imaging in medical diagnosis and therapy ; 25
Identifiers: LCCN 2017013424| ISBN 9781498736343 (hardback ; alk. paper) |
ISBN 1498736343 (hardback ; alk. paper)
Subjects: LCSH: Cancer--Radiotherapy. | Diagnostic imaging.
Classification: LCC RC271.R3 B43 2017 | DDC 616.99/40642--dc23
LC record available at https://lccn.loc.gov/2017013424

Visit the Taylor & Francis Web site at
http://www.taylorandfrancis.com

and the CRC Press Web site at
http://www.crcpress.com

Contents

Series Preface

Since their inception over a century ago, advances in the science and technology of medical imaging and radiation therapy are more profound and rapid than ever before. Further, the disciplines are increasingly cross-linked as imaging methods become more widely used to plan, guide, monitor, and assess treatments in radiation therapy. Today, the technologies of medical imaging and radiation therapy are so complex and computer driven that it is difficult for the people (physicians and technologists) responsible for their clinical use to know exactly what is happening at the point of care, when a patient is being examined or treated. The people best equipped to understand the technologies and their applications are medical physicists, and these individuals are assuming greater responsibilities in the clinical arena to ensure that what is intended for the patient is actually delivered in a safe and effective manner.

The growing responsibilities of medical physicists in the clinical arenas of medical imaging and radiation therapy are not without their challenges, however. Most medical physicists are knowledgeable in either radiation therapy or medical imaging and expert in one or a small number of areas within their disciplines. They sustain their expertise in these areas by reading scientific articles and attending scientific talks at meetings. In contrast, their responsibilities increasingly extend beyond their specific areas of expertise. To meet these responsibilities, medical physicists periodically must refresh their knowledge of advances in medical imaging or radiation therapy, and they must be prepared to function at the intersection of these two fields. How to accomplish these objectives is a challenge.

At the 2007 annual meeting of the American Association of Physicists in Medicine in Minneapolis, Minnesota, this challenge was the topic of conversation during a lunch hosted by Taylor & Francis Group and involving a group of senior medical physicists (Arthur L. Boyer, Joseph O. Deasy, C.-M. Charlie Ma, Todd A. Pawlicki, Ervin B. Podgorsak, Elke Reitzel, Anthony B. Wolbarst, and Ellen D. Yorke). The conclusion of this discussion was that a book series should be launched under the Taylor & Francis banner, with each volume in the series addressing a rapidly advancing area of medical imaging or radiation therapy of importance to medical physicists. The aim would be for each volume to provide medical physicists with the information needed to understand technologies driving a rapid advance and their applications to safe and effective delivery of patient care.

Each volume in this series is edited by one or more individuals with recognized expertise in the technological area encompassed by this book. The editors are responsible for selecting the authors of individual chapters and ensuring that the chapters are comprehensive and intelligible to someone without such expertise. The enthusiasm of volume editors and chapter authors has been gratifying and reinforces the conclusion of the Minneapolis luncheon that this series of books addresses a major need of medical physicists.

The series *Imaging in Medical Diagnosis and Therapy* would not have been possible without the encouragement and support of the series manager, Lu Han, of Taylor & Francis Group. The editors and authors, and most of all I, are indebted to his steady guidance of the entire project.

William R. Hendee
Founding Series Editor

Preface

The introduction of flat-panel detectors for clinical imaging in radiation therapy in the mid-1990s was the beginning of the age of computerized image guidance in this field. The early electronic portal imaging devices (EPIDs) provided fast, easy clinical *beam's eye view* (*BEV*) imaging for accurate pretreatment patient positioning. With the introduction of on-board kilovoltage (kV) imaging systems in the early 2000s, megavoltage (MV) portal imaging has mostly fallen out of favor with most research and development favoring the newer low-energy systems. Recently, however, there has been a resurgence of interest in MV imaging for quality assurance, portal dosimetry, cone-beam computed tomography (CBCT), and real-time tumor tracking. The purpose of this book is to bring all of this information together in one place so that a reader may derive a clear understanding and appreciation for the current clinical applications and future opportunities afforded by beam's eye view MV imaging.

IMAGE GUIDANCE IN RADIATION THERAPY

Image guidance is key to the precision and accuracy of radiation therapy. Modern radiation therapy equipment enables the delivery of very precisely calculated amounts of radiation in complicated shapes in order to maximize target conformality and minimize the damage to healthy tissues. This exquisitely delivered therapy is for naught unless the patient anatomy is located in the correct location during every treatment fraction. In fact, poorly controlled localization can lead to harmful effects such as toxicities in organs at risk and even debilitating or lethal injury. On the flip side, highly precise and accurate localization can not only prevent harm but also enable radiation dose escalation to the target providing greater tumor control and extending survival.

There are several approaches to image guidance in radiation therapy, each with advantages and disadvantages. Current clinical technologies include optical surface imaging, kV X-rays, magnetic resonance imaging, positron emission tomography, and MV X-ray imaging. Among these options, the last (MV imaging) is the most prevalent while also the most undervalued. Nearly every new clinical linear accelerator comes equipped with MV imaging as a standard feature. Given the high "penetration" in the marketplace, innovations in MV imaging can translate to a large, widespread impact. Therefore, the cost/benefit ratio for improvements in MV imaging is highly favorable.

MV IMAGING

External beam radiation therapy is most often performed with clinical linear accelerators providing a photon beam with a peak energy in the 6–25 MV range. The photons that are not absorbed in the patient pass through unattenuated and are lost in the shielding of the treatment room. But there is valuable information contained in the distribution of these photons, which can be collected by deploying an imaging panel on the exit side of the patient. This is sometimes called *beam's eye view* imaging because the resulting images are formed by the treatment beam, showing exactly what is being irradiated. In addition to the anatomical information, the beam intensity and field shape as a function of time data can be acquired to assure the accuracy and precision of treatments and/or provide an input for real-time adaptation. Although this technique has some limitations, the main benefits include the lack of additional imaging dose, target visualization during treatment, minimal localization degeneracy, and, in the case of CBCT, more accurate dose calculation.

ORGANIZATION OF THE BOOK

This book has been written to provide a comprehensive introduction to the history, current state of the art, and future prospects for MV imaging. Chapters 1 through 3 introduce the development of portal imaging, the construction of modern flat-panel imagers, and the modeling of these devices with Monte Carlo techniques. Chapters 4 through 7 present the current clinical applications of EPIDs and MV imaging from the beam's eye view. Chapters 8 through 12 provide a glimpse of the future: novel techniques and innovative detector designs that can enable real-time motion management, adaptive radiotherapy, and automatic quality assurance.

In this book, we have assembled the experts in the field of MV imaging to provide a complete assessment of the current and future capabilities of MV imaging in image-guided radiation therapy. By bringing together the history, technical basis, clinical uses, and future possibilities in a single book, it is anticipated that the reader will gain a wide range of appreciation for the strengths and weaknesses of MV imaging. As long as the vast majority of radiation therapy procedures continue to be administered with X-ray radiation, there will be a role for beam's eye view imaging to ensure the precision and accuracy. It is hoped that this book will inspire researchers to develop further innovations, not heretofore imagined, to improve the quality, safety, and efficacy of radiation therapy.

Editor

Ross I. Berbeco, PhD, is a board-certified medical physicist, associate professor of Radiation Oncology at Harvard Medical School and director of Medical Physics Research at the Brigham and Women's Hospital and Dana–Farber Cancer Institute, Boston, Massachusetts. He earned his PhD in high-energy experimental physics at the University of Michigan, before transitioning to medical physics. Dr. Berbeco began working with beam's eye view (BEV) imaging during his postdoctoral fellowship at the Massachusetts General Hospital, Boston, Massachusetts, and has since continued to research BEV imaging for tumor localization to facilitate applications such as delivered dose reconstruction, adaptive radiation therapy, and tumor tracking.

Contributors

Eric Ford
Department of Radiation Oncology
University of Washington
Seattle, Washington

Olivier Gayou
Lexington Clinic
Department of Radiation Oncology
John D. Cronin Cancer Center
Lexington, Kentucky

Peter Greer
Calvary Mater Newcastle Hospital
University of Newcastle
Newcastle, New South Wales, Australia

John H. Lewis
Department of Radiation Oncology
University of California, Los Angeles
Los Angeles, California

Ben Mijnheer
Division of Radiotherapy
Netherlands Cancer Institute
Antoni van Leeuwenhoek Hospital
Amsterdam, the Netherlands

Daniel Morf
Varian Medical Systems
Baden, Switzerland

I. Antoniu Popescu
Department of Medical Physics
British Columbia Cancer Agency
Vancouver, British Columbia, Canada

James L. Robar
Department of Radiation Oncology
Dalhousie University
Halifax, Nova Scotia, Canada

Joerg Rottmann
Department of Radiation Oncology
Brigham and Women's Hospital
Harvard Medical School
Dana–Farber Cancer Institute
Boston, Massachusetts

Jeffrey V. Siebers
Department of Radiation Oncology
University of Virginia School of Medicine
Charlottesville, Virginia

Josh Star-Lack
Varian Medical Systems
Palo Alto, California

Marcel van Herk
Division of Molecular and Clinical Cancer Sciences
University of Manchester
Manchester, United Kingdom

Philip Vial
Department of Radiation Oncology
Sydney South West Area Health Service
Sydney, New South Wales, Australia

PART 1

FUNDAMENTALS

1

History of MV imaging

MARCEL VAN HERK

1.1 MOTIVATION

The purpose of radiotherapy is to eradicate a tumor with high-energy X-rays while sparing the surrounding organ-at-risks (OAR) as much as possible. Typically, there will always be some discrepancy between the planning situation and treatment. Treatment margins are employed to make the treatment robust against these discrepancies. A misplacement of the tumor with respect to the beams that exceeds the treatment margins can cause a geometric miss. The importance of this effect has been recognized for some time. Marks (1974) compared field placement errors (FPE) with and without immobilization. Byhardt in 1978 reported FPE per site, whereas Brahme in 1984 demonstrated the effect of field shifts on estimated tumor control probability (TCP), and Kinzie et al. (1983) showed that recurrences occurred more frequently when inadequate margins are applied (Marks 1974, Byhardt et al. 1978, Brahme 1984, Kinzie et al. 1983).

Rabinowitz et al. (1985) related FPE to clinical outcome where major protocol variations were observed frequently and were shown to affect the survival. The only available method to measure FPE at that time was portal film, although early attempts had been made to integrate X-ray on linear accelerators (LINAC) or cobalt sources (Figure 1.1) (Lokkerbol and Smit 1961, Biggs et al. 1985). In addition, a rare report exists of remote visual evaluation of setup based on a fluorescent screen mirror system *without* a camera (https://www.historad.com/en/#!/en/100-years-radiotherapy-netherlands-cancer-institute-rebuilding/image-guided-rotational-therapy/). These approaches were not widely disseminated, mainly due to a lack of means to digitize the images and insufficient computing power to quickly process and analyze the films. A complicating factor was the difference in perspective between the imaging and treatment beamlines. This made it very difficult to interpret the images, as computed tomography (CT) and digitally reconstructed radiographs (DRRs) were not yet available.

At this time (mid-1980s), X-ray film was used for megavoltage (MV) imaging; however, the film was expensive, cumbersome, and error prone due to the low contrast. Investigators such as

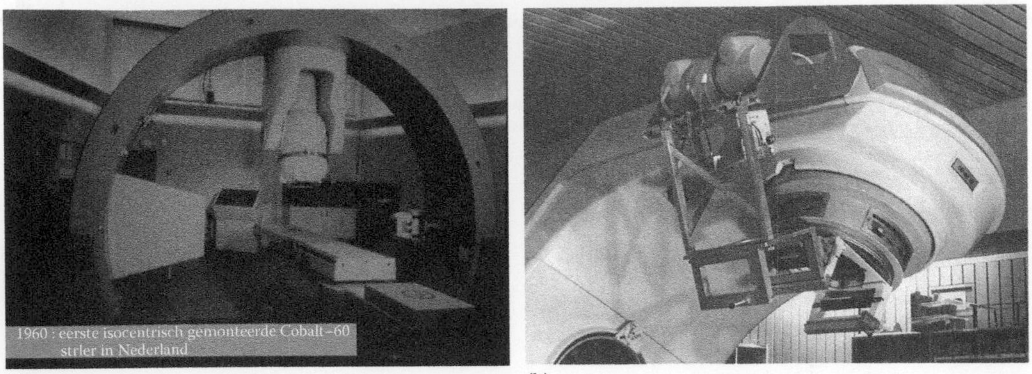

(a) (b)

Figure 1.1 (a) Early reported integration of imaging on a radiotherapy machine often utilized diagnostic X-rays. An X-ray tube and image intensifier-based X-ray detector are placed perpendicular to the beamline of this early isocentric Co60 machine developed at the Netherlands Cancer Institute. The device can be seen in action here: https://www.youtube.com/v/9_B8DfvBvKY&hl=en_US&feature=player_embedded&version=3. (b) Biggs et al. reported in 1985 the integration of an X-ray imaging chain (film-based) on a linear accelerator. For both systems, image analysis was hampered by the use of a different imaging and treatment beamline.

Galbraith (1989) pursued optimization of the film-screen cassettes. In 1985, Meertens showed the feasibility of digital image processing to enhance contrast and sharpness of MV films (Meertens 1985b). However, the home-built laser scanner employed in that study took several hours to digitize one film and the limited computing power available (e.g., Digital PDP-11) further increased the time needed to process images, limiting these approaches to research only. Grayscale display devices were not common, so processed data needed to be printed on film for review. The necessary film printers for digital images appeared with CT scanners in the early 1980s. In those days, most common computer equipment (terminals and printers) supported text only. Analysis of portal films was done by manual comparison with 2D kilovoltage (kV) radiographs or digital reconstructed radiographs using, for example, rulers or templates (Figure 1.2). Around this time, it became clear that some digital solution would be highly preferable. The introduction and availability of microprocessors facilitated this technological shift. The Intel 8086 processor, one of the first 16 bits microprocessors with sufficient computing power for image processing, had appeared in 1978 enabling digital image processing on small and low cost computer systems. The time was ripe for *electronic* portal imaging devices (EPIDs).

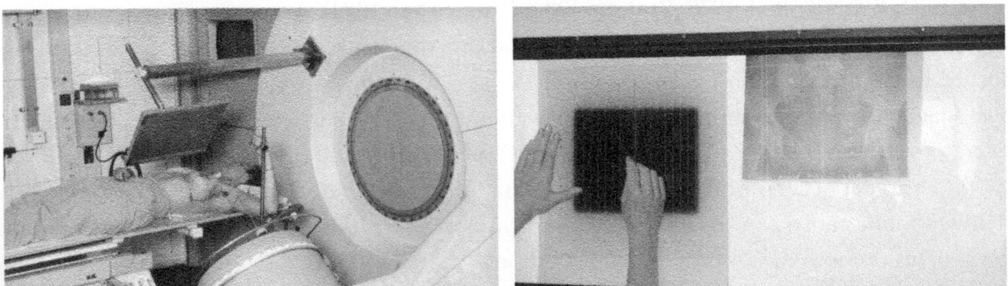

(a) (b)

Figure 1.2 (a) Portal film-screen cassette mounted on a linear accelerator. (b) Portal films were visually compared with reference images (e.g., simulator images) often with the aid of templates or rulers. The hands shown are of Dr. J. Lebesque. (Courtesy of the Harm Meertens, the Netherlands Cancer Institute.)

Coinciding with the initiation of the EPID development, a number of workshops were held that eventually evolved into the recurring *Electronic Portal Imaging (EPI)* conferences. This meeting series is still continuing to date—every other year, mostly in Europe, United States, or Australia. The memorable first workshop (not counted in the series though) was held in Chapel Hill, North Carolina in 1987 (Figure 1.3a), organized by George Sherouse. Norman Bailey and Arthur Boyer organized the subsequent meeting in Las Vegas in 1989. At this meeting, physicist Shlomo Shalev organized an image quality competition and *enhance-off*. The *Las Vegas phantom* was introduced there, which is still commonly used, and was later combined with commercial analysis software, called PIPS (Rajapakshe 1996). Subsequent meetings were held in Newport Beach, San Francisco (Figure 1.3b), Amsterdam, Houston, Brussels, Vancouver, Brighton, Melbourne, Leuven, Sydney, Aarhus, and St. Louis. In its current form, the workshop is now called *Electronic Patient Imaging* and covers all forms of in-room patient imaging.

Over the years, the meetings changed focus from detector development to image analysis, clinical application, setup error correction protocols, treatment margins, use of implanted markers, portal dosimetry, tracking, and magnetic resonance (MR) guidance. Most important, the meetings always have had a multidisciplinary character, with physicists, physicians, and radiographers in the organizing committee, presenting and in attendance. Noteworthy is the 2006 meeting chaired by Kay Hatherley, the radiographer greatly responsible for the early introduction of EPID in Australia. She opened the meeting giving a welcome speech from inside the shark tank of the Melbourne aquarium.

Workshop on Megavoltage Imaging and Image Processing

The Division of Radiation Oncology of the University of North Carolina School of Medicine is pleased to announce the 1987 Workshop on Megavoltage Imaging and Image Processing to be held in Chapel Hill February 25, 26, and 27. Topics will include:

- the role of portal films in precision radiotherapy
- optimal technique for portal film exposure
- state of the art in film design for radiotherapy
- portal film contrast enhancement (digital and analog methods)
- megavoltage "fluoroscopy" and filmless imaging
- medical-legal issues in the use of filmless and enhanced portal images
- THE PORTAL FILM ENHANCE-OFF

The Workshop will bring together a group of physicists, physicians, computer scientists, engineers, and industry specialists to foster an open discussion of the current state of megavoltage image acquisition and processing. In addition, a set of test images will be sent to participating institutions for enhancement and the resulting enhanced images will be presented for comparison and discussion.

Approximately 15 CME credits will be available. We anticipate a registration fee for the two and one half days of approximately $50.00. A proceedings will be published.

For more information on registration, to submit a paper for presentation, to volunteer to participate in the enhance-off, or to purchase the proceedings please contact:

George W. Sherouse, M.S.
or
Julian Rosenman, Ph.D., M.D.
Radiation Oncology
UNC School of Medicine
Chapel Hill, NC 27514
(919) 966-1101

(a)

(b)

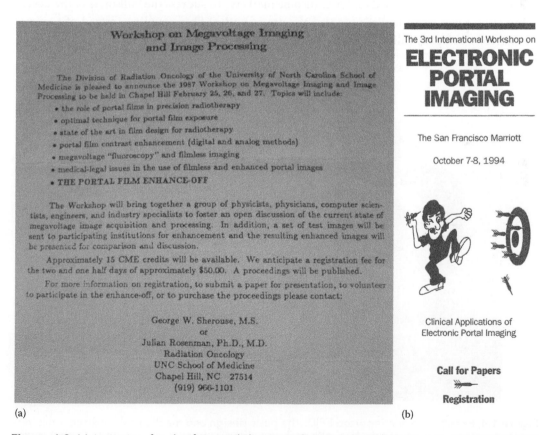

Figure 1.3 (a) Invitation for the first workshop on *Electronic Portal Imaging* in Chapel Hill that brought together about 25 pioneers in the field. (b) The 3rd workshop was visited by around 200 people. The logo, designed by Shlomo Shalev, very aptly visualized the goal of electronic portal imaging: hitting the target. (Courtesy of Marcel van Herk.)

1.2 EARLY DEVICES

The most obvious design for an EPID is to combine a fluorescent screen with a video camera similar to the devices used for low-energy X-rays. For compactness and to keep the camera out of the high-energy radiation beam, a tilted mirror is typically placed below the fluorescent screen, while the camera is looking in a direction parallel to the screen. One of the earliest reported systems using this technology is by Benner et al. (1962) (Figure 1.4).

However, video systems have several drawbacks. The relatively low light output of the fluorescent screen, in combination with the poor light collection due to the lens aperture leads to a low detective quantum efficiency (DQE) of well below 1%, while the devices are very bulky. Larger lens apertures would lead to poor resolution due to spherical aberrations. Yet, because of their simple design, it was these kinds of systems that were first adopted as commercial solutions for companies such as Elekta and Siemens (Benner et al. 1962, Baily 1980, Leong 1986, Visser 1990). Add-on systems using this technology were developed by Eliav and Cablon (de Boer 2000, Odero 2009). The problem of low light collection was addressed by using amplified cameras and later cooled CCD cameras (Franken 2004). Cooled cameras also alleviated issues with radiation damage to the CCD chip. Cablon produces such a system, with software developed in collaboration with the Erasmus Medical Center in Rotterdam, the Netherlands, although Elekta also produces a CCD camera device intended for lower income markets. To address the bulkiness of the device, Wong et al. (then in St. Louis) proposed the use of optical fiber coupling instead of a mirror and started a company called FiberVision (Figure 1.4) (Wong et al. 1990). However, interestingly

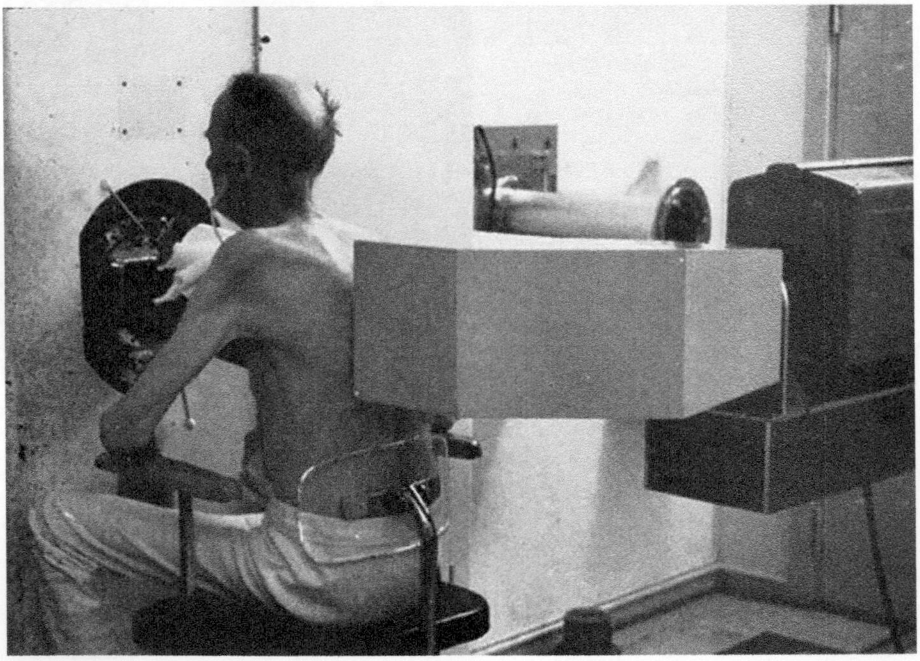

Figure 1.4 Possibly the first reported EPID. The basic design and mechanical construction is very similar to later screen/mirror devices, although the video camera (gray box on the right) is very bulky. Here the device is used in the fixed beamline of a 30 MV Betatron. (Courtesy of Radium Hospital, Oslo, Norway, 16 December 1961; From Benner S. et al., *Phys Med Biol.*, 7, 29–34, 1962.)

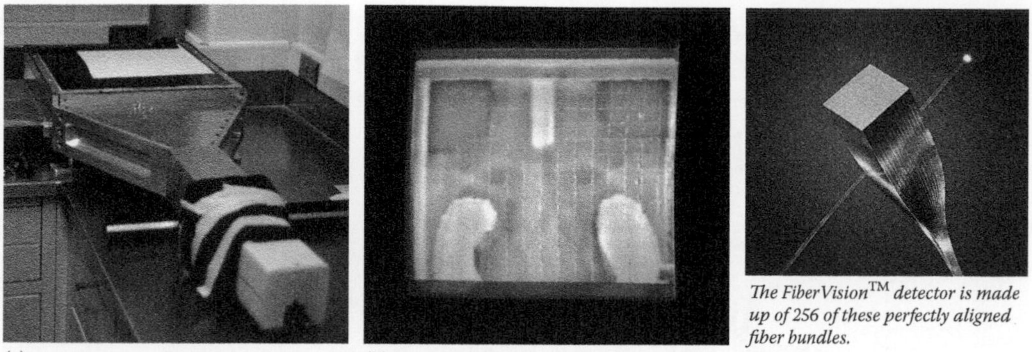

The FiberVision™ detector is made
up of 256 of these perfectly aligned
fiber bundles.

Figure 1.5 The FiberVision EPID is a screen/video camera system, where the mirror has been replaced by a tapered bundle of optical fibers. Images (a) and (b) were made by Marcel van Herk during a visit to St. Louis in 1988, image (c) was provided by John Wong.

the fiber technology did not overcome low DQE because the light-acceptance angle of the tapered fibers was similar to a lens system (Boyer et al. 1992) and the difficulty of manufacturing and large weight made this approach impractical (Figure 1.5).

The pursuit of compact solutions led to several subsequent innovations. The first truly compact system was based on a scanning line array of diodes and was introduced by Lam et al. (1986). However, due to the scanning approach and the low DQE of the applied diodes, this device had poorer quantum efficiency than video-based systems (~0.01%). Morton and Swindell combined photocells with heavy tungsten alloy scintillators bringing the DQE of such a scanning device to practical levels of about 0.5% (Morton et al. 1991), and first clinical use of the device was reported, in particular in breast (Gildersleve et al. 1994).

However, low sampling efficiency still limited the DQE, as the detectors are located only for a very short time at any array location. Ionization chambers were an alternative detector but these suffer from an ever lower efficiency due to the low density of the ionization medium. Liquid or high-pressure gases were therefore considered a more suitable ionization medium for imaging detectors. However, due to the large number of required pixels, it was not feasible to associate each pixel with its own electronics to amplify and integrate the signal (a method that would be employed much later in amorphous silicon [aSi] devices). Instead, some form of multiplexing was required to limit the required number of amplifiers to one of the dimensions of the imaging array. A mechanical solution for this multiplexing is, so called kinetistatic charge detection, proposed by Dibianca and Barker (1985). However, rather than using pixel detectors, the idea was to use strip detectors filled with high-pressure gas with a carefully controlled ion drift speed. By moving the detector system with exactly the same speed as transfer of ion speed in the gas, ionization in the liquid would be static in space, and each location in the medium acts as its own signal collector, realizing a sampling efficiency close to 100%. As a result, the detector promised a very high DQE, up to 25% depending on the thickness and density of the liquid or gas layer, while maintaining a relatively simple design. However, 30 years later, this proposed device is still in the prototype stage, with a recent iteration presented in 2006 (Samant and Gopal 2006). Another form of multiplexing can be achieved by in-plane rotation of an array of liquid-filled strip detectors (Bova et al. 1987). This detector uses tomographic reconstruction to obtain a 2D image using only a one-dimensional array of detectors and amplifiers (Figure 1.6a shows a prototype with a single detector that could be translated).

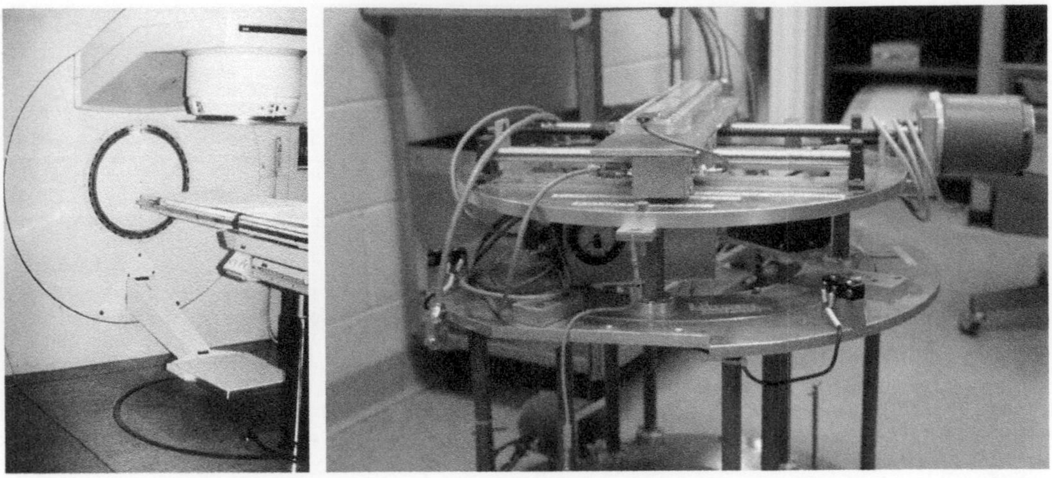

(a) (b)

Figure 1.6 (a) Scanning scintillator detector array reported by Morton and Swindell in 1991. (Courtesy of Will Swindell.) The device consists of 128 zinc tungstate scintillation crystals individually coupled to photodiodes. The array is scanned in 4–10 s over the image plane to obtain a 128 × 128 pixel portal image. (Courtesy of Swindell.) (b) Prototype rotating detector developed by Bova et al., University of Florida. (Courtesy of Marcel van Herk [during his visit to Gainsville].)

This idea was abandoned, likely due to the limited processing power available at that time. The use of ultrapure liquids in a wire chamber-like configuration was pursued at the University of Michigan, Ann Arbor, Michigan following its use at the European Center for Nuclear Research (CERN) for high-energy detectors. However, it was difficult to achieve sufficient sampling efficiency if no detectors were read out simultaneously as the ion lifetime would be very short (Boyer et al. 1992). No practical device was ever demonstrated. The only commercially implemented liquid-filled detector used electronic multiplexing, and was successfully marketed as the Varian PortalVision, with two main detector versions released around 1990 and 1997. The authors started their own development of the scanned detector using air-filled chambers (van Herk 1985). The detector array consists of two parallel plates with perpendicular electrode strips, where one electrode set is connected to a switching high-voltage power source per column, whereas the other electrode set is connected to a single electrometer per row. With air-filled chambers, images were successfully acquired but the device suffered from a very low efficiency (due to the short sampling per chamber, and extremely small signal due to the small chamber size and density), as well as a low resolution due to the long lateral range of ionization charge through the ionization medium—initially air. The low resolution could be counteracted by using a Lucite spacer with cylinder-shaped holes for each electrode intersection. However, the major breakthrough arrived when the chamber was filled with a liquid, isooctane, as the ionization medium. The first prototypes built had 30 rows and 30 columns (Meertens 1985a), and later prototypes 128 rows and 128 columns (Figure 1.7) (van Herk and Meertens 1988). The 128 × 128 system was used clinically for a short time, initially as a general purpose EPID and later as a portable EPID to use during total body irradiation (TBI) (e.g., Gladstone et al. 1993). The major benefit of this system for application in TBI was its relatively high radiation resistance due to the use of discrete components.

The original 128 × 128 pixel device sampled each chamber row for 20 ms, giving a total scan time of about 3 s. This corresponded to an expected sampling efficiency of less than 1%. To improve

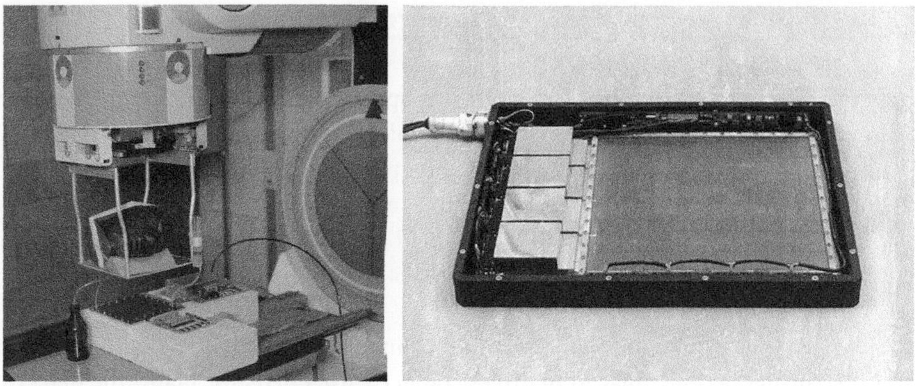

(a) (b)

Figure 1.7 Two prototypes of the liquid-filled EPID. (a) The device (30 × 30 pixels) demonstrated early feasibility of liquid-filled detectors but was too small for clinical use. (b) The device (128 × 128 pixels, 32 × 32 cm^2 active area), with a carbon fiber housing, was used briefly clinically prior to commercialization and later for verification of TBI placing the detector on the patient bed, imaging the lung region during TBI to check the location of lung shields. (Courtesy of author.)

the sampling efficiency, elaborate sampling schemes based on Hadamard matrices were attempted but were unsuccessful in demonstrating major benefits of this approach as these only reduced electronics noise, not quantum noise (van Herk et al. 1992). However, later analysis showed that the sampling efficiency was actually much higher than 1%, which could be attributed to an unexpected long ion lifetime in the liquid. This prolonged lifetime was due to water pollution of the liquid slowing down ion mobility, and as a result greatly reducing recombination (van Herk 1991). The incoming radiation therefore forms a type of latent image in the liquid while the high voltage is switched off, and during that time, ion loss is due to recombination only. However, the recombination causes a distinct nonlinear dose-response curve, which is very close in form to a square root function (Boellaard et al. 1996). In typical radiotherapy use, the ion lifetime was about 200 ms (depending on the dose rate). Sampling efficiency therefore depended mostly on the readout speed and improved from 3% in the prototype and Mark-I detector to about 20% in Mark-II devices with faster readout. The corresponding DQE ranged from 0.03% to 0.2%. The liquid-filled device was first commercialized by Brown–Boveri, whose radiotherapy section was later taken over and continued by Varian, albeit focusing on detector development and software only. All commercial liquid-filled ionization chamber devices had a detector array with 256 × 256 pixels of 1.27 × 1.27 mm. One noteworthy development that coincided with the development of the first clinical prototype in Netherlands Cancer Institute (NKI) was the introduction of carbon fiber composite to create light-weight housings with good transparency for photons (Figure 1.7b). This technology was subsequently adopted for new fixtures for setup and immobilization of patients.

In addition to electronic detectors, the development of systems for computed radiography should also be mentioned. At first, this technology was based on selenium plates similar to the ones used in analog photocopiers. Later, laser-stimulated phosphors were used, a technology that is still employed. The advantage of these plates is that they provide very high resolution, while avoiding the use of expensive films. However, as readout is off-line it does not solve one of the major obstacles of film, the need for development, making online application impractical (Astapov et al. 1981, Gur 1989, Geyer 2006).

1.3 AMORPHOUS SILICON DEVICES

The benefit of active pixel technology for EPIDs was apparent for a very long time before its first practical implementation. The work of Larry Antonuk, working at the University of Michigan should be highlighted. As early as 1992, Dr. Antonuk presented the design of amorphous hydrogenated silicon arrays at the second EPI workshop in Newport Beach, California (Figure 1.8) (Antonuk et al. 1992). In their abstract they presented a device of 10 cm × 11 cm and 240 × 250 pixels, but indicated that devices with sizes of 25 to 50 cm edge could be realized in the future with real-time readout.

In the design, which has not fundamentally changed since its introduction, the detector consists of a scintillation plate placed in direct contact with a detector plate. The detector plate defines each pixel by a large photodiode, in which the inherent capacity is used to integrate the light signal. The pixel is combined with a switching field-effect transistor that connects each detector row to the readout electronics to extract the residual charge, and recharge each pixel for a next readout. The major advantage of the active pixel technique is that integration of the light signal takes place in the photodiodes of each pixel separately, giving it a close to 100% sampling efficiency. The tight coupling between the fluorescent screen and the photodiode array also improves the light utilization from below 1% in camera systems to about 30%. However, it would take almost a decade until such devices would become mainstream and replace other EPID technologies. The main limitations that needed to be overcome were the development of fabrication plants that could handle the required size of detector plates, the development of suitable readout electronics, and the electrical connection between the detector array and the electronics, nowadays using a rubber material with anisotropic electrical conductivity. The development of these technologies was driven by innovations in flat-panel display devices for TV and computer applications. Current flat-panel detectors have more than one million pixels on detector arrays of larger than 40 × 40 cm. Innovations compared to the original reported device in 1992 include faster readout, amplifiers with programmable gain to reduce electronics noise, and improved radiation hardness of the readout electronics. Radiation hardness of the panels has always been very good due to the amorphous nature of the semiconductor material (Boudry and Antonuk 1996).

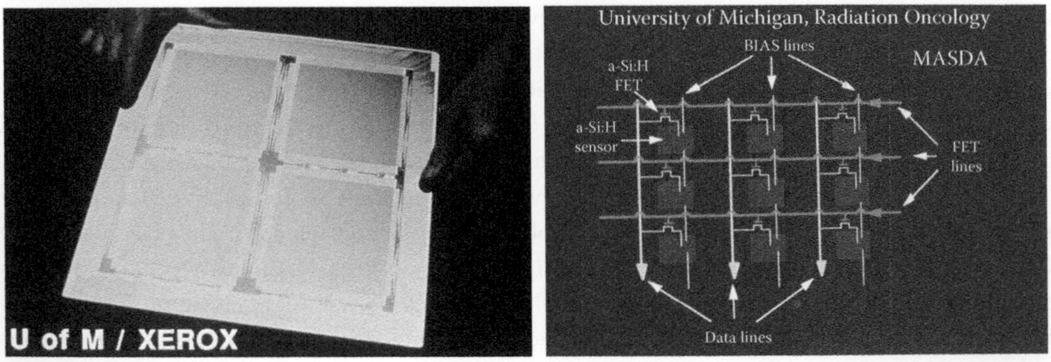

(a) (b)

Figure 1.8 (a) Early amorphous silicon detector plates. (b) Electronic schematic diagram of an aSi panel. (Courtesy of Larry Antonuk, University of Michigan.)

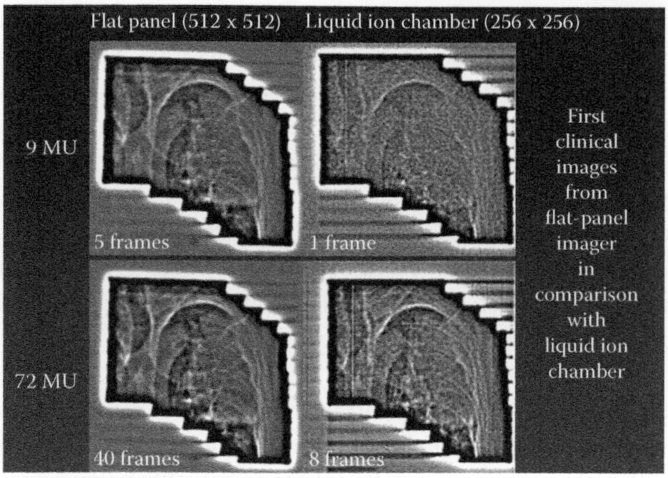

Figure 1.9 Images from an early aSi device (Heimann detector later used in Elekta IviewGT EPID), compared to images acquired on a liquid-filled device with the same dose. The visual image quality suggests about a factor 8 difference in quantum efficiency between the devices, mostly related to the scanning efficiency. For visual comparison, images were processed with equivalent unsharp mask filters. (Courtesy of Marcel van Herk, 2000.)

A comparison of the image quality of the then mainstream liquid-filled ionization chamber array and an early aSi detector provided by Elekta, made by the author in the year 2000 is shown in Figure 1.9. The relative image quality suggests an appropriate factor of 8 improvement in quantum efficiency; i.e., the top left image on an aSi detector has similar quality as the bottom right image taken on a liquid-filled detector with 8 times more dose.

1.4 EARLY CLINICAL EXPERIENCE

Initially, megavoltage (MV) radiographs were mainly used for weekly patient imaging, which provided an ad hoc style of setup error correction that was not very effective in reducing geometric uncertainties. With the clinical introduction of the first EPIDs, the development of image processing and analysis, data analysis, correction protocols, and margin calculations could start in earnest. In 1988, the first EPID was introduced in clinical practice at the NKI. The *simultaneous boost* technique for prostate cancer (Lebesque and Keus 1991) originated from this early EPID work. This technique is a good example of an advanced margin-reducing treatment enabled by innovative technology. Initial work on portal image analysis focused on detection of the radiation's field edge in relation to patient anatomy (e.g., Bijhold et al. 1991a). At that time, EPIDs were often not rigidly mounted on the gantry. Therefore, the detected field shape, in combination with a field-edge matching algorithm provided an essential step in the analysis of treatment quality (Bijhold et al. 1991b). In addition, field-shape analysis was developed to validate that the actual delivered field shape coincided with the planned one (Bijhold et al. 1992a, Leszczynski 1993, Dong and Boyer 1996). The next step involved automatic detection and registration of the, mostly bony, anatomy, and later radiopaque markers. In addition, 3D–2D registration was pioneered in the EPID field (Gilhuijs et al. 1996, Murphy 1997). One of the latest developments in this line is the use of kV-MV image pairs for ultrafast setup correction of markers (Mutanga et al. 2012). Independent of the analysis methods, correction protocols need to be in place. Initially, off-line correction protocols were proposed and used by several groups, mainly because of the time requirements for a table move that had to be performed manually after reentering the treatment bunker. Bijhold et al. (1992b) first reported that the use of a strict correction protocol reduced the amount of required corrections and increased the accuracy compared to intuitive assessment. The decision protocol developed by Bel et al. (1993) was implemented very early in three clinics and demonstrated a large increase in setup accuracy for prostate treatments (Bel et al. 1996) with very limited workload. A natural extension from off-line correction protocol rules was the introduction of adaptive radiation therapy by Yan et al. (1995). With automatic couch movement, the need for limiting the number of corrections reduced resulting in no action-level protocols (de Boer 2001) and daily online corrections. Both the procedures are still widely used today.

Field of image analysis for electronic portal imaging (EPI) reached maturity in the mid-1990s, at which time the application in Europe really took off. This shift was apparent at the 2nd International Workshop on *Electronic Portal Imaging* in 1992. The first clinical online use of EPIDs was reported by de Neve et al. (1992). The cable of a table hand pendant was extended, allowing remote control of the table to correct for incorrect patient positioning detected using a Siemens Beamview device. They reported results of 21 patients, where around 90% of images (taken with relative high fraction of the treatment dose) were of sufficient quality for evaluation. Treatment was interrupted for roughly 10% of fields, followed by remote adjustment. Gilhuijs and van Herk reported on the development of their algorithms and software system for detection of setup errors based on chamfer matching of field edge and anatomy (Gilhuijs and van Herk 1993). The last step was to extend the system with software for automatic preparation of reference images (mostly simulator images), where field edges and anatomical features were extracted automatically. Later the same algorithms were used to automatically detect anatomy on DRRs. This system was partly integrated in the Varian PortalVision device and the specific software developed at NKI is still used to date at that institution, rather than other available commercial solutions. Other automatic and semiautomatic image-matching algorithms based on image regions, moments, landmarks, multiscale medial axes, and curves were presented by the authors Boyer, Moseley, Munro, Jaffray, Shalev, Fritsch, and Balter (Moseley and Munro 1994, Radcliffe et al. 1984, Fritsch et al. 1995, Dong and Boyer 1995). Many of these early algorithms are still widely used in commercial systems. Early clinical application results based on the analysis of hundreds of images were discussed by radiographers from Winnipeg and by physicist Mike Herman from Baltimore (Herman et al. 1994). Authors Jaffray, Bisonnette, and Yu presented physical characterizations of EPID systems and proposed optimizations related to scatter (e.g., optimizing build-up plates and magnification). Also early attempts were made to quantify dosimetric benefits of online and off-line correction procedures and derive treatment margins. These developments inspired the pursuit of margin *recipes*. Not in the EPI meeting, but in the same year (1992), NKI author Bijhold was the first to separate random and systematic errors (Bijhold et al. 1992b), which was an essential step that eventually led to the development of margin recipes that are still widely used. In the following years, radiopaque markers were introduced that enabled more accurate patient setup, in particular in the prostate (Nederveen 2003). EPIDs were then further studied to validate multileaf collimator (MLC) motion, and eventually for portal dosimetry. At the *EPI96* workshop, half of the talks were about clinical implementation and there were several talks about using markers, and cone-beam computed tomography (CBCT) was presented by David Jaffray. Even though clinical application of EPIDs in Europe started in the mid-1990s, it took quite a long time before the application became widespread in the United States. For a large part, this discrepancy may be attributed to billing rules that did not cover EPI for a long time.

With the wide-scale implementation of CBCT for patient setup verification, EPIDs are less and less frequently used for setup verification. Currently there is a large interest to use the EPID signal for quality assurance (QA) of the delivery, by means of EPID dosimetry, using a variety of methodologies (van Elmpt 2008).

1.5 SUMMARY

Over a period of about 20 years (Table 1.1), EPI has become an indispensable component of virtually every medical LINAC. The field initially focused on detector development, but eventually covered all aspects of image-guided radiotherapy such as image processing, decision protocols (adaptive radiotherapy), and treatment margins. With the introduction of secondary imaging systems on LINACs, such as CBCT and magnetic resonance imaging (MRI), the application of these detectors is changing from patient setup toward treatment QA for dose and patient motion.

Table 1.1 Timeline of historical EPID developments

Year	Topic	
1960	Cobalt with perpendicular X-ray unit	Lokkerbol and Smit (1961)
1961	First use of video EPID	Brenner (1963)
1974	Importance of setup errors published	Marks (1974)
1978	First 16 bit microprocessors	https://en.wikipedia.org/wiki/Intel_8086
1980	First modern video EPID	Baily (1980)
1980	CT scanners introduced in RT	
1985	First X-ray unit on LINAC	Biggs et al. (1985)
1985	Digital film processing	Meertens (1985b)
1986	Scanning diode array EPID	Lam et al. (1986)
1988	Liquid-filled matrix ionization chamber	van Herk and Meertens (1988)
1990	Fiber vision	Wong et al. (1990)
1990	First commercial EPIDs released	
1991	Edge detection algorithms	Bijhold et al. (1991a)
1992	Prototype aSi arrays	Antonuk et al. (1992)
1992	EPID image analysis methodology	van Herk (1993)
1993	First off-line correction protocols	Bel et al. (1993)
1995	Prostate motion measured	van Herk (1995)
1996	First 3D/2D matching algorithms	Gilhuijs et al. (1996) (Murphy 1996)
1998	First EPID dosimetry applications	Boellaard et al. (1998)
2000	First CBCT reported	Jaffray (2000)
2000	First aSi detectors commercial	
2002	aSi used in CBCT system	Jaffray (2002)
2005	First commercial CBCT systems	

REFERENCES

Antonuk LE, Boudry J, Huang W, McShan DL, Morton EJ, Yorkston J, Longo MJ, Street RA. Demonstration of megavoltage and diagnostic x-ray imaging with hydrogenated amorphous silicon arrays. *Med Phys.* 1992; 19(6):1455–1466. Erratum in: *Med Phys.* 1993; 20(3):825.

Astapov BM, Mamontov VV, Atkochius VB. Electrogammagraphy: A rapid method in aligning and controlling the irradiation in radiotherapy. *Med Radiol (Mosk).* 1981; 26(3):43–47.

Baily NA. Video techniques for x-ray imaging and data extraction from roentgenographic and fluoroscopic presentations. *Med Phys.* 1980; 7(5):472–491.

Bel A, van Herk M, Bartelink H, Lebesque JV. A verification procedure toimprove patient set-up accuracy using portal images. *Radiother Oncol.* 1993; 29(2):253–260.

Bel A, Vos PH, Rodrigus PT, Creutzberg CL, Visser AG, Stroom JC, Lebesque JV. High-precision prostate cancer irradiation by clinical application of an offline patient setup verification procedure, using portal imaging. *Int J Radiat Oncol Biol Phys.* 1996; 35(2):321–332.

Benner S, Rosengren B, Wallman H, Netteland O. Television monitoring of a 30 MV x-ray beam. *Phys Med Biol.* 1962; 7:29–34.

Biggs PJ, Goitein M, Russell MD. A diagnostic X ray field verification device for a 10 MV linear accelerator. *Int J Radiat Oncol Biol Phys.* 1985; 11(3):635–643.

Bijhold J, Gilhuijs KG, van Herk M, Meertens H. Radiation field edge detection in portal images. *Phys Med Biol.* 1991a; 36(12):1705–1710.

Bijhold J, Gilhuijs KG, van Herk M. Automatic verification of radiation field shape using digital portal images. *Med Phys.* 1992a; 19(4):1007–1014.

Bijhold J, Lebesque JV, Hart AA, Vijlbrief RE. Maximizing setup accuracy using portal images as applied to a conformal boost technique for prostatic cancer. *Radiother Oncol.* 1992b; 24(4):261–271.

Bijhold J, van Herk M, Vijlbrief R, Lebesque JV. Fast evaluation of patient set-up during radiotherapy by aligning features in portal and simulator images. *Phys Med Biol.* 1991b; 36(12):1665–1679.

Boellaard R, Essers M, van Herk M, Mijnheer BJ. New method to obtain the midplane dose using portal in vivo dosimetry. *Int J Radiat Oncol Biol Phys.* 1998; 41(2):465–474.

Boellaard R, van Herk M, Mijnheer BJ. The dose response relationship of a liquid-filled electronic portal imaging device. *Med Phys.* 1996; 23(9):1601–1611.

Boudry JM, Antonuk LE. Radiation damage of amorphous silicon, thin-film, field-effect transistors. *Med Phys.* 1996; 23(5):743–754.

Bova FJ, Fitzgerald LT, Mauderli WM, Islam MK. Real-time megavoltage imaging. *Med Phys.* 1987; 14: 707.

Boyer AL, Antonuk L, Fenster A, van Herk M, Meertens H, Munro P, Reinstein LE, Wong J. A review of electronic portal imaging devices (EPIDs). *Med Phys.* 1992; 19(1):1–16.

Brahme A. Dosimetric precision requirements in radiation therapy. *Acta Radiol Oncol.* 1984; 23(5):379–391.

Byhardt RW, Cox JD, Hornburg A, Liermann G. Weekly localization films and detection of field placement errors. *Int J Radiat Oncol Biol Phys.* 1978; 4(9–10):881–887.

de Boer HC, Heijmen BJ. A protocol for the reduction of systematic patient setup errors with minimal portal imaging workload. *Int J Radiat Oncol Biol Phys.* 2001; 50(5):1350–1365.

de Boer PBM, de Boer JC, Heijmen BJ, Pasma KL, Visser AG. Characterization of a high-elbow, fluoroscopic electronic portal imaging device for portal dosimetry. *Phys Med Biol.* 2000; 45(1):197–216.

De Neve W, Van den Heuvel F, De Beukeleer M, Coghe M, Thon L, De Roover P, Van Lancker M, Storme G. Routine clinical on-line portal imaging followed by immediate field adjustment using a tele-controlled patient couch. *Radiother Oncol.* 1992; 24(1):45–54.

DiBianca FA, Barker MD. Kinestatic charge detection. *Med Phys.* 1985; 12(3):339–343.

Dong L, Boyer AL. A portal image alignment and patient setup verification procedure using moments and correlation techniques. *Phys Med Biol.* 1996; 41(4):697–723.

Dong L, Boyer AL. An image correlation procedure for digitally reconstructed radiographs and electronic portal images. *Int J Radiat Oncol Biol Phys.* 1995; 33(5):1053–1060.

Franken EM, de Boer JC, Barnhoorn JC, Heijmen BJ. Characteristics relevant to portal dosimetry of a cooled CCD camera-based EPID. *Med Phys.* 2004; 31(9):2549–2551.

Fritsch DS, Chaney EL, Boxwala A, McAuliffe MJ, Raghavan S, Thall A, Earnhart JR. Core-based portal image registration for automatic radiotherapy treatment verification. *Int J Radiat Oncol Biol Phys.* 1995; 33(5):1287–1300.

Galbraith DM. Low-energy imaging with high-energy bremsstrahlung beams. *Med Phys.* 1989; 16(5):734–746.

Geyer P, Blank H, Alheit H. Portal verification using the KODAK ACR 2000 RT storage phosphor plate system and EC films. A semiquantitative comparison. *Strahlenther Onkol.* 2006; 182(3):172–178.

Gildersleve J, Dearnaley D, Evans P, Morton E, Swindell W. Preliminary clinical performance of a scanning detector for rapid portal imaging. *Clin Oncol (R Coll Radiol).* 1994; 6(4):245–250.

Gilhuijs KG, van de Ven PJ, van Herk M. Automatic three-dimensional inspection of patient setup in radiation therapy using portal images, simulator images, and computed tomography data. *Med Phys.* 1996; 23(3):389–399.

Gilhuijs KG, van Herk M. Automatic on-line inspection of patient setup in radiation therapy using digital portal images. *Med Phys.* 1993; 20(3):667–677.

Gladstone DJ, van Herk M, Chin LM. Verification of lung attenuator positioning before total body irradiation using an electronic portal imaging device. *Int J Radiat Oncol Biol Phys.* 1993; 27(2):449–454.

Gur D, Deutsch M, Fuhrman CR, Clayton PA, Weiser JC, Rosenthal MS, Bukovitz AG. The use of storage phosphors for portal imaging in radiation therapy: Therapists' perception of image quality. *Med Phys.* 1989; 16(1):132–136.

Herman MG, Abrams RA, Mayer RR. Clinical use of on-line portal imaging for daily patient treatment verification. *Int J Radiat Oncol Biol Phys.* 1994; 28(4):1017–1023.

Jaffray DA, Siewerdsen JH. Cone-beam computed tomography with a flat-panel imager: Initial performance characterization. *Med Phys.* 2000;27(6):1311–1323.

Jaffray DA, Siewerdsen JH, Wong JW, Martinez AA. Flat-panel cone-beam computed tomography for image-guided radiation therapy. *Int J Radiat Oncol Biol Phys.* 2002;53(5):1337–1349.

Kinzie JJ, Hanks GE, MacLean CJ, Kramer S. Patterns of care study: Hodgkin's disease relapse rates and adequacy of portals. *Cancer.* 1983; 15;52(12):2223–2226.

Lam KS, Partowmah M, Lam WC. An on-line electronic portal imaging system for external beam radiotherapy. *Br J Radiol.* 1986; 59(706):1007–1013.

Lebesque JV, Keus RB. The simultaneous boost technique: The concept of relative normalized total dose. *Radiother Oncol.* 1991; 22(1):45–55.

Leong J. Use of digital fluoroscopy as an on-line verification device in radiation therapy. *Phys Med Biol.* 1986; 31(9):985–992.

Leszczynski KW, Shalev S, Gluhchev G. Verification of radiotherapy treatments: Computerized analysis of the size and shape of radiation fields. *Med Phys.* 1993; 20(3):687–694.

Lokkerbol, H, Smit JW. A new apparatus for telecobalttherapy. *Transactions IXth International Congress of Radiology*, vol. 1l. Stuttgart, 1961, p. 1498.

Marks JE, Haus AG, Sutton HG, Griem ML. Localization error in the radiotherapy of Hodgkin's disease and malignant lymphoma with extended mantle fields. *Cancer.* 1974; 34(1):83–90.

Meertens H, van Herk M, Weeda J. A liquid ionisation detector for digital radiography of therapeutic megavoltage photon beams. *Phys Med Biol.* 1985a; 30(4):313–321.

Meertens H. Digital processing of high-energy photon beam images. *Med Phys.* 1985b; 12(1):111–113.

Morton EJ, Swindell W, Lewis DG, Evans PM. A linear array, scintillation crystal-photodiode detector for megavoltage imaging. *Med Phys.* 1991; 18(4):681–691.

Moseley J, Munro P. A semiautomatic method for registration of portal images. *Med Phys.* 1994; 21(4):551–558.

Murphy MJ. An automatic six-degree-of-freedom image registration algorithm for image-guided frameless stereotaxic radiosurgery. *Med Phys.* 1997; 24(6):857–866.

Mutanga TF, de Boer HC, Rajan V, Dirkx ML, van Os MJ, Incrocci L, Heijmen BJ. Software-controlled, highly automated intrafraction prostate motion correction with intrafraction stereographic targeting: System description and clinical results. *Med Phys.* 2012; 39(3):1314–1321. doi:10.1118/1.3684953.

Nederveen AJ, Dehnad H, van der Heide UA, van Moorselaar RJ, Hofman P, Lagendijk JJ. Comparison of megavoltage position verification for prostate irradiation based on bony anatomy and implanted fiducials. *Radiother Oncol.* 2003; 68(1):81–88.

Odero DO, Shimm DS. Third party EPID with IGRT capability retrofitted onto an existing medical linear accelerator. *Biomed Imaging Interv J.* 2009; 5(3):e25. doi:10.2349/biij.5.3.25.

Rabinowitz I, Broomberg J, Goitein M, McCarthy K, Leong J. Accuracy of radiation field alignment in clinical practice. *Int J Radiat Oncol Biol Phys.* 1985; 11(10):1857–1867.

Radcliffe T, Rajapakshe R, Shalev S. Pseudocorrelation: A fast, robust, absolute, grey-level image alignment algorithm. *Med Phys.* 1994; 21(6):761–769.

Rajapakshe R, Luchka K, Shalev S. A quality control test for electronic portalimaging devices. *Med Phys.* 1996; 23(7):1237–1244.

Samant SS, Gopal A. Analysis of the kinestatic charge detection system as a high detective quantum efficiency electronic portal imaging device. *Med Phys.* 2006; 33(9):3557–3567.

van Elmpt W, McDermott L, Nijsten S, Wendling M, Lambin P, Mijnheer B. A literature review of electronic portal imaging for radiotherapy dosimetry. *Radiother Oncol.* 2008; 88(3):289–309. doi:10.1016/j.radonc.2008.07.008.

van Herk M. Development of an imaging system for patient set-up monitoring during radiotherapeutic treatment with high energy photons. Report on the experimental work for the partial fulfilment of the Master of Science degree in Experimental Physics at the University of Amsterdam, Amsterdam, the Netherlands, March 1985.

van Herk M. Physical aspects of a liquid-filled ionization chamber with pulsed polarizing voltage. *Med Phys.* 1991; 18(4):692–702.

van Herk M, Bel A, Gilhuijs KG, Vijlbrief RE. A comprehensive system for the analysis of portal images. *Radiother Oncol.* 1993; 29(2):221–229.

van Herk M, Bijhold J, Hoogervorst B, Meertens H. Sampling methods for a matrix ionization chamber system. *Med Phys.* 1992; 19(2):409–418.

van Herk M, Bruce A, Kroes AP, Shouman T, Touw A, Lebesque JV. Quantification of organ motion during conformal radiotherapy of the prostate by three dimensional image registration. *Int J Radiat Oncol Biol Phys.* 1995; 33(5):1311–1320.

van Herk M, Meertens H. A matrix ionisation chamber imaging device for on-line patient setup verification during radiotherapy. *Radiother Oncol.* 1988; 11(4):369–378.

Visser AG, Huizenga H, Althof VG, Swanenburg BN. Performance of a prototype fluoroscopic radiotherapy imaging system. *Int J Radiat Oncol Biol Phys.* 1990; 18(1):43–50.

Wong JW, Binns WR, Cheng AY, Geer LY, Epstein JW, Klarmann J, Purdy JA. On-line radiotherapy imaging with an array of fiber-optic image reducers. *Int J Radiat Oncol Biol Phys.* 1990; 18(6):1477–1484.

Yan D, Vicini F, Wong J, Martinez A. Adaptive radiation therapy. *Phys Med Biol.* 1997; 42(1):123–132.

2

Detector construction

DANIEL MORF

2.1 GENERAL REQUIREMENTS FOR IMAGING SYSTEMS

2.1.1 Introduction

Until the early 1990s, film dominated the imaging procedures in radiation oncology treatment environments. Working with film was a cumbersome and time-consuming process not suitable for much of today's imaging procedures in radiation therapy. The use of film was primarily limited to quality assurance (QA) and, in certain situations, verifying the patient's position. There was minimal integration with the treatment machine and interfacing with various treatment techniques was limited. Along with the advancement in digital imaging technology and, with it, the ability to provide images quickly or even in real time, the requirement for a tight integration with the treatment machine, and in particular, with the beam-delivery subsystem, became crucial for acquiring images with superior quality.

Section 2.1 will briefly address the treatment machine components relevant for imaging and how they influence the image quality and detector design. The focus here will include the detector design, key detector components, and the operation of flat-panel image detector. Section 2.5 will explain advantages and limitations of megavoltage (MV) imaging.

2.1.2 Principles of beam generation (linear accelerator)

Most radiation therapy machines utilize a linear accelerator (LINAC) to generate the high-energy X-ray beam. Treatment machines with radioactive sources will not be considered in this book, as they are seldom equipped with digital imaging systems.

The two principle accelerator types are traveling wave and standing wave accelerators. Both types of accelerators do not deliver the radiation as a continuous beam; rather it is delivered in discrete pulses. The pattern is not relevant for radiation treatment. However, for image acquisition, it is essential to synchronize with the beam pulse pattern to achieve good image quality. The actual electron acceleration takes place in roughly 5 μs. After delivery of each beam pulse, there is a delay of several milliseconds (2.5–200 ms) until the next beam pulse occurs. As a result, the pulse repetition frequency (PRF) can vary between 5 and 400 Hz. The different PRFs are used to control the dose rate for a given energy. Figure 2.1 shows a timing diagram of a typical beam delivery pulse pattern. Each of the shown beam pulses has a microstructure, due to the inherent phase focusing, and consists of roughly 2×10^4 micropulses, 330 ps apart and 30 ps wide (3 GHz accelerator system). This microstructure can be ignored for image acquisition synchronization.

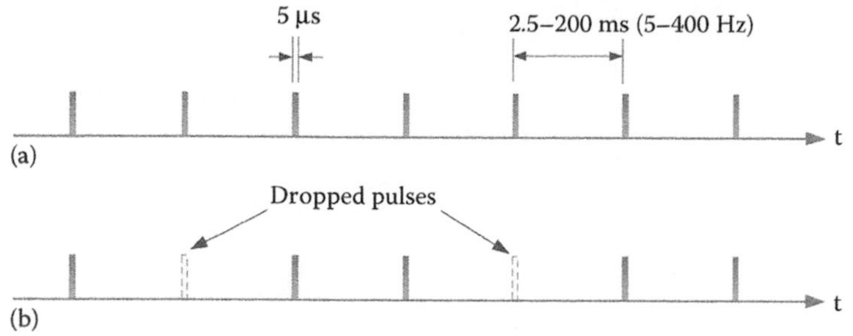

Figure 2.1 (a) Timing diagram of a typical beam pulse pattern generated by a linear accelerator. (b) *cut-out* pulses (Some accelerator types do not only vary with the frequency for regulating a particular dose rate, they *cut-out* pulses and generate in this way an irregular pulse pattern).

The beams of the accelerated electrons have a relatively narrow energy spectrum (± 3%–5%). However, when these narrow band electrons are converted to photons, a spectrum is generated that starts at low keV values and spans up to the energy of the accelerated electrons. Figure 2.2 (solid line) shows a typical photon spectrum of a 6 MV treatment beam.

When passing through the patient, the photons get absorbed. Due to the energy dependence of the X-ray cross section in a body, the low-energy part of the beam spectrum gets absorbed more heavily than the high-energy photons. This changes the spectrum as shown in Figure 2.2 (dashed line).

In addition to these spectrum changes, there is a radial dependency of the energy. When the beam is passing through a flattening filter (FF) in the head of the treatment machine, the photons near the central axis have to pass through more materials (typically copper and tungsten) in order to achieve a flat beam profile. There is a disproportionate absorption of low-energy photons near the central axis. The further away they can pass through the filter, the less material they have to pass through. As a consequence, the hardening of the beam spectra has a radial dependency. The spectrum is hardened near the central axis. This is important to note, since the detector and, in particular, the scintillator has an energy dependent response. Applications using the MV detector as a dose verification device have to correct for this energy dependence (Figure 2.3).

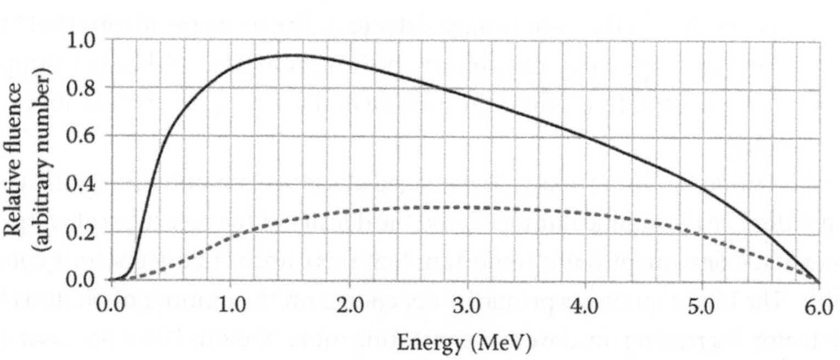

Figure 2.2 6 MV photon beam spectrum. For an unattenuated 6 MV beam (solid line) and spectrum after passing through 20 cm of soft tissue and 3 cm of bone (dashed line). Note that the scatter is not considered in this diagram.

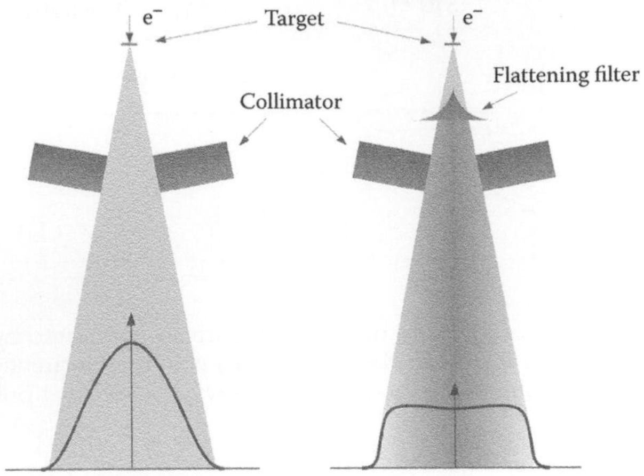

Figure 2.3 (a) Beam profile of a flattening filter-free (FFF) beam and (b) a beam profile with flattening filter (FF). There is also a radial symmetric energy falloff from the central axis toward the field edge, as indicated by the gray shades of the beam. The darker shading corresponds to a higher average energy (beam hardening).

2.2 DIGITAL IMAGING AND DETECTOR TYPES FOR X-RAY IMAGING

What is a digital X-ray image? What kind of information does the image content represent? It is important to understand that an X-ray image is not comparable to an analog film image. An analog image, basically, represents an analog value—no discrete steps—of an object being projected onto a film. There is also no discretization in the spatial domain; no spatial bins such as pixel with a given area are present. An X-ray image is, by nature, a distribution of quanta that are registered into pixel bins. The X-ray image is formed by individual photons that basically carry binary information. It is either there (not absorbed in the object being imaged) or not there (absorbed in the object).

Imagine a man standing in the rain with an umbrella. In this analogy, raindrops represent the photons and the umbrella represents the object to be imaged. When the rain starts, the contour of the umbrella is barely visible on the floor (image detector). The more raindrops that fall, the sharper the contour gets. The raindrops carry only binary information; they either get stopped (absorbed) by the umbrella or they can fall freely onto the ground and form an *image*. The more raindrops, the better the *image quality*.

The same holds true for an X-ray image. The imaged object will potentially absorb a certain number of photons, depending on the atomic number Z and the density of the object, producing different signal levels in the detector. Consequently, it is important that there are as many photons contributing to an image as possible. The image quality is primarily dependent on the number of photons being captured by the image detector. Increasing the dose and generating more photons is not necessarily the preferred method, since this increases the dose to the patient. Therefore, it is important to capture all the photons impinging on an image detector. Contemporary MV image detectors capture only 1% to 2% of the photons. The majority of the photons travel through the detector without interacting with it. Some physical constraints make efficiency improvements difficult. In order to understand the physical limitation, we need to have a closer look at the detector design and X-ray conversion. We will focus here on high-energy detectors, which rely primarily on the indirect detector principle (see Section 2.2.3).

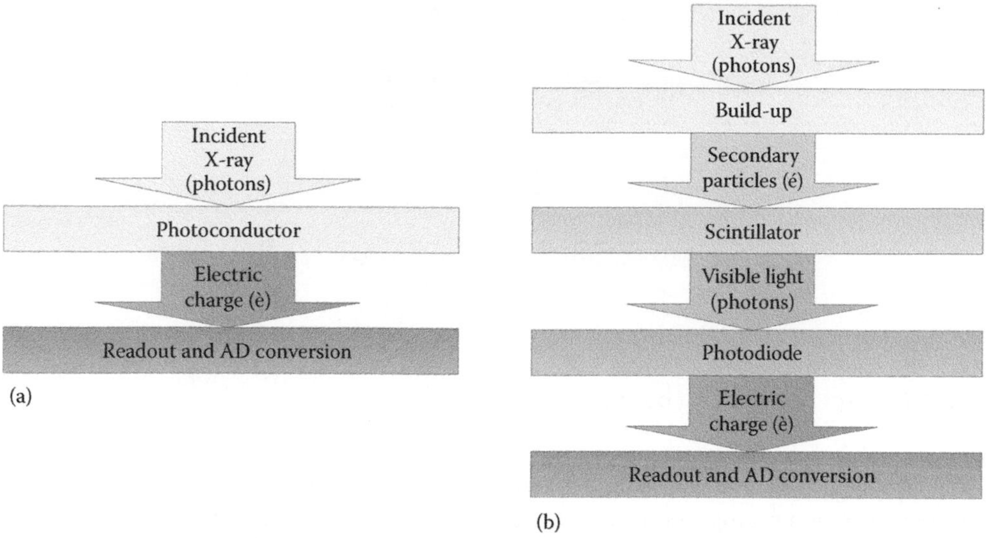

Figure 2.4 Physical processing steps of (a) a direct converter and (b) an indirect converter.

Almost all of today's radiation therapy systems utilize a flat-panel detector for imaging and QA tasks. These latter applications, including dosimetric and machine parameter verification, are becoming important functions of modern flat-panel detectors.

2.2.1 Amorphous silicon flat-panel detectors

There are two basic types of amorphous silicon (aSi) flat-panel detectors.

1. Direct converter type, which utilizes a thick layer of photoconduction material for X-ray detection.
2. Indirect converter type that incorporates a phosphor layer to produce visible light photons after interaction with an X-ray photon.

Figure 2.4 compares the different physical processing steps of a direct and an indirect converting flat panel.

Direct converting flat panels are not well suited for high-energy imaging application and are therefore discussed here only briefly. All flat-panel detectors currently used on LINAC are based on the principle of indirect conversion.

2.2.2 Direct converter

The incident high-energy photon is converted into an electron-hole pair in the photoconduction layer. A high voltage is applied between the top electrode and the pixel electrode in order to generate an electric field in the photoconductor. This electric field is responsible for separating the electron and the hole. The electron travels to the top electrode and the hole to the pixel electrode (bottom) and causes a change in electric potential on the pixel capacitor. The electric potential change is reset when the pixel is read out and the necessary charge is measured and digitized in the readout electronics (Figure 2.5).

The photoconducting material used in many direct conversion detectors is amorphous Selenium (aSe). This material is not well suited for absorbing high-energy photons due to its low atomic

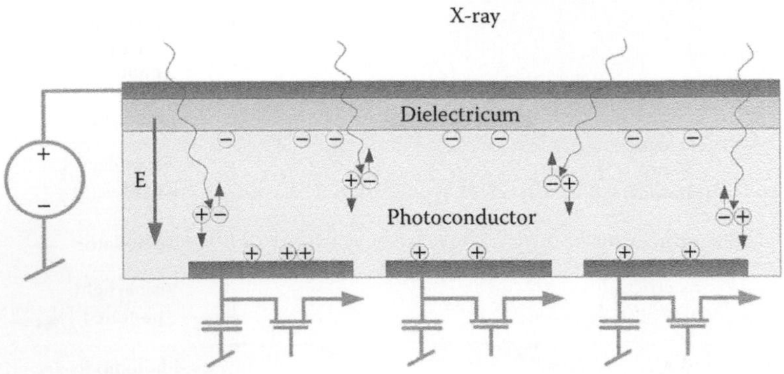

Figure 2.5 Direct converter flat panel.

number (Z = 34). Materials with a higher atomic number are better suited the cross section for Compton interaction is proportional to Z per unit volume (see Table 2.1). Compton interaction is the dominant interaction for the therapy beam-energy spectrum. The only way to increase the efficiency would be to use a material with a higher atomic number or to thicken the photoconducting layer. Due to technical difficulties of controlling the fabrication of thick and stable layers over large areas, the maximum thickness of aSe is limited to approximately 1 mm.

The use of photoconducting material with a higher atomic number has also been investigated. In particular, the use of mercuric iodide (HgI_2) with an effective atomic number of Z = 66 and lead iodide (PbI_2) with an effective atomic number of Z = 62.7 have been studied. The reliable and uniform deposition of these materials over a large area has not been resolved. Due to the high Z, both materials would have been very efficient for detection of high-energy photons of a therapy beam.

2.2.3 Indirect converter

Image detectors based on the indirect converter principle (Figure 2.6) convert the incident X-ray photons in multiple steps (Figure 2.4b). First, the photons are converted to secondary electrons;

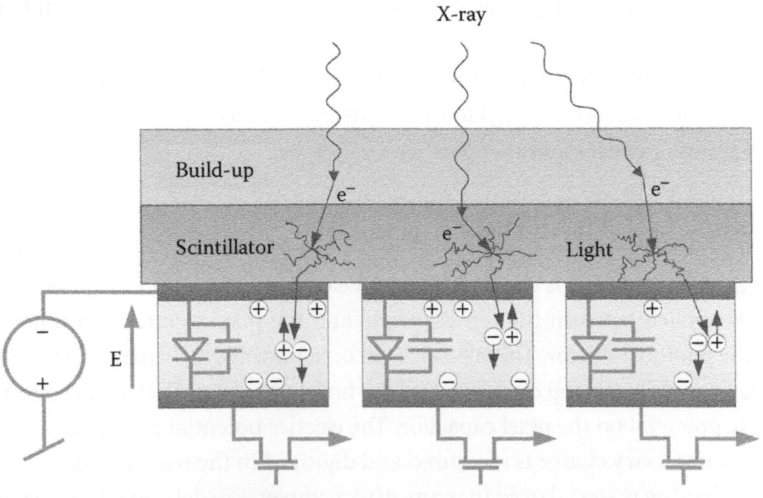

Figure 2.6 Indirect converter flat panel.

these electrons deposit their energy by either ionization or by atomic excitation. The latter is the preferred effect in scintillators, because atoms produce the energy difference as a light photon when they revert to their unexcited state (see Figure 2.9). Thus, one electron will excite hundreds or even thousands of atoms and generate a corresponding number of light photons. These light photons are captured by the aSi photodiode. The light creates electron-hole pairs in the aSi diode, which are then stored in the intrinsic capacitance of the photodiode. This accumulation of charge changes the electric potential at the photodiode. When reading out the detector, the initial potential will be reestablished by transferring a charge into the photodiode. The amount of charge that is transferred to the photodiode is measured and digitized. This digitized value represents the pixel signal and hence the number of photons interacting with the pixel.

2.2.4 Advanced technologies

Flat-panel imager technologies have been around for the past 15 years. Can we expect new technologies that will gain importance in the future? Certain trends are now visible in the diagnostic market; manufacturers have added complementary metal-oxide-semiconductor (CMOS)–based imagers to their product portfolio. However, it is unlikely that they will replace the aSi-based imager. Rather, they will extend the range of applications a digital imager can be used for, especially where high resolution, low-system noise, and fast readout speed is required. Radiation therapy imaging systems using a high-energy treatment beam as a source, would not profit from the more expensive CMOS technologies, since the CMOS technology does not address the most important shortcoming of aSi-based flat-panel detectors, namely, poor detection efficiency. About 98% of the photons incident on the detector do not interact with the detector and, therefore, do not contribute to the image signal. There are several paths being pursued to improve the performance. One idea is to use high Z material to increase the cross section for high-energy photons. Alternatively, adding more scintillation material to the detector would also increase the efficiency. However, increasing the thickness of the scintillator also increases the blurring and reduces the spatial resolution. A third approach to address this limitation is to stack up multiple detection layers (including the pixel array) within the same image detector. Apart from the efficiency improvement, there are other developments under consideration to improve the performance of current flat-panel detectors. Efforts are being made to increase the readout speed, reduce the imager lag, enlarge dynamic range, and improve radiation hardness of the detector. These are important steps toward better image quality and will be the basis for enabling new applications.

2.3 FLAT-PANEL DETECTOR DESIGN

2.3.1 X-ray conversion

2.3.1.1 HOW TO DETECT PHOTONS

Photons can only be detected if they interact with matter. In the energy range used for radiation therapy, two types of interactions dominate: Photoelectric and Compton. Both interaction types are briefly explained in order to understand specific detector properties. Pair production, which starts at photon energies above 1.02 MeV, does not play a significant role in high-energy imaging. A detailed explanation of these effects can be found in other sources.

2.3.1.2 PHOTOELECTRIC EFFECT

Photoelectric absorption takes place when a photon interacts with an electron of the inner shell of an atom. The entire energy of the photon will be transferred to the electron, the secondary particle. Part of the energy removes the electron from the shell (=binding energy) and the remaining energy is transferred to the electron as kinetic energy. Thus, the photon is completely absorbed. The electron with the binding energy closest to the energy of the incident photon, but not less than that, has the highest probability of being ejected. The ejected electron leaves a vacancy that will be filled from an electron of a shell further out that has less binding energy. The energy difference when moving from an outer shell to an inner shell will be radiated either as characteristic X-ray or as Auger electrons. These tertiary particles have an isotropic distribution. For imaging we are primarily interested in the secondary particle (the emitted electron) as these particles carry enough energy to undergo further reactions in the scintillator and eventually generate the signal that form the image.

Probability of interaction: The probability of a photon undergoing a photoelectric interaction depends mainly on (1) the energy of the photon and (2) the atomic number, Z, of the material. Photoelectric interaction probability for a given energy of the photon per unit volume is approximately proportional to Z^4. For low atomic numbers, the proportionality is closer to $Z^{4.5}$ than Z^4. For a given Z, the interaction probability is roughly proportional to $1/E^3$. These values reduce to 1/E for incident photon energies above the rest mass of the electron (0.511 MeV).

2.3.1.3 COMPTON EFFECT

Compton scattering (Compton effect) is the dominant interaction of X-ray photons in the therapeutic energy range (4–25 MV) with tissue and with image detectors. The Compton effect is an inelastic interaction of a photon with a weakly bound electron of the outer shell of the absorber material. The interacting photon transfers part of its energy to the electron that will be ejected from the atomic shell. The incident photon is scattered and emitted with the rest energy, that is, the incident energy minus the energy that was transferred to the ejected electron. As with all particle interactions, energy and momentum must be conserved. There are limits to the maximum scattering angle and the maximum energy that can be transferred to the scattered electron. The maximum angle of the scattered electron cannot exceed 90° from the direction of the incident photon. The photon can scatter at any angle. However, for the scattered photon, there are scatter angle dependent limitations of the energy. The energy of a photon scattered at 90° is always less than or equal to 511 keV and the maximum energy for a 180° scattered photon (backscatter) cannot exceed 255 keV. However, a scattered high-energy electron can undergo a bremsstrahlung event and generate a photon with higher energy then allowed by the above constraints.

Probability of Interaction: The probability of Compton interactions depends on the electron density (electrons/cm^3) of the material. The probability is independent of the atomic number Z per unit mass and proportional to the density and to the atomic number Z per unit volume, with the exception of hydrogenated material. In hydrogenous material, the electron density is almost doubled compared to anhydrogenous materials. Hence, the probability of Compton interaction is higher in hydrogenous material with the same mass as in anhydrogenous material. The fraction of the energy transferred to the electron is dependent on the energy of the incident photon. If the energy of the incident photon is well below the electron's

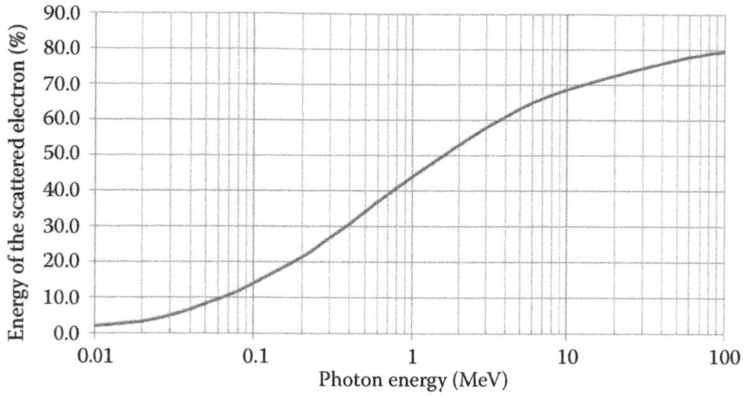

Figure 2.7 Energy of secondary electrons.

residual energy (<<511 keV), then most of the energy remains with the scattered photon. The energy of the electron is then small. If the incident photon energy is much higher than the rest energy of the electron, then most of the energy is transferred to the electron. Figure 2.7 shows the fraction of the energy transferred to the secondary electron as a function of the incident photon energy. The values are averaged over all scatter angles.

2.3.1.4 SUMMARY OF PHOTON INTERACTION

Table 2.1 illustrates the relationship of the interaction probability to atomic number Z and beam energy E for different photon interaction types. Note that the table is a simplified illustration of the interaction–probability relationships.

2.3.2 IMAGE-DETECTION LAYERS

Photons that are not detected in the imager represent lost information and, depending on the application, wasted radiation dose to the patient. Thus, the goal of every detector is to capture as many primary photons as possible. As outlined earlier, the efficiency of a high energy detector is relatively poor, due to the small cross section of the high-energy photons. In today's image detectors, less than 2% of the photons interact and contribute to the image signal. It is important to understand that flat-panel detectors are energy integrating devices and do not count or accumulate photons. The majority of high-energy image detectors are based on the indirect conversion principle. Figure 2.8 shows the basic components of a high energy flat-panel detector. Each component will be described further in detail.

Table 2.1 Qualitative overview of dependence of interaction probability on atomic number Z and energy E for different interaction type

Type of interaction	Relationship with atomic number (Z) per unit volume	Relationship with incident photon energy (E)
Photo effect	Z^4–$Z^{4.5}$	$E^{-3.5}$ (E << 0.5 MV), E^{-1} (E >> 0.5 MV)
Compton effect	Z	$E^{-0.5}$–E^{-1}
Pair production	Z^2	Log(E)

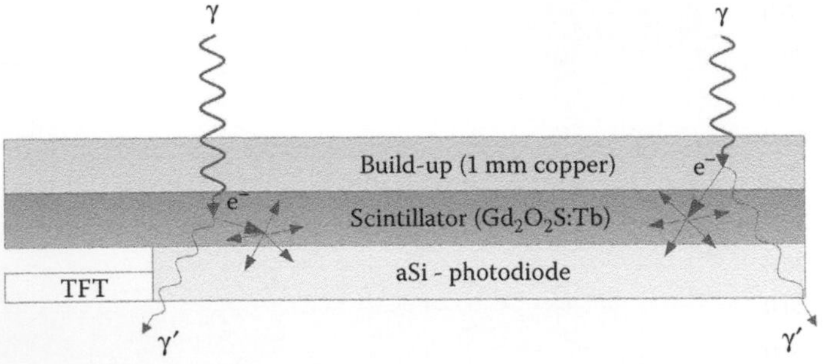

Figure 2.8 High-energy (megavolt) flat-panel detector cross section.

2.3.2.1 COPPER BUILD-UP (DOSE BUILD-UP EFFECT)

As we have seen previously, the majority of the photons will interact with the detector by Compton scattering. A high-energy electron is ejected and will travel a certain distance, before it deposits its energy. The energy deposition of high-energy charged particles (electrons) takes place through multiple collisional losses. Due to the high kinetic energy of these charged particles, they can travel a substantial distance before they have deposited all of their energy through ionization, excitation, and radiation losses. Thus, maximum energy is not deposited at the surface of the detector. Instead, the peak energy deposition is shifted by approximately the average range of the electrons. In order to maximize detection of photons or their secondary electrons, the ideal point to collect the signal is where the deposition peak is located. That is why a 1 mm copper plate is placed on top of the scintillator. The copper acts as a build-up and ensures that the area of maximum energy deposition lies in the scintillator, the place where the light is generated.

2.3.2.2 SCINTILLATOR (LIGHT GENERATION)

The primary function of the scintillator is to convert the photons and the secondary electrons to light. The Compton interactions occur not only in the build-up plate but also in the scintillator. Indirect flat-panel imagers use a phosphor as scintillation material. These phosphors are typically doped with an element that creates an activation state (luminescence centers) in the forbidden band. Most scintillators for high-energy flat panels use gadolinium oxysulfide (Gd_2O_2S) as a scintillation material. It is deliberately doped with terbium (Tb), which determines the wavelength of the emitted light photons. In a Gd_2O_2S:Tb scintillator, the peak wavelength is around 545 nm (green light). Figure 2.9 shows a simplified illustration of the scintillation process. When the secondary particle (Compton electron) deposits part of its energy, it excites electrons of the atom with which it interacts. These electrons are excited into the conduction band; from there they fall back through the activation states to the valence band. When moving from the excited activator state to the activator ground state, the energy difference is released as a light photon. In materials doped with terbium, the energy difference of the two activator states is such that it equals the energy of a green light photon. Thus, the released photon has a wavelength of 545 nm or about 2.3 eV. However, some electrons return to the valence band without going through the activator center and do not generate light.

The secondary electron releases its energy through multiple collision losses. This results in many light photons being generated along the track of the electron. Each interaction that passes

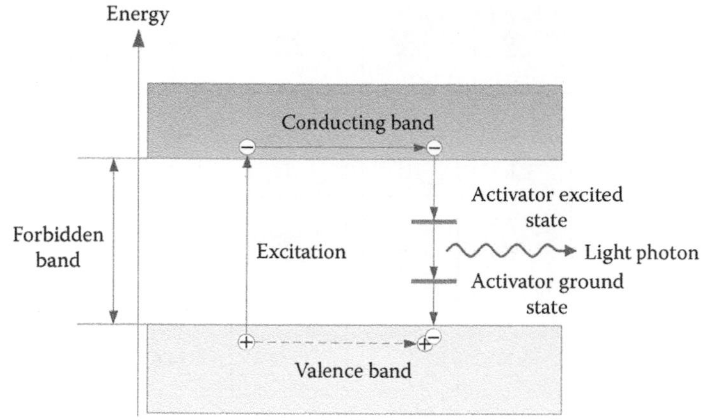

Figure 2.9 Light generation in the scintillator.

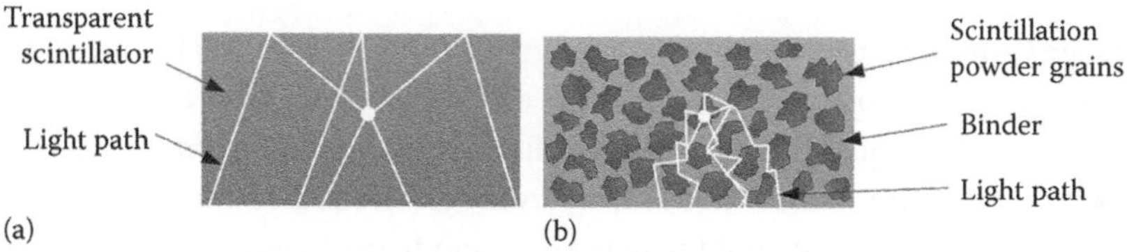

Figure 2.10 Light scatter in (a) transparent scintillator and (b) scintillation powder in a binder.

through the activator state consumes about 2.3 eV, which is transferred to the light photon. For example, a 2 MeV photon ejects a 500 keV secondary electron. This electron undergoes multiple collisions and could release more than 220,000 light quanta. Due to competing energy loss processes, only around 25,000 light quanta are released. Note that one initial photon, in our example a 2 MeV photon, has now generated around 25,000 light photons, each carrying redundant information. This is known as quantum amplification or conversion gain of the phosphor. It is not advisable to create a phosphor that is transparent, because the isotropic light generated could travel a long distance within the phosphor and cause blurring. Constructing the phosphor out of small light-reflective particles will scatter the light and reduce blurring (Figure 2.10). These small phosphor particles are mixed into a binder. The index of reflection of the binder is deliberately chosen to be different from the index of refraction of the phosphor particle. This enhances the reflectivity of the particle such that the light does not move through the particle; instead, it tends to reflect off the surface of neighboring particles. Thus, the spread of the light is reduced. Most scintillators used for high-energy image detectors are constructed as described in this chapter.

2.3.2.3 AMORPHOUS SILICON PIXEL ARRAY

The task of the aSi pixel array is to convert the visible light (545 nm for gadolinium oxysulfide [GOS] scintillator) to an electric charge that can subsequently be read out by the electronics. A pixel consists of a photodiode that converts light to an electric charge and a thin film transistor (TFT) switch, which can be turned on to read out the captured charge. The photodiode also acts as a charge storage device. All of the accumulated charge will remain on the photodiodes intrinsic capacitor, until it is read out by the electronics. The pixels (photodiode and TFT) are arranged in a matrix as shown in Figure 2.11.

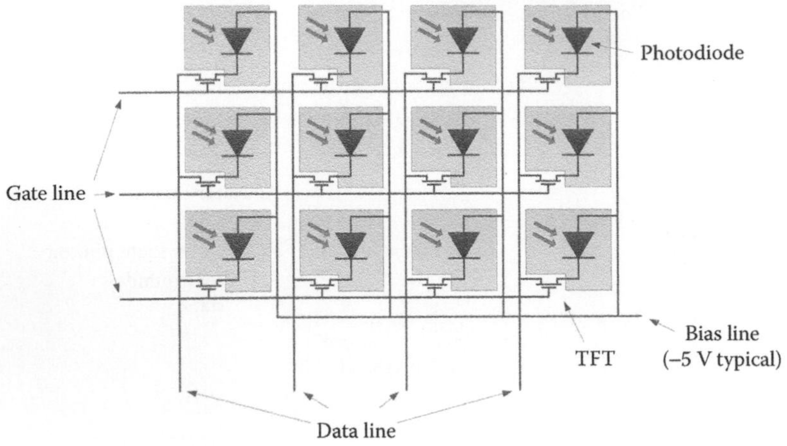

Figure 2.11 Simplified aSi pixel array for a detector with 3 × 4 resolution.

Light-to-Charge Conversion (Photodiode): Converting the light photons to an electric charge is the task of the photodiode. The photodiode is an aSi PIN diode with a light-transparent top metal layer. When light quanta enter the photodiode electron-hole pairs are generated. The electric field of the reverse biased diode separates the electrons and the holes. The electrons travel to the cathode (n-doped) and the hole to the anode (p-doped) of the diode (Figure 2.12). The structure of a PIN diode forms an intrinsic capacitor that is able to store the generated charge. As long as the PIN diode is not read out, the charge is accumulated in this intrinsic capacitor. Hence, the photodiode can be thought of as a charge-integrating device.

Charge Readout: After a reset and before radiation is turned on, the voltage across the photodiode equals the VBias voltage (in our example 5 V). The anode is at −5 V and the cathode is at 0 V. When the photodiode is exposed to light, as a consequence of the interactions described earlier in this chapter, charge is generated and the cathode potential drops (Figure 2.13). The anode potential cannot move since it is tied to the VBias potential. Readout of the charge by turning the TFT-on establishes the initial potential at the anode as shown in Figure 2.13. There is no dedicated reset cycle as reading out the charge resets the pixel. The charge that is read out to reestablish the voltage level at the cathode of the

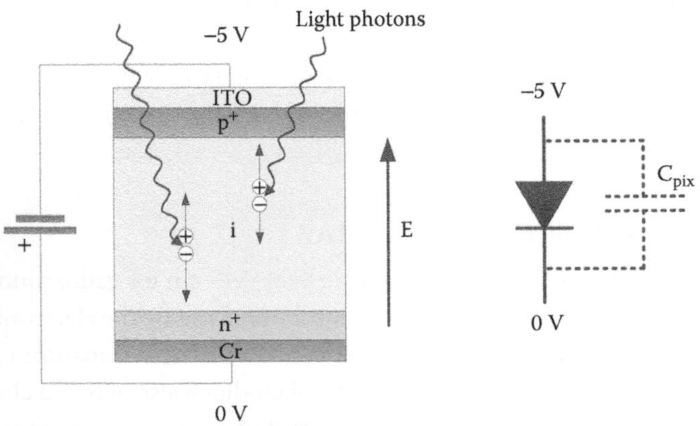

Figure 2.12 Charge generation in the pixel diode.

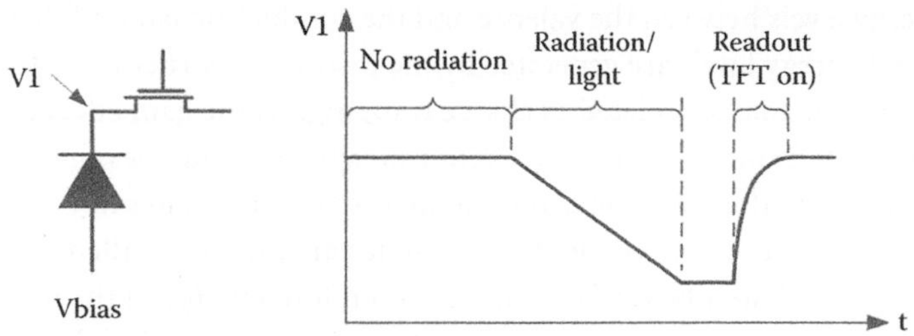

Figure 2.13 Voltage level at the photodiode. Note the timing is not to scale. In practice, the exposure time is generally much longer than the readout (TFT-on) cycle.

photodiode (V1 in the above figure) is measured in an external charge integration amplifier. The charge is directly proportional to the integrated light incident on the photodiode and therefore also proportional to the X-ray photons interacting with the detector. This results in a linear relationship between the fluence incident on the detector and the measured pixel signal for a given energy spectrum.

Readout Time Constant: The time to read out the charge stored in the photodiode is limited by physical properties of the pixel. The intrinsic capacitance and the resistance, when the TFT is switched on, form a resistor–capacitor (RC) network. The time constant $\tau = R \times C$ of this network limits how fast the charge can be read out. Currently used aSi pixel arrays have time constants on the order of a couple of μs. If the TFT were switched on for time equal to τ, only 67% of the charge would be read out. An incomplete charge readout would cause similar image artifacts as with the lag effect (described in Unwanted Pixel Array Properties). For this reason, it is common to set the readout time to be at least 4τ, which enables a transfer of approximately 99% of the charge to the integration amplifier. These physical properties are the primary cause for the readout time restriction, and hence limit the maximum achievable frame rate.

Unwanted Pixel Array Properties: As with any physical device, there are certain real phenomena that degrade or limit the actual performance of flat-panel detectors. One such phenomenon is the dark current associated with the photodiode. The dark current is present in all diodes and is similar to the reverse bias leakage current in nonoptical devices. The dark current has the same direction as the photocurrent and can therefore generate a signal in the absence of radiation. This dark current is responsible for part of the offset signal that is seen in aSi-based flat-panel detectors. However, the major part of the offset signal originates in the electronics. Since the aSi photodiode array is an integrating device, the dark current is integrated between two readout cycles and is therefore dependent on the frame readout speed. As with any diode leakage current, the dark current is also a function of the temperature. As a consequence, the offset calibration or dark-field calibration of the imager is dependent on temperature and frame readout speed. It is therefore recommended to perform dark or offset field calibration on a regular basis. Some systems perform this calibration in the background without any user interaction. Another characteristic of aSi arrays is charge trapping in the pixel diode. It appears to the user as image lag or gain loss. Lag effect, sometimes referred to as ghosting or memory effect, is primarily caused by charge trapping in the photodiode. Lag effects originating in the scintillator are typically much smaller. Charge trapping occurs if there are additional energy levels between the valence and the conducting band of the photodiode. These additional energy levels are generated by the disordered structure of the aSi. There are two noticeable effects that are related to charge trapping: (1) the gain effect and (2) the lag effect. The first 5–20 frames acquired after beam-on appear to have less gain than the subsequent frames. Before the radiation is turned on, the trapping states in the photodiode are empty. With the start of radiation, the trapping states are gradually filled up. These trapped charges cannot immediately be read out, and are therefore *missing* in the first few frames of the readout. After some time, the trapping states are filled and all of the newly generated charge can be read out by the electronics. The loss of charge at the beginning of the sequence appears as if the detector would have less gain, hence the name gain effect. When the radiation is turned off, the trapped state releases the charge as a function of time. Even though there is no radiation present, a charge that was caught in the trapped state can be read out and will appear as lag or as a ghost image. Charge trapping and charge releasing are exponential functions and are continuous processes, which reach an equilibrium during radiation.

2.4 OPERATION OF A FLAT-PANEL IMAGER

The previous sections detailed the detection and conversion of X-ray to an electronic signal and explained the function and properties of the different detector components. This section describes the detector operation and how it is integrated and synchronized with the beam delivery system, the LINACs.

A flat-panel image detector incorporates a two-dimensional pixel array. Each pixel consists of a photodiode and a TFT switch. The TFT gate connections of an entire row are tied to one gate line. The TFT switches in a particular row will be turned on simultaneously when the gate line is activated (Figure 2.14). Only one gate line is turned on at a time. As soon as the TFTs of one row is activated,

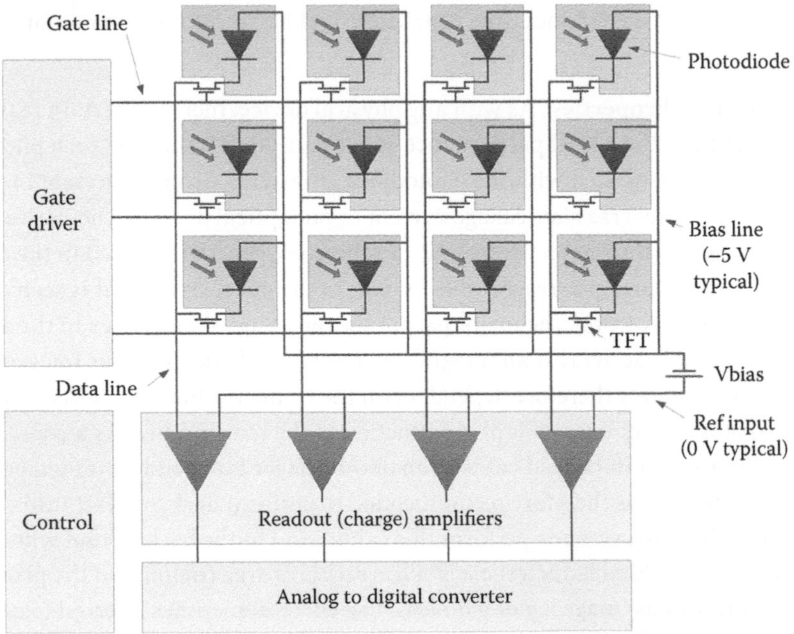

Figure 2.14 Schematic representation of a 3 × 4 aSi array with associated electronics.

the charges are transferred via the data line to the integration amplifier. Each pixel in a particular row has an individual data line and charge amplifier. However, all pixels in the same column are connected to the same data line and therefore share the same charge amplifier (Figure 2.14). The readout scheme described reduces the number of contacts to the panel and simplifies the design.

A frame readout sequence starts either at the top or at the bottom of the panel by activating the first gate line. After the entire charge has been read out and transferred to the charge amplifier, the gate line is turned off and the adjacent gate line is turned on. This process continues until all rows have been read out. The gate lines are typically addressed in sequence, although they could be addressed in any order. After a short pause, the readout starts again and the next frame is read out. All integrating flat-panel detectors are continuously read out even if there is no radiation being delivered and no image is requested. The integrating behavior of this detector design continuously accumulates signals such as dark current even in the absence of a radiation source. These accumulated background signals, together with other high-precision analog voltages, can only be kept constant when the readout timing of the detector is also constant. This is achieved with the continuous readout scheme. Error signals that are constant can be corrected with a calibration scheme as it is outlined in Section 2.4.3.

2.4.1 Design and Electronics

2.4.1.1 GATE DRIVER ELECTRONICS

The gate driver electronics consist mainly of a very long shift register with an output stage that is able to drive the TFT gates. Typical TFT off-voltages are –10 V. TFT on-voltages are in the order of 12 V. In most designs, the output stage of the shift register can be switched on and off to control the gate-on time, the time the charge is transferred to the readout amplifier.

2.4.1.2 READOUT ELECTRONICS

Each data line is connected to a dedicated charge integration amplifier. The integration amplifiers have very sensitive input stages in order to convert charges as small as a couple of femto coulomb into voltages that can be further processed. It is common to use a technique called correlated double sampling (CDS) to cancel out any offset and low-frequency noise signals. The CDS circuit captures a first sample just before the TFT is turned on and establishes a *base* signal level. The TFT is then turned on for a predefined period that allows the charge to be transferred to the amplifier and establishes a second signal level. The difference of the two signal levels represents the readout pixel charge. Only this signal difference is further processed and eventually converted to a digital signal. Modern design incorporates the analog-to-digital converters (ADC) in the same application specific integrated circuit (ASIC) as the charge amplifier. Older designs perform the analog-digital conversion in separate chips or even outside of the detector. Typical pixel value quantizations are 14 or 16 bit.

2.4.1.3 CONTROL ELECTRONICS

The control electronics have two main functions. First, they generate the signal used to control the timing of the gate driver and the readout circuit. The timing may vary depending on the different readout modes, the selected dose rate, and the requested dose per image. The control electronics also format the digital pixel data such that they can be sent to the host for further processing.

2.4.2 Detector readout and synchronization techniques

There is a restriction on the minimum time an aSi flat-panel detector can be read out. One fundamental limitation is the RC time constant arising from the TFT-on resistance and the pixel capacitance. Assuming a typical TFT-on resistance of 0.7 mOhm and a pixel capacitance of 10 pF, the time constant will be up to 7 μs. In order to read out 99% of the pixel charge, the TFT-on time should be at least 4 time constants or about 30 μs. Some spare time is required for the readout and other delays, so a minimal pixel time of 40 μs would be used in this case. This is the time that is required to read out a complete row, since all pixels in one row are being read out simultaneously. Assuming a detector with a resolution of 1280 × 1280, the minimum row time has to be multiplied by the number of rows in order to get the frame time. Hence, the time for reading out the entire detector is roughly 50–100 ms. Reading the detector during one of the 5 μs beam pulses would create uncorrectable image artifacts and has to be avoided as outlined in Section 2.1.2.

The time between two consecutive beam pulses can be as short as 2.5 ms; therefore, it is not possible to read out a complete frame in-between two beam pulses. Three basic detector readout methods are implemented by the system integrator and are chosen depending on the imaging use case. The three detector readout methods are described in Sections 2.4.2.1 through 2.4.2.3. Depending on the manufacturer, the name of the method might vary, but the principles remain the same. The timing diagram is drawn for illustrative purposes for a 32 × 32 detector array (32 columns and 32 rows). The accumulated charge is shown for one pixel in row 1; however, all pixels in the same row will demonstrate similar behavior.

2.4.2.1 Projection radiography

Projection radiography is primarily used for single pre- and posttreatment imaging. The beam delivery system synchronizes with the image acquisition system, so that the dose is delivered in-between two consecutive frame readouts. This method prevents any artifacts induced by the beam pulse, but requires the beam delivery to be paused for at least one frame readout time (50–100 ms), which is not desirable for in-treatment imaging (Figure 2.15).

If a radshot image is acquired during treatment, one or two reset frames (frames without beam) are required to clear the detector prior to the desired image acquisition. This adds substantially more beam-pause time and is typically undesirable during treatment delivery.

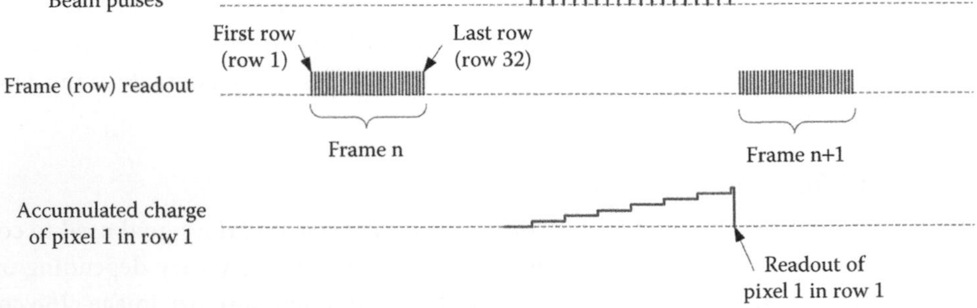

Figure 2.15 Radshot mode timing diagram for a frame with 32 rows.

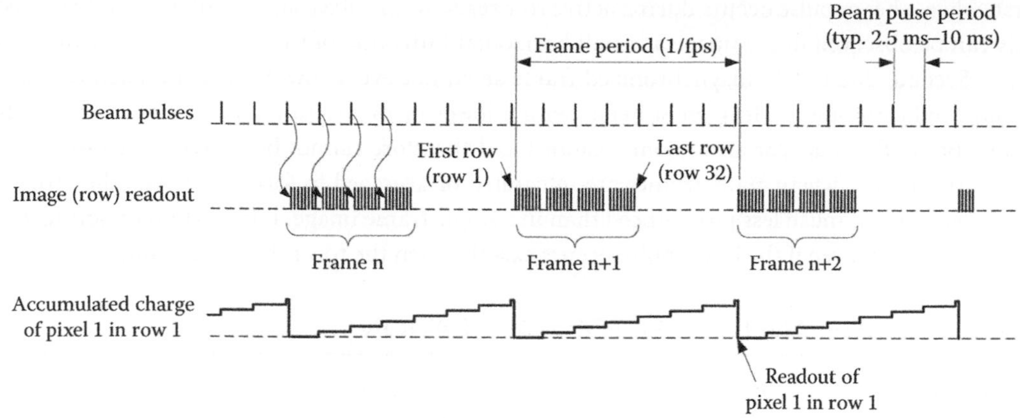

Figure 2.16 Synchronized mode timing diagram for a frame with 32 rows.

2.4.2.2 SYNCHRONIZED IMAGING

Synchronized image acquisition coordinates the row readout with the beam pulse delivery such that it never reads out at the same time as a beam pulse is being delivered. Based on the selected machine parameters (dose rate, energy, etc.), the imaging system calculates the number of rows that can be read in-between two consecutive beam pulses. Instead of reading a complete frame at once, the frame is partitioned in groups of rows that fit in-between beam pulses. After the readout of the last row, a delay is added before the start of a new frame readout. This delay is adjusted such that the required amount of dose is accumulated between two consecutive readout cycles of the same row (Figure 2.16). Note that the pixel signal is the charge that is accumulated between frame n and frame n+1.

This mode is best used for beam's eye view (BEV) imaging, since it can run in the background and does not influence beam delivery. However, for systems or treatment modes where beam pulses are randomly dropped, *banding artifacts* can occur due to differential accumulation of beam pulses.

2.4.2.3 UNSYNCHRONIZED IMAGING

The unsynchronized image acquisition mode reads out frames completely independent of the beam pulse timing (Figure 2.17). The imager is *free running*. This causes two types of image artifacts.

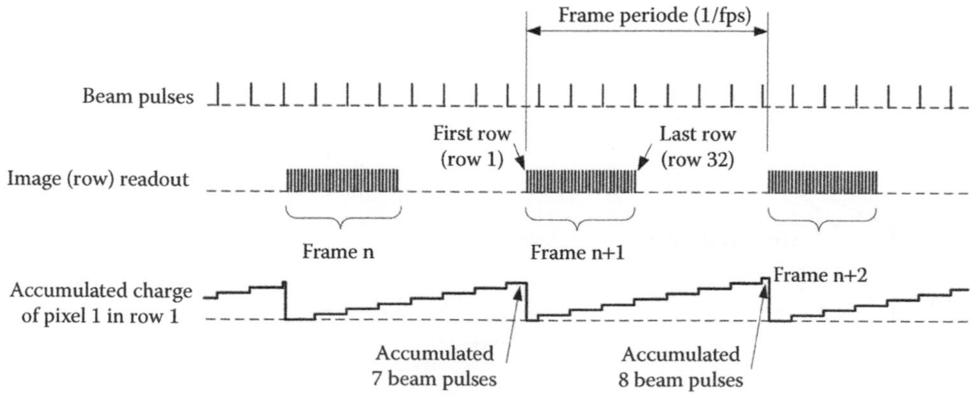

Figure 2.17 Unsynchronized mode timing diagram for a frame with 32 rows.

First, when a beam pulse occurs during active row readout, the pixel signal will be slightly altered. This can be observed in the image as small horizontal (direction of the row) lines that exhibit an offset. Second, due to the unsynchronized frame start, not every row has accumulated the same amount of beam pulses. Horizontal bands with a lower or higher signal can be observed in the image. Both effects appear at random locations and therefore cannot be corrected. However, for dosimetry modes, where many frames are integrated or averaged to form an image, the artifacts smear out and are much less pronounced than in a single frame image. It is worth mentioning that no charge is lost, even if the beam pulse occurs exactly when the row is being read out.

2.4.3 Flat-Panel Detector Calibration

The goal of flat-panel detector calibration is to correct each pixel, so that all of them appear to have the same properties in terms of offset and gain. If you look at the pixel-response curve in Figure 2.18, the calibration makes the two pixel-response curves (pixel a and pixel b) look identical.

Due to inconsistent manufacturing, there are pixels that do not function at all or exhibit unstable behavior. The task of the defective pixel correction function is to replace these defective pixels with an estimated pixel value.

2.4.3.1 Offset Calibration (Dark Field Calibration)

The offset calibration enables the correction of any offset present in the system. The inherent pixel offset value consists of two main components: (1) the pixels diode dark current as described in Section 2.3.2.3 and (2) the various analog electronics offset along the signal path. The offset values are readout timing and temperature dependent; hence, regular offset calibration is essential for maintaining good image quality.

Reading the detector in the absence of the X-ray beam provides the offset value of all pixels. In order to reduce the random noise associated with every image, the offset image is read several times and then averaged. The offset value of each pixel is then stored in the offset image. During normal image acquisition, the individual offset of each pixel is read and subtracted from the acquired value. As a result, the individual pixel offset is removed.

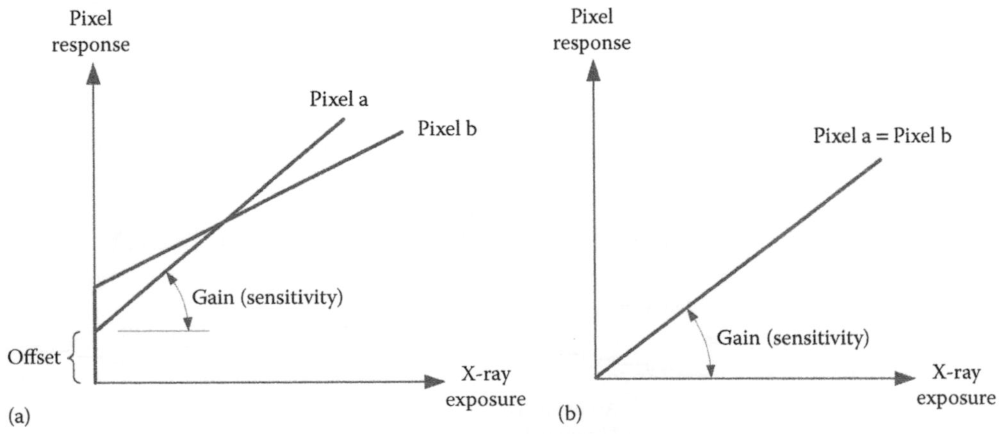

Figure 2.18 Pixel-response curves for two pixels: (a) Before calibration and (b) after calibration.

2.4.3.2 GAIN CALIBRATION (FLOOD FIELD CALIBRATION)

Each pixel has a slightly different sensitivity (signal gain), as depicted in Figure 2.18. In addition to the pixel gain variation, there is a multiplicative gain component that originates in the readout electronics. The gain correction adjusts the gain of each pixel such that all pixels appear to have the same gain. The gain image that is used for correction is acquired with a full-field exposure (flood field). Multiple flood fields are averaged to prevent image degradation due to the presence of random noise. The acquired gain image is normalized with the average (or median) value of the acquired image.

The full-field exposure is not a uniform field (the beam profile is not flat), so not every pixel is exposed to the same amount of radiation. It is not possible to distinguish whether the difference in signal level is due to different pixel sensitivity or different *local* exposure levels. Therefore, values in the gain correction image correct both sensitivity differences and beam nonuniformities. The resulting image (without an object) is always flat. This is a desirable effect when the image is acquired for standard imaging use cases. However, if the image is used for verifying correct dose delivery—as is the case for dosimetry use cases—this will definitely not yield the desired result. Algorithm and correction mechanisms exist to separate the sensitivity correction component from the beam nonuniformity correction component in the gain image. This primarily allows for correcting the pixel sensitivity while maintaining the beam non-uniformity.

2.4.3.3 DEFECTIVE PIXEL CORRECTION

Every flat-panel detector has pixels and sometimes lines or columns that do not respond correctly. These defective pixels have to be corrected. Typically, the systems replace defective pixels by an average or the median pixel value of the nearest neighbor pixels. The pixel correction is performed after the offset and gain correction.

2.5 MV IMAGE QUALITY

The fact that the high-energy flat-panel detector is in the axis of the beamline allows and enables the capture of images without additional radiation during treatment delivery. It also provides information of the two important dosimetric axes (x and y). The depth (along the beam axis) information is less important, since there is no rapid dose falloff in the direction of the beam axis. A unique advantage of the high-energy beam is the ability to penetrate through high atomic number material such as hip implants or teeth fillings.

High-energy (MV) images suffer in contrast and sharpness when compared to kilovoltage (kV) images. The contrast that appears in the image is mainly influenced by two physical processes:

1. The ability of the object being imaged to create absorption difference when the beam passes through anatomy with different density. The ability to absorb X-rays is strongly dependent on the energy of the photons and the type of interaction that takes place. The photon interaction type is mainly a function of the atomic number Z (see Section 2.3.1) and the beam energy. These dependencies are best visualized in the energy-dependent attenuation coefficients. Figure 2.19 shows the linear attenuation coefficients for bone and soft tissue. The linear attenuation coefficient is chosen because it visualizes the ability of different materials to absorb X-rays for the same path length. For imaging purposes, this is more illustrative than using density normalized mass attenuation coefficients.

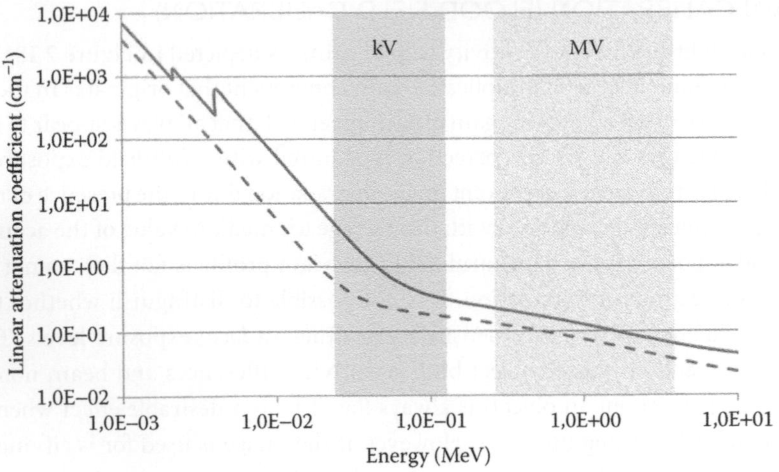

Figure 2.19 Linear attenuation coefficients; solid line: bone; dashed line: soft tissue. (*Source*: www.nist.gov/pml/data/xraycoef/). The dark gray area marks the energy range of diagnostic kV imaging and the light gray area marks the range of MV imaging.

As can be seen in Figure 2.19, the differences in the attenuation coefficients between bone and soft tissue are smaller for MV energies than for kV energies. This results in smaller attenuation differences and consequently less signal differences between bone and soft tissue when MV energies are used for imaging.

2. The second process reflects the ability to efficiently absorb photons in the flat-panel detector. The absorption efficiency is a function of the atomic number Z, the density, and the thickness of the absorber material. As outlined in Section 2.3.2, the absorber material consists of a build-up and a scintillator layer. The scintillator uses a material with a relatively high Z (gadolinium oxysulfide: Z = 64). The dominant interaction process is photoelectric absorption up to 400 keV. Above this energy, the Compton interactions dominate. As outlined in Section 2.3.1, the probability of interaction for the photoelectric effect is roughly proportional to Z^4 and for Compton it is roughly proportional to Z (Table 2.1). The dominance of the Compton interactions in MV flat-panel imagers is one important factor for the relatively poor quantum detection efficiency of high-energy flat-panel detectors. Current flat-panel detectors used for treatment imaging have a quantum detection efficiency of around 1%–2%, so only a small fraction of the photons are converted to an image signal. For comparison, diagnostic (kV) imaging systems have a similar detector construction, but reach a detection efficiency of 50%–70%, due to the dominance of the photoelectric effect.

Theoretically, any subtle difference in contrast could be made visible (assuming sufficiently small pixel value quantization), if they were not limited by the noise that is present in all X-ray images. The majority of today's detectors are described as *quantum limited*. That means the only noise source that can be observed in the image is a consequence of the statistical spatial distribution of the X-ray quanta. This noise is sometimes referred to as quantum noise. Any other noise source such as system noise is negligible and therefore not visible in the image. The quantum noise can be described by a Poisson distribution. As a consequence, the noise becomes a function of the signal amplitude itself: $\sigma = \sqrt{N}$, where N is the number of quanta and σ is the quantum noise. The only way to reduce the noise is to detect more quanta. Assuming the imaging dose is constant, the detection efficiency of the detector had to be improved to capture a bigger fraction of the photons. Reduced noise will then improve the visibility of the object contrast. Concepts to increase detector efficiency are described in Section 2.2.4.

Another disadvantage of imaging with a high-energy photon beam is the electron shower that is created by Compton interactions. One photon can create electrons with relatively high energy (Figure 2.7). These electrons release their energy through multiple ionization and excitation processes. The secondary photon can interact with matter and generate electrons that then deposit their energy some distance away from the incident interaction. As a result, the scatter component of an MV beam is not negligible. Due to the relatively high energy, scatter cannot be reduced by an antiscatter grid. These mechanisms, together with blurring in the detection layers, contribute to the degradation of the spatial resolution, and hence the loss of sharpness.

3

Monte Carlo simulation of EPIDs

JEFFREY V. SIEBERS AND I. ANTONIU POPESCU

3.1 INTRODUCTION

Evaluating electronic portal imaging device (EPID) detector design alternatives and understanding its performance for imaging and dosimetry applications is enhanced by having a detailed understanding of how the EPID responds as a radiation detector. A primary tool for evaluating expected detector response and performance is Monte Carlo (MC) radiation transport simulations. This chapter describes MC as applied for EPID imaging and dosimetry, utilizes MC to evaluate basic detector performance to incident radiation, and demonstrates the use of MC to compute EPID images.

MC simulations have been used extensively in EPID studies as they have undergone several generations of evolution, from liquid ionization chambers to camera-based systems, to aSi-based flat-panel detectors. Studies range from investigating the image-formation process (Bissonnette et al. 2003), detector design (Wowk et al. 1994; El-Mohri et al. 1999; Cho et al. 2001; Schach von Wittenau et al. 2002; Ko et al. 2004; Radcliffe et al. 2009), simulation of detector characteristics such as detective quantum efficiency (DQE), noise power spectrum (NPS), and modulation transfer function (MTF) (Bissonnette et al. 2003; Star-Lack et al. 2014), detector optical transport (Liaparinos et al. 2006; Michail et al. 2010; Blake et al. 2013), investigating alternative detector designs (Monajemi et al. 2006; Sawant et al. 2006; Teymurazyan and Pang 2012), considering accelerator target design to optimize imaging (Flampouri et al. 2002), analyzing detector spectral response (Yeboah and Pistorius 2000; Laure Parent et al. 2006), energy-deposition kernel generation for use with other calculation algorithms (Keller et al. 1998; McCurdy et al. 2001; Kirkby and Sloboda 2005; Li et al. 2006; Wang et al. 2009), to generating simplified calculation models for clinical use (Jung et al. 2012).

Although further evolution in EPID design is anticipated, this chapter concentrates on current aSi flat-panel detectors.

3.2 EPID MONTE CARLO

EPID MC simulations are typically used to (1) understand detector response; (2) predict images either with or without the patient; and (3) compute energy-deposition kernels to aid alternative calculation algorithms. Depending on the calculation, the source of incident particles can range from a treatment head simulation and/or source model, to parallel beam of monoenergetic particles incident on the EPID surface. In the case of predicting a through-patient image, a patient model, typically based on CT scan of the patient, is placed between the source and the EPID with ensuing particle transport through the patient. The MC simulation through the EPID is typically based on detailed geometric descriptions provided by the manufacturer. In addition to the typical radiation interactions in the detector, the simulation may include transport of optical photons within the detector as this affects the signal recorded by the detector.

3.2.1 Geometric modeling

For accurate response simulation, the geometric model for an EPID MC simulation should be based on the manufacturer design specifications. It has been shown that for some imagers, backscatter from materials downstream of the EPID contribute to the detector signal (Ko et al. 2004), meriting their inclusion in the simulation geometry. In some circumstances, minor detector elements that contribute, for example, <~0.5% of the signal can be neglected from the model, but at the risk of degraded model accuracy. Similarly, backscattering materials can often be adequately modeled with a simplified model.

The schematic illustration of a typical imager along with materials included in the MC geometric model is shown in Figure 3.1. The MC geometric model includes the front cover and the detector case, as well as a detailed model of the internal image detection unit (IDU) (not all materials are shown). The materials downstream of the IDU need to be included in the MC simulation only if radiation interactions in those materials contribute to energy deposition in the screen. Materials downstream of the IDU are required for simulation of Varian's aS500 and aS1000 imagers (Ko et al. 2004), but not for the aS1200 imager or for Elekta iViewGT imagers (Laure Parent et al. 2006).

For the MC, the geometric model consists of multiple parallel planes of uniform materials. This geometry is straightforward to implement in Geant, GATE, MCNP (Schach von Wittenau et al. 2002), Penelope, EGSnrc, and the BEAMnrc, DOSXYZnrc, and DOSRZ user codes. The MC scored energy in the screen serves as a surrogate for the detector response. The screen can be geometrically subdivided into individual pixel elements for scoring, or a virtual scoring mesh

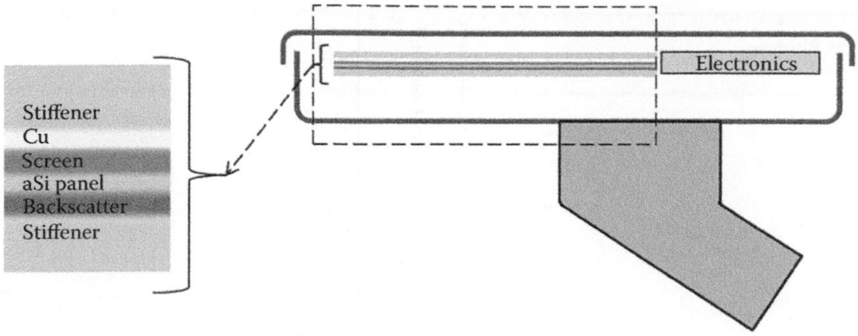

Figure 3.1 Schematic drawing of a typical EPID. Items included in the dashed box are included in the MC simulation model. The stiffener, screen, and backscatter are simplified representations of the multiplane components used in simulations. Figure is not to scale.

can be overlaid on the screen for scoring (Figure 3.5). This later method is further discussed in Section 3.3.

3.2.2 Light transport

Although flat-panel imagers yield a signal when operated in direct measurement mode without an intensifying screen (El-Mohri et al. 1999; Sabet et al. 2010), intensifying screens are used to increase sensitivity in all commercial EPIDs. A typical intensifying screen is the Lanex Fast Back screen (Carestream Health, Rochester, NY), which increases the detector sensitivity by a factor of ~8 (Vial et al. 2008). The intensifying screen dominates the detector energy response (Section 3.2.3) and contributes to signal blurring from the spread of optical photons in the imager (Kirkby and Sloboda 2005; Blake et al. 2013). MC simulations of detector response including optical photon scatter have demonstrated that it can be ignored without compromising simulation results for most scenarios. As a result, energy deposition in the screen serves as a suitable surrogate for EPID response in MC simulations (Blake et al. 2013).

3.2.3 Detector response

Figure 3.2 shows the MC computed energy deposited per monoenergetic source particle for photons and electrons incident on a Varian aS1200 EPID imager. The photon scale has been multiplied by a factor of 10 for easier comparison of the responses. The photon energy response is high at low energies due to high photoelectric cross section of the Gd_2O_2S:Tb screen. As the photoelectric cross section decreases with increasing energy, so does the EPID's photon response until the signal increases due to Compton interactions.

Electrons with $E<3$ MeV have insufficient energy to traverse the materials upstream of the phosphor screen; therefore only the bremsstrahlung from these electrons contribute to the EPID signal above 3 MeV, the signal from electrons rapidly rises with electron energy. The electron-response curve looks similar to a backward electron depth-dose curve, as expected when a fixed-depth detector is irradiated with electrons of increasing energy. Above 4 MeV, the signal per electron is more than 10 times the signal per photon.

As indicated in Figure 3.2, the EPID energy response is substantially different from water-equivalent detectors, particularly for $E < 1$ MeV photons. This is not an overresponse of the EPID,

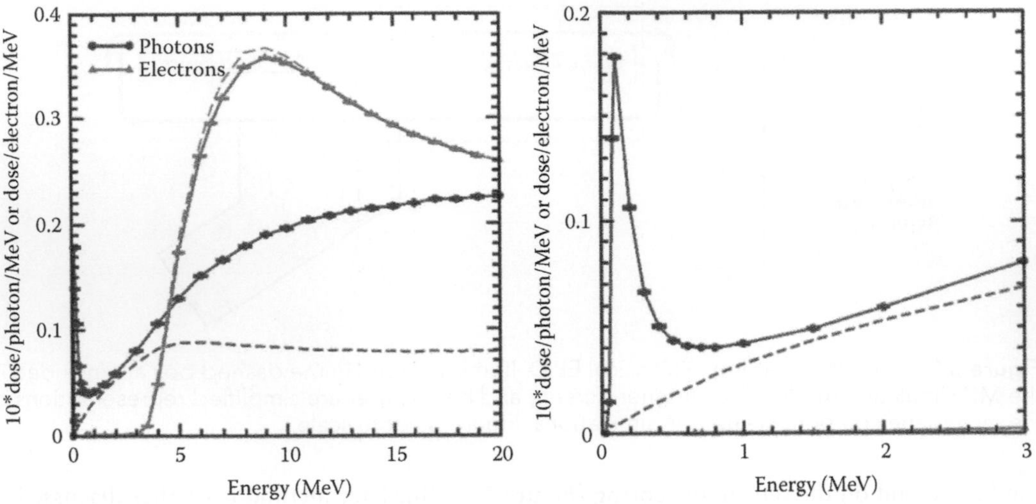

Figure 3.2 Dose deposited in the screen per source particle as a function of particle energy for monoenergetic photons and monoenergetic electrons incident on a Varian aS1200 EPID. The photon dose scale has been multiplied by 10. Above 5 MeV, dose/photon is <1/10th the dose/electron. The dashed curve is the relative dose deposited per source particle at 1.5 cm depth in water. The energy response of the screen differs from water, particularly at energies below 1 MeV.

but merely the natural effect of photoelectric interactions in the higher effective Z than water detector. This nonwater-like energy response challenges most dose calculation algorithms that use water-based energy deposition kernels.

Using MC, the photon response as a function of energy can be subdivided by the materials that generate the secondary electrons that interact in the phosphor screen as shown in Figure 3.3.

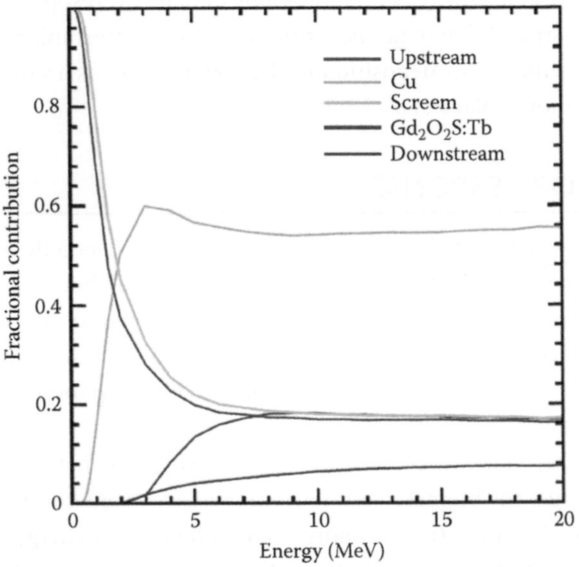

Figure 3.3 Fractional contribution of secondary electrons contributing to energy deposition in the screen for the Varian aS1200 imager as a function of photon energy. The different curves represent different detector materials. Note, totals may not add to 1 due to contributions from the aSi glass panel (max 2.5%@ 6 MeV) and other materials between the Cu build-up and the aSi panel (max 3%@1 MeV). Upstream signifies detector elements traversed before the Cu build-up, whereas downstream detector elements after the aSi glass panel.

Contributions from materials upstream of the Cu build-up, the entire screen, and the Gd_2O_2S:Tb screen layer and materials located downstream of the aSi glass panel are shown. At low energies, all signal is from photons interacting in the screen, with Cu build-up components starting to contribute at E>0.5 MeV. By 3 MeV, the Cu build-up levels off to contribute 0.56–0.60 of the total dose. The components upstream of the Cu screen begin contributing at the same energy that external electrons begin contributing (Figure 3.2), 3 MeV. As energy increases, the dose contribution from electrons generated upstream of the screen increases to nearly 20% of the total dose, whereas the detector unit backscatter components contribute up to 8% of the signal.

3.2.4 Energy-deposition kernels

Example MC computed energy-deposition kernels for photons and electrons are shown in Figure 3.4. These kernels are computed by simulating monoenergetic particles incident perpendicular to the imager at the imager center and scoring the EPID response. It is interesting to note the response kernel is narrowest at intermediate photon energies; the 1.0 MeV kernel is narrower than those at 0.15 and 10 MeV. This is due to the transition from photoelectric to Compton, to pair-production dominated events, and contributions from backscattered electrons. The electron kernels are substantially broader than the photon kernels due to multiple Coulomb scattering in materials upstream of the imager screen.

3.2.5 Monte Carlo calibration to detector response

The output from typical EPID MC simulations is the dose deposited in the Gd_2O_2S:Tb intensifying screen, which converts electrons produced in the Cu build-up plate into visible light, which is then detected by the aSi glass panel. The light output is considered to be directly proportional to the energy deposited in the screen. Due to slight variations in flat-panel manufacturing processes, real detectors have pixel-to-pixel sensitivity variations. Similarly, readout electronic variations result in further pixel-to-pixel sensitivity variations and a nonzero background signal. To account for these variations, dark-field and energy-dependent flood-field calibrations are typically performed to

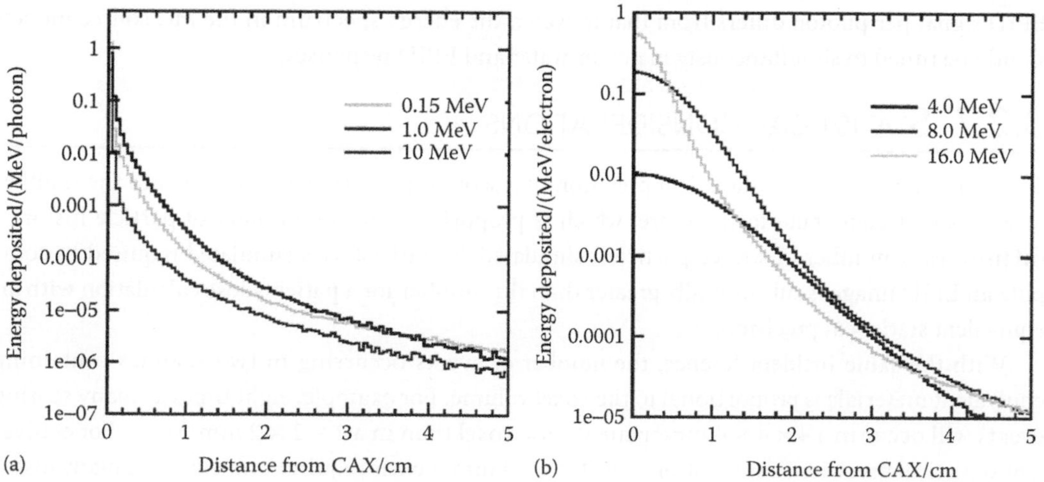

Figure 3.4 Monte Carlo computed energy-deposition kernels for monoenergetic (a) photons and (b) electrons for the Varian aS1200 EPID imager. The photon kernel has a minimum width near 1 MeV, whereas the electron kernel gets narrower with increasing energy.

normalize measured pixel responses. This normalization, however, suppresses the true beam profile. Flattened beams typically have *horns* at shallow depths, whereas flood-field corrected images will have a flat profile. Therefore, the true beam profile must be restored to measured images prior to intercomparing them with MC calculations.

Processes used to restore the true beam profile to a flood-field flattened measured image include (a) applying an empirical (Siebers et al. 2004) or measured (Van Esch et al. 2004) correction matrix, (b) ensuring that the flood-field image is uniform (Siebers et al. 2004), or (c) direct calibration of the relative pixel response (Greer 2005; Sun et al. 2015; Boriano et al. 2013). Since the differential energy response of the EPID differs from that of water, film, and other typical beam-profile detectors, care must be taken when the calibration is based on matching another detector, since the detector response will deviate with respect to such a calibration when the energy spectrum incident on the EPID changes. Only direct calibration of the relative pixel response, without reference to a non-EPID—measured beam profile will result in a calibration that is robust to energy spectrum changes, with the response being proportional to the dose to the EPID imaging phosphor.

The conversion of measured EPID pixel-intensity per monitor unit to MC computed dose per treatment head source particle can be accomplished by equating the measured and computed responses under a single set of common conditions, for example, a 10×10 cm^2 field with the imager at 105 cm source-detector distance (SDD).

For a properly commissioned MC source model, no additional calibration should be required to account for output dependencies, for example, due to field-size dependencies of output factors.

3.2.6 EPID source-model considerations

For pretreatment and patient-EPID image calculations, accuracy is limited by the accuracy of the source model used in the calculations. Although MC source-model tuning typically matches depth dose and lateral beam profiles for phantoms located near isocenter, users must be aware that the EPID is usually positioned at a highly extended source to surface distance (SSD) (up to ~180 cm) compared with typical patient-dose calculations (~90–110 SSD). Therefore, one must ensure that the model remains valid at these extended distances. Furthermore, as the energy dependence of the EPID signal per photon differs from that in water, the energy spectrum in the MC source models should be tuned to simultaneously match in-water and EPID responses.

3.2.7 Statistical considerations

In an MC simulation, the statistical precision of a scored quantity is proportional to the number of events that contribute to the score, which is proportional to the number of particle histories (nHistories = number of source particles) simulated. The nHistories simulated required to compute an EPID image is substantially greater than the number for a patient-dose calculation with an equivalent statistical precision.

With the same incident fluence, the number of events occurring in two volumes containing equivalent materials is proportional to the voxel volume. For example, eight times as many scoring events will occur in a $4 \times 4 \times 4$ mm^3 patient-dose voxel than in a $2 \times 2 \times 2$ mm^3 voxel. For equivalent statistical precision, a simulation with $4 \times 4 \times 4$ mm^3 voxels requires one-eighth as many histories as one with $2 \times 2 \times 2$ mm^3 voxels. A similar scaling can be performed to determine the number of histories for an EPID simulation of equal statistical precision. Consider a 400×300 mm^2 EPID containing 512 × 384 0.78 × 0.78 mm^2 pixels. The sensitive region of the EPIDs phosphor screen is ~0.3 mm thick; therefore, the EPID voxel volume is ~1/45th that of a 2 × 2 × 2 mm^3 voxel. For a 1024 × 768 or 1024 × 1024 imager, the voxel-volume ratio is 1/180th. If this were the only factor, when aiming for equivalent statistical precision, a single-beam MC calculation with the EPID located at the plane of the patient would require simulation of at least 45 times as many histories as required for a patient-dose calculation.

For nonequivalent materials, the ratio of the material interaction cross sections must also be accounted for when determining the ratio of the number of interactions per unit volume. Near 2 MeV, the dose per photon for the EPID and an equivalent water detector are about equal (Figure 3.2). Therefore, there are nearly equivalent events per unit volume in this Compton-dominated energy region. At both low energies and high energies, the EPID signal is larger in comparison to water, but always within a factor of ~2.5. For a 6 MV beam with mean photon energy near 2 MeV, the material differences have little to no effect on the required number of histories. Similarly, the beam-hardening caused by the patient will have little effect on nHistories.

For simulations of in-treatment imaging, the EPID is placed downstream of the patient at an SDD of 150–180 cm. Compared with the patient at 100 cm source-to-axis distance (SAD), the particle fluence is reduced by 2.25 (SSD = 150) to 3.3 (SSD = 180); hence, the NPS for these simulations must be increased by these factors for equivalent statistical precision.

An additional factor that affects nHistories for EPID calculations is the number of beams used in the treatment delivery. For patient-MC dose calculations, the nHistories required for a given statistical precision is independent of the number of beams because the beams converge in the high-dose region at isocenter and the quantity of interest is the integrated patient dose from the treatment. Unfortunately, this independence does not carry over for the unique EPID images formed by each beam. When per-beam EPID images are desired, the nHistories for each beam must be sufficient to reach the desired statistical precision. Over an entire treatment, this multiplies the total nHistories by the number of beams. This becomes especially burdensome when cine-EPID images are desired for arc-therapy verification, with images at, for example, 180 control points, or continuous cine images at 10 Hz yielding 600 independent EPID frames for a 60 s, 360° gantry rotation.

The overall nHistories required for EPID image creation can be staggering when a statistical precision similar to that which is needed for patient-dose calculations. The increase in nHistories and computation time can be as low as a factor of 45 when a 512 × 384 imager is simulated for a single beam with the EPID at isocenter, but as high as 350,000 for a 1024 × 768 imager located at 180 cm SDD for a 600 frame cine acquisition. The time required to compute these EPID images can be reduced with multiple fast multicore CPU or GPUs, and utilizing variance reduction techniques and approximations in calculations (Sections 3.3 and 3.4).

3.3 VARIANCE REDUCTION AND APPROXIMATIONS

Many publications discuss variance-reduction techniques for particle transport. The interested reader can find detailed discussion and references with respect to variance reduction in Jenkins et al. (1988) and Siebers et al. (2005).

Variance reduction refers to techniques to reduce the computer CPU time T necessary to determine a quantity R within a specified statistical uncertainty σ_R without changing the computer hardware.

T is proportional to N, the number of statistically independent particle tracks simulated. The variance, σ_R^2, decreases with $1/N$, the product $T\sigma_R^2$ is constant, and the efficiency ε of s simulation is given as

$$\varepsilon = \frac{1}{\sigma_R^2 T}$$

Variance-reduction techniques play an essential role in radiotherapy treatment head MC simulations and patient-dose calculations. Techniques such as bremsstrahlung splitting, Russian roulette, history repetition, simultaneous transport of particle sets, and fictitious cross sections have enabled research on delivery systems, detectors, and dose deposition in heterogeneous patients and phantoms. Variance reduction has also enabled clinical patient-dose assessments.

For EPID simulations, application of variance-reduction techniques during the treatment head and patient transport portions of a simulation are critical to ensure that enough particles are incident on the EPID to achieve a statistically meaningful image (Section 3.2.7). The above variance-reduction techniques can also be applied to particles transported through the imager; however, the greatest benefit is achieved by techniques that ensure that the incident particle contributes to the image creation as opposed to transporting through the imager without interacting. As such, an effective variance-reduction technique for EPID MC simulations is forced interactions of incident photons with the imager components followed by implicit capture (survival biasing) in which the incident particle survives each collision, but with its weight reduced by its survival probability. With this technique, multiple progeny of each incident particle can contribute to the EPID image. The efficiency improvement of this process can be enhanced by sampling multiple different particle cascades from a single incident particle to further sample the energy deposition distribution. Additional gains can be made with application of source-particle history repetition, which will result in multiple identical particles incident at different locations on the imager. Here, full reapplication of the imager signal at multiple locations can be applied, greatly improving calculation efficiency. Reapplication of the energy depositions at multiple locations is similar to MC kernel convolution, discussed in Section 3.5.

True variance-reduction techniques introduce no bias in the quantity being computed. As the goal of MC is to compute the quantity of interest with the lowest total uncertainty, it is often prudent to trade-off small systematic errors for large efficiency gains, thereby enabling reduction in the random, and more importantly, total uncertainty. Many approximations inherently exist in an MC simulation, for example, condensed history electron transport is an approximation; however, extensive comparisons with experiments have demonstrated that condensed history techniques can be used without introducing detectable bias. Other typical approximations utilized include geometry simplifications, energy cutoffs, physics models, and others. Ideally, all approximations are backed up with a sensitivity study to determine the systematic error introduced.

When a particle is transported in an MC code, a substantial fraction of the CPU time in a given particle history can be consumed by the boundary crossing algorithm used to transport a particle across material or scoring region boundaries. Although the boundary crossing algorithm used can affect both the speed and accuracy of the MC simulation, for energy deposition scoring in the EPID phosphor screen, eliminating the boundaries between adjacent scoring regions, as shown in Figure 3.5, can substantially reduce the simulation time without introducing significant bias. The method separates the particle transport geometry and its associated

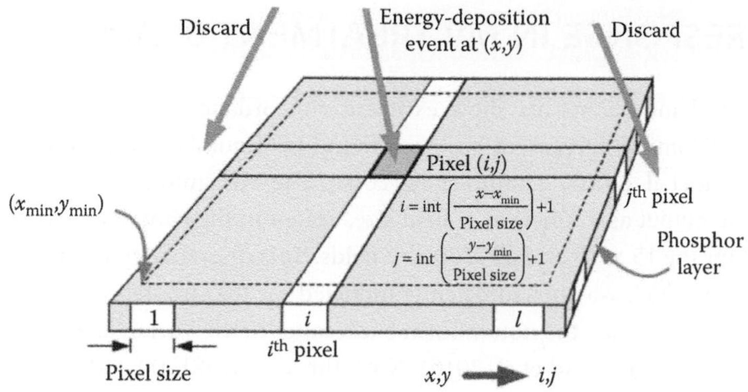

Figure 3.5 Schematic representation of the direct conversion of the energy-deposition event position to the pixel index. Instead of requiring the MC to transport particles through individual pixels at the imager screen level, energy-deposition events at a position (x, y) are directly translated to pixel coordinates (i, j). This virtual pixelization removes the need for the MC to perform interpixel boundary crossings, and, therefore, reduces the MC computation time. (From Siebers J. V. et al., *Med. Phys.*, 31, 2135–2146, 2004. With permission.)

boundaries from the energy-deposition grid, which is boundary-less. The condensed-history step electron-energy deposition event locations are directly translated to pixel coordinates. Eliminating the scoring-plane boundaries reduces calculation time by a factor of >4 for a typical EPID MC simulation. However, bias is theoretically introduced by the boundary artifact indicated in Figure 3.6. This effect has been described by Bielajew et al. (2000). In practice, however, the bias introduced is negligible and is blurred out by the spread of the optical photons within the phosphor screen.

Other geometric simplifications have also been used to reduce MC calculation time for EPIDs. Frauchiger et al. (2007) investigated several geometric EPID models and concluded that eight-layer geometry results in similar EPID dose distributions compared with the accurate 24-layer geometry, yet reduces the CPU time by about a factor of 6.

When computing through-patient EPID images with 6 MV beams, electron transport in the patient can be neglected, as these electrons have no direct contribution to the EPID signal (Figure 3.2). The in-direct bremsstrahlung contribution from electrons in the patient is negligible.

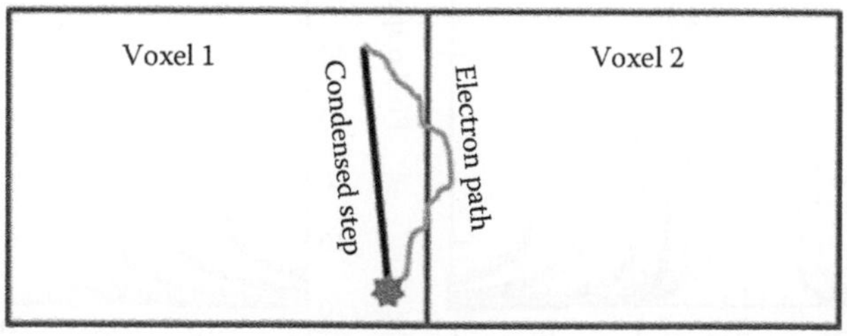

Figure 3.6 Theoretical bias introduced eliminating interscoring region boundaries in the EPID screen in an EGSnrc simulation. Without boundaries, the condensed history step length can be large with respect to the voxel size and all energy deposition would be scored in Voxel 1, whereas the electron path simulated with the intervoxel boundary would consider the distance to the scoring region boundary, choose a shorter step length, and deposit some of the energy in Voxel 2 along the electron path. In practice, this bias is negligible.

3.4 EPID RESPONSE IN MV TREATMENT BEAMS

MC simulations and measurements show excellent concordance for 6 MV beams as shown in Figures 3.7 and 3.8. Similar agreement has been found by a number of authors Such as Spezi and Lewis (2002); Parent et al. (2007); Cufflin et al. (2010). The MC simulation reproduces both profile shape and absolute output as a function of field size, except in the gantry-couch profile at distances of −5 to −15 cm for the 15 × 15 and 20 × 20 cm² fields. This discrepancy is due to backscatter from the EPID mounting, which was not sufficiently modeled for the simulation. Techniques have since been introduced to account for the nonuniform backscatter of the Varian aS500 and aS1000 imagers (Wang et al. 2009; Rowshanfarzad et al. 2010). Note, the Varian aS1200 and Elekta iViewGT do not have the asymmetric profile problem since they have sufficient uniform backscatter material.

An example of good agreement between MC and EPID measurement is shown in Figure 3.8 for pretreatment intensity-modulated radiation therapy (IMRT) quality assurance (QA). Visually, the measured and computed profiles are nearly indistinguishable. Profiles shown in Figure 3.8c and the EPID dose-difference histogram in Figure 3.8d demonstrate the accuracy achievable with MC for 6 MV beams.

MC-based studies of image formation for 6 MV photons show that electrons originating in the treatment head or in a patient can be safely omitted from MC calculations for most imagers since few electrons have sufficient energy (~3 MeV) to reach the EPID screen. For higher energy beams, however, electrons contribute a substantial fraction of the signal, as is shown in Figure 3.9, which compares 18 MV measured and simulated images as a function of field size. A calibrated MC computed image excluding electrons will fail to reproduce the beam profile as a function of field size. For large field sizes, the beam horns are exaggerated and the out-of-field dose is underestimated. The right panel shows that the electron contribution is a broad-profile without noticeable beam edges. For the 25 × 25 field size, the signal from electrons for an 18 MV beam is ~18% of the photon signal.

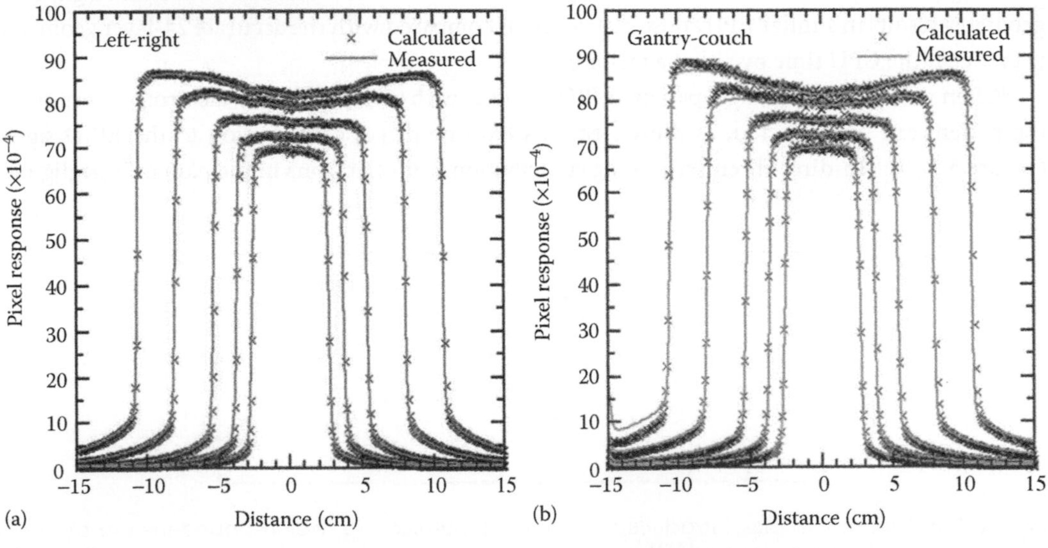

Figure 3.7 Comparison between measured and MC computed beam profiles as a function of field size for 6 MV beams on an aS500 imager. Comparisons are in terms of absolute pixel response with the measurement and MC cross calibrated in a 10 × 10 cm² field. The discrepancies at distances −5 to −15 cm for the 15 × 15 and 20 × 20 cm² field in the gantry-couch direction are attributed to inadequate modeling of the backscatter from components downstream of the detector unit. (From Siebers J. V. et al., *Med. Phys.*, 31, 2135–2146, 2004.)

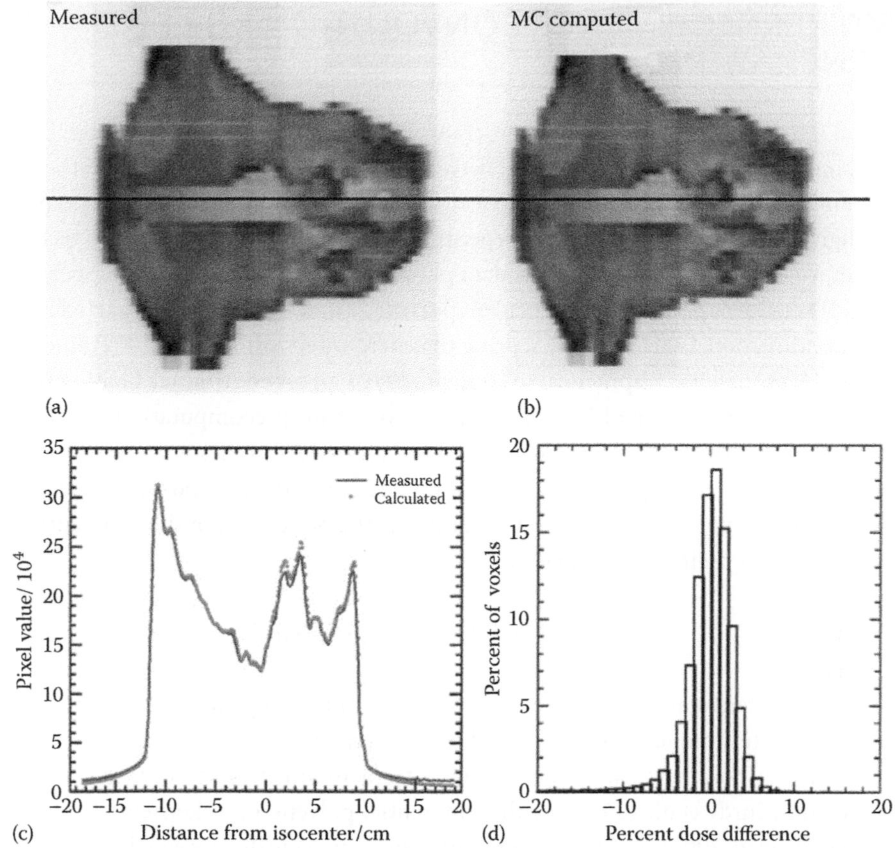

Figure 3.8 Measured and MC computed pretreatment aS500 EPID images for a sliding-window head-and-neck IMRT field. The MC simulation included transport through the treatment head, the moving MLC leaves, and the portal imager. The MC image accurately reproduces the intensity modulation and the tongue-and-groove effect from the MLC delivery. The average pixel difference was 0.5% and the gamma passing rate for this field was 99.7% with a 2%, 2 mm criterion. (Adapted From Siebers J. V. et al., *Med. Phys.*, 31, 2135–2146, 2004.)

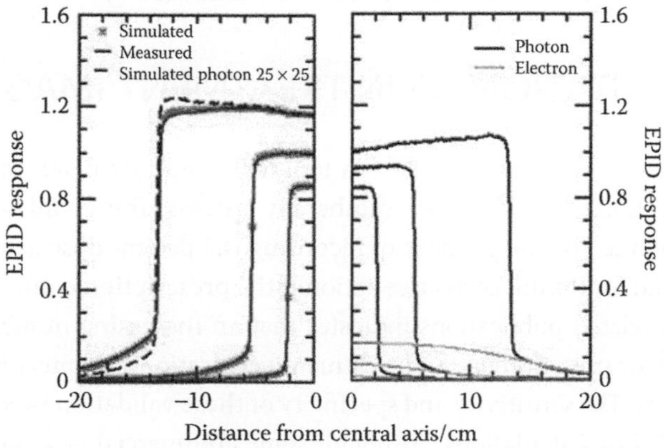

Figure 3.9 (a) Comparison of measured and MC simulated 18 MV beam profiles with and without contributions from treatment-head generated electrons and (b) photon and electron contributions to the EPID response. Exclusion of the electron component for high-energy beams affects the calibration such that the large field beam horns are exaggerated and the out of field dose is underestimated if they are neglected in the simulation.

3.5 MC KERNEL FLUENCE CONVOLUTION FOR EPID IMAGE GENERATION

An EPID is a fixed geometry device that can change its location with respect to the beam. From the EPID's point of view, the only difference between one beam configuration and the next is the differential energy fluence incident on the detector, which analog MC simulates one particle at a time. Although forced interactions, history repetition, and other variance reduction techniques are utilized to improve the signal per EPID-incident particle with some success, the upper limit of the signal per EPID-incident particle for an incident particle $P(E, u)$ where E is the particle energy and u is the particle direction, is achieved by scoring the energy-deposition kernel $K(P, E,u)$ at the particle's location. Although it is impractical to compute $K(P, E,u)$ for continuous floating point values of E and u, careful discretization of E and u along with one-time precomputation of $K(P, E,u)$ make the problem tractable. This is the basis for kernel convolution techniques.

Kernel convolution techniques utilize MC computed EPID dose-response kernels along with particle fluence to compute the dose to an imager pixel. The basic form of the convolution to compute dose to position $\vec{r}$ on the EPID panel is

$$D(\vec{r}) = \iint \Phi(E,\vec{r}')EK(E,\vec{r}-\vec{r}')dEd\vec{r}'$$

that convolves the differential energy fluence $E\Phi(E,\vec{r}')$ with the energy differential kernel $K(E,\vec{r}-\vec{r}')$. Kernel convolution techniques have been used extensively for predicting images from MC computed fluence using polyenergetic kernels derived from monoenergetic kernels (McCurdy and Pistorius 2000; McCurdy et al. 2001), radially-differential polyenergetic kernels to account for off-axis hardening (Li et al. 2006), and direct summation of images created by weighted monoenergetic kernels (Wang et al. 2009). Although each of these algorithms discretizes the problem and neglects directional dependence (u) of the incident particle on the EPID (kernel tilting is not implemented), they reproduce measured EPID images with an accuracy that is similar to that of full analog EPID MC simulations. With sufficient kernels, kernel-based methods reproduce analog MC computations within the statistical precision of the analog MC computation, thereby making them a reasonable time-saving substitute for full analog EPID MC simulations.

3.6 MC COMPUTATION OF IN-TREATMENT IMAGES

MC simulations can play an important role as a tool for independent verification of patient dose in every treatment fraction in an adaptive radiotherapy process. This could be ideally done online, using fast codes, or offline, for the purpose of recording the patient dose accumulated during the course of treatment and for taking corrective action if the prescription constraints are not met.

A survey of EPID related publications indicates that an increasing number of institutions are currently using EPID for pretreatment and treatment verification. Chapters 6 and 11 of this book cover these applications. The sensitivity and specificity of these validation tasks is dependent on the accuracy of the EPID dose calculations. Although some commercial treatment planning systems (TPSs) provide EPID dose or image predictions, as noted, the differential energy response of aSi EPIDs and TPS water-kernel–based dose calculations limit the accuracy to which non-MC or non-EPID specific dose response kernels can reproduce EPID signals. MC simulations are ideally suited for treatment QA purposes and allow users to perform verification tasks in an independent manner, in a framework outside the TPS. Besides their proven advantage in terms of dose calculation accuracy, MC simulations allow one to investigate the influence of source components for beams of a particular type and their contaminant particles. For independent treatment verification, some MC codes present the advantage of being open-source and publicly available. This is a feature that allows for user customization in ways that would be impossible with proprietary commercial QA solutions.

A truly *patient-specific* QA protocol should be relevant to the actual patient treatment and be capable of providing patient-pose specific dosimetry. Although phantom-based commercially available QA solutions are widely used for pretreatment verification, they would obviously not be able to catch during treatment errors, such as patient mispositioning, errant multileaf collimator (MLC) position, and so on. In contrast, patient 3D dose calculation using real-time linear accelerator (LINAC) log information and transit EPID dosimetry (in conjunction with cone-beam computed tomography (CBCT) data acquired in every treatment fraction) can provide the information needed for accurately assessing the patient-dose accumulation over the entire course of treatment.

EPID can be used for both pretreatment and transit (*in vivo*) dose verification. Pretreatment dose verification allows for detection of errors made during the transfer of treatment parameters between the TPS and the treatment unit and for detection of malfunctioning equipment (machine output variations, flatness and symmetry errors, or leaf positioning errors, e.g., due to incorrect MLC calibration). Transit dose verification allows for detection of errors related to patient treatment (setup errors, organ motion, anatomy changes, or changes related to different patient arm position, e.g., or to the fact that the beam passes through a different part of the couch or restraining device). An EPID-based adaptive radiotherapy workflow has been described in detail by Persoon et al. (2013), although their dose calculation algorithm was Varian's Acuros, rather than MC. For transit dosimetry, a model of the treatment couch (e.g., Teke et al. [2011]) should be included, besides the patient CT dataset, in the MC simulation.

The types of errors that can be detected by comparing measured and MC predicted EPID images depend on the coupling between the MC image prediction algorithm and the patient-dose prediction algorithm. If the simulated source particles for the patient-dose calculation are used for the pretreatment EPID verification and particles exiting the patient-dose calculation algorithm are used for transit dose verification, there is little opportunity for the delivery errors to be introduced and go undetected. Comparison of measured and predicted pretreatment images confirms the beam delivery, including congruence of the information transferred to the delivery system and that used for the MC patient-dose computation. Comparison of the MC predicted and measured transit images confirms the patient attenuation. Patient shifts through homogeneous regions could be difficult, if not impossible to detect; however, the LINAC beam delivery can be rigorously verified. On the other hand, if there is no coupling between the MC predicted images and the patient-dose calculation, comparison between measured and such stand-alone MC predicted EPID images might not reveal dosimetry errors that may originate in the TPS or TPS data transfer. Improper conversion to DICOM (or transferring the wrong beams) would influence the MC prediction algorithm and the LINAC delivery, making such errors undetectable. Similarly, errors in the patient-dose calculation (e.g., wrong output factors or MLC characteristics such as the dosimetric leaf gap) would not be detected by stand-alone MC predictions. These, error if present, would be uncovered by a direct TPS to MC *patient*-dose comparison that simultaneously computes the EPID image. Some authors have accounted for deviations between TPS predicted and delivered fluence by utilizing pretreatment EPID images to back-project the beam fluence and reconstruct a phase space that can then be sampled to calculate the MC patient-dose distribution and predict the transit EPID dose (van Elmpt et al. 2006).

Although the earlier EPID MC studies typically focused on integrated field images to verify the accumulated field shape and dose, for dynamic treatments such as volumetric modulated arc therapy (VMAT), it is desirable to also predict and verify images per control point. Podesta et al. (2014) provided a calibration method for time-dependent pretreatment dose verification for both flattened and flattening filter free (FFF) beams, and showed that sufficient data are present in the LINAC log files and EPID frame headers to reliably synchronize and resample portal image frames to correspond to treatment plan control points.

Asuni et al. (2013) described a method for simultaneously providing MC dose deposition in both the patient CT dataset and any type of planar (entrance or exit) detector for VMAT plans.

Su et al. (2015) developed an efficient simulation technique and demonstrated the feasibility of including transit EPID cine-mode MC simulations as part of a comprehensive patient-specific VMAT QA process. They developed a novel tool within the BEAMnrc/DOSXYZnrc environment that is capable of providing transit EPID distributions, per control point, or over any user-defined time interval, in a single simulation, for any VMAT plan delivered to a patient, as described in Section 3.7. Simultaneously, the total MC dose distribution is generated within the patient CT phantom.

A meaningful dose reconstruction requires simulations through an updated anatomy dataset, since there may be patient anatomy variations during a fractionated course of treatment. McVicar et al. (2015) have proposed a workflow to assess daily dose distributions using a patient CBCT data acquired before each fraction and MC simulations. All image and contour registrations are performed automatically with a deformable image registration software suite and MC simulations are performed daily on the anatomy-updated deformed CT set, using the automated MC QA system described in Popescu et al. (2015). An indirect validation of the dose reconstruction is performed by comparing daily EPID images acquired during patient treatment with MC predicted ones (using the Su et al. [2015] technique mentioned above), based on the updated patient anatomy. The LINAC performance is intrinsically captured by EPID images. The dosimetric effects of LINAC performance can be capture by using LINAC log files for MC inputs.

At present, there is no general consensus on the thresholds for action level, when differences between actual and predicted EPID doses are noted, since it is difficult to interpret these differences in terms of their implications for tumor control probability and normal tissue complication probability.

3.7 ADVANCED CLINICAL APPLICATION: 4D EPID IMAGE PREDICTION

Popescu and Lobo (2013) have introduced a new class of phase spaces in international atomic energy agency (IAEA) format (Capote et al. 2006), called *4D phase spaces*, in which a time-like variable (most commonly, the MU index) is part of the record of each particle, along with its position, energy, charge, momentum vector direction, LATCH, and so on. A similar feature was proposed much earlier by Spezi et al. (2001), but it was limited to the native (planar and static) BEAMnrc phase space format and thus not amenable to VMAT simulations in which the patient geometry is time-dependent. The 4D phase space takes advantage of the fact that the IAEA format permits storage of nonplanar phase spaces; the 4D phase space is stored at the patient CT boundary. This scoring geometry allows for the capturing of all particles that exit the patient, regardless of their direction, and can be saved in either the LINAC (BEAMnrc) or patient (DOSXYZnrc) coordinate systems (Figure 3.10).

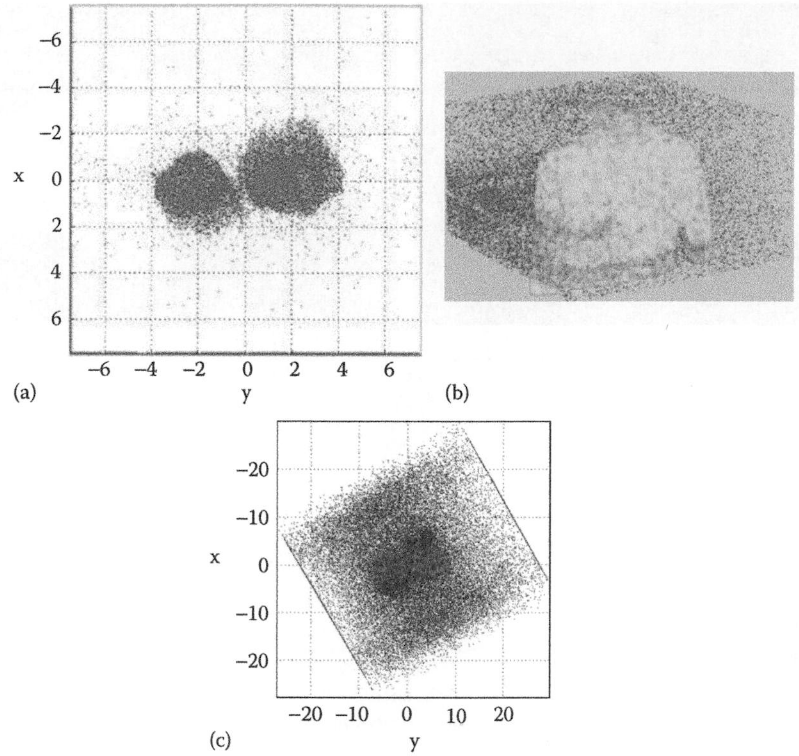

Figure 3.10 An entrance phase space, scored by BEAMnrc at the bottom plane of the LINAC, shown in beam's eye view (a) is used as the input for a DOSXYZnrc VMAT simulation. The simulation provides a dose distribution in the patient CT phantom (not shown) and an exit phase space, which can be visualized in either DOSXYZnrc coordinates (b) or BEAMnrc coordinates (c). Each particle in the exit phase space is labeled with the appropriate MU index. The exit phase space can further be used as input for EPID simulations.

The 4D exit phase space is used as input to EPID MC simulations, which are performed in the LINAC coordinate system since, in this system, the EPID position is invariant as the gantry rotates around a patient.

In comparison with discrete per-control-point simulations, an advantage of the 4D phase space approach is that the continuity of the beam delivery between control points is considered in the same manner as it occurs for the beam delivery. Even for large control point spacing, beam delivery (hence patient dose) is accurately simulated.

The time/MU index permits computation of EPID images over arbitrary user-defined gantry angles, including between adjacent control points, partial arcs, or integrated over the entire treatment delivery (Popescu and Lobo 2013; Su et al. 2015). Figure 3.11 shows MC simulated EPID dose images for a small (20% of the total MU delivered), arbitrarily selected, MU interval of a two-arc VMAT treatment for prostate and pelvic nodes. Image (a) resulted from a simulation through the original planning CT of the patient, whereas image (b) resulted from a simulation through a rescan, two weeks after the beginning of the treatment. Each simulation has relatively large (~9%) standard deviations due to the statistical issues discussed in Section 3.2.7. Although typical methods of comparing images with this much uncertainty are not applicable, systematic changes can be detected by taking advantage of the fact that the MC reports both the dose and the uncertainty on a pixel-by-pixel basis. If the patient was identical in the two simulations, the differences of the

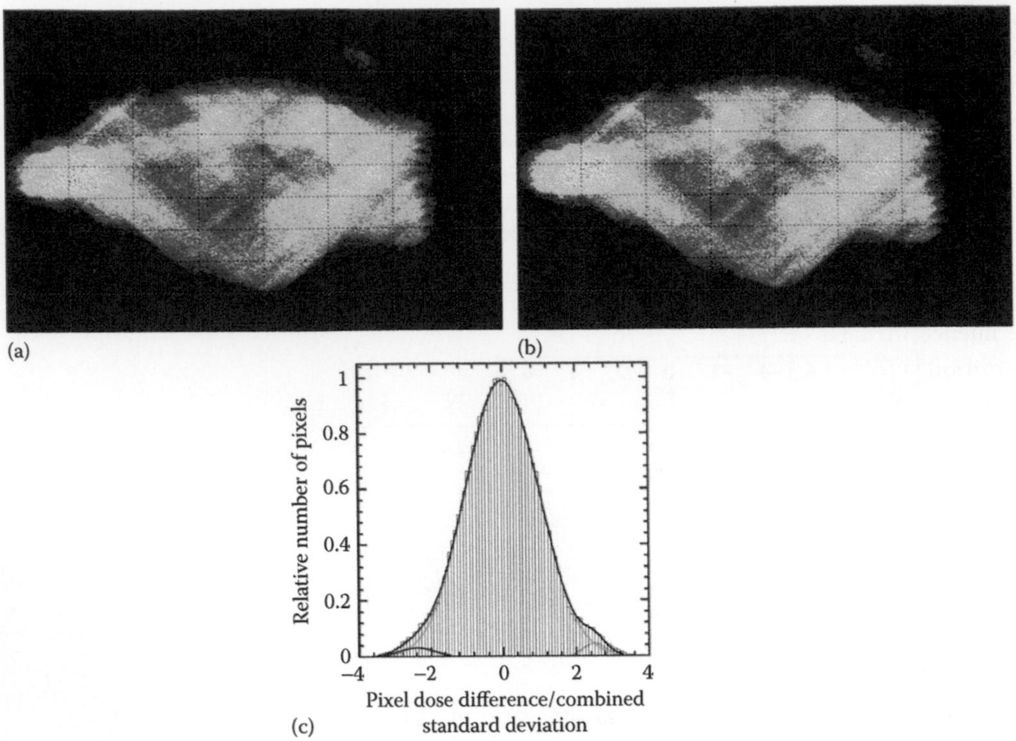

Figure 3.11 MC simulated EPID dose images for a 20% of the MUs from a VMAT delivery. (a) was simulated through the original CT of the patient, whereas (b) was simulated through a rescan, two weeks after the beginning of the treatment. (c) shows the histogram of the pixel dose difference divided by the combined statistical standard deviation from the two simulations. Fitting the histogram with normal distributions reveals three distributions: the 96.6% of pixels with a systematic offset of $\mu_M = -0.01$, $\sigma_M = 1.0$; 4% of pixels with $\mu_H = +2.5$, $\sigma_H = 0.3$; and 1.6% of pixels with $\mu_L = -2.4$, $\sigma_L = 0.5$.

doses divided by the standard deviation of the difference would yield a normal distribution with a mean of zero and standard deviation of one. When the patient differs, systematic differences in the geometry change the distribution shape. Fitting with a series of normal distributions can be used to infer the changes. In this case, direct differences of pixel doses with respect to the combined standard deviation from the two MC simulations reveals three normal distributions that describe the resulting change. The majority of pixels (96.9%) are fit with a normal distribution offset by $\mu_M = -0.01$ with $\sigma_M = 1.0$. The remaining pixels come from two distributions: 1.4% with $\mu_H = +2.5$, $\sigma_H = 0.3$ and 1.6% with $\mu_L = -2.4$, $\sigma_L = 0.5$. These changes are due to changes in beam transmission through the patient.

3.8 SUMMARY

EPID MC simulations can be used for simulating and studying the imaging and dosimetric properties of the imager, to compute energy-deposition kernels to aid alternative EPID calculation algorithms, and to compute images for patient treatment QA. Nearly every major MC code package has been used to simulate EPIDs. Geometric modeling of most EPIDs is straightforward, and can be simulated by a series of uniform material slabs. It is best to base the geometric modeling on vendor provided specifications to ensure accurate response modeling. EPIDs in clinical use have a minimum of 1 mm of Cu build-up prior to a scintillating screen. The build-up material suppresses the EPID response to incident electrons with energy less than ~3 MeV. Compared with water, MC simulations show that clinical EPIDs have a greater response to low-energy photons due to photoelectric interactions with the high Z materials in the scintillation screen.

Due to their small pixel size, large distance from the source, and when there is a need to compute an independent image for each beam or beam segment, the number of particles required for statistically relevant MC simulations is substantially greater than the number required for patient-dose calculations. As such, variance-reduction techniques play an important role in EPID simulations. An effective method for EPID computation is to use MC to compute the fluence incident on the imager, then convolving that fluence with MC computed energy-deposition kernels.

MC computed EPID images accurately reproduce measurements when an accurate source representation is used. Simulations are useful to predict pretreatment and during-treatment images integrated over the entire treatment delivery, or images for a portion of a given beam's delivery.

REFERENCES

Asuni G, van Beek TA, Venkataraman S, Popescu IA and McCurdy BMC (2013) A Monte Carlo tool for evaluating VMAT and DIMRT treatment deliveries including planar detectors, *Phys. Med. Biol.* 58: 3535–3550.

Capote R, Jeraj R, Ma C-M, Rogers DWO, Sanchez-Doblado F, Sempau J, Seuntjens J and Siebers JV (2006) Phase-space database for external beam radiotherapy, Report INDC(NDS)-0484, Nuclear Data Section, International Atomic Energy Agency, Vienna, Austria.

El-Mohri Y, Antonuk LE, Yorkston J, Jee KW, Maolinbay M, Lam KL. et al. (1999) Relative dosimetry using active matrix flat-panel imager (AMFPI) technology. *Med. Phys.* 26: 1530–1541. doi:10.1118/1.598649.

Flampouri S, Evans PM, Verhaegen F, Nahum AE, Spezi E and Partridge M (2002) Optimization of accelerator target and detector for portal imaging using Monte Carlo simulation and experiment. *Phys. Med. Biol.* 47: 3331–3349. doi:10.1088/0031-9155/47/18/305.

Frauchiger D, Fix MK, Frei D, Volken W, Mini R, Manser P (2007) Optimizing portal dose calculation for an amorphous silicon detector using Swiss Monte Carlo Plan, *J. Phys.: Conf. Ser.* 74(01): 21005.

Greer PB (2005) Correction of pixel sensitivity variation and off-axis response for amorphous silicon EPID dosimetry, *Med. Phys.* 32(12): 3558–3568.

Jenkins TM, Nelson WR, Rindi A (Eds.) (1988) *Monte Carlo Transport of Electrons and Photons.* Boston, MA: Springer US.

Jung JW, Kim JO, Yeo IJ, Cho Y-B, Kim SM, DiBiase S (2012) Fast transit portal dosimetry using density-scaled layer modeling of aSi-based electronic portal imaging device and Monte Carlo method. *Med. Phys.* 39: 7593–7602. doi:10.1118/1.4764563.

Keller H, Fix M and Rüegsegger P (1998) Calibration of a portal imaging device for high-precision dosimetry: A Monte Carlo study. *Med. Phys.* 25: 1891–1902. doi:10.1118/1.598378.

Kirkby C, Sloboda R (2005) Comprehensive Monte Carlo calculation of the point spread function for a commercial a-Si EPID. *Med. Phys.* 32: 1115–1127. doi:10.1118/1.1869072.

Ko L, Kim JO, Siebers JV (2004) Investigation of the optimal backscatter for an aSi electronic portal imaging device. *Phys. Med. Biol.* 49: 1723–1738. doi:10.1088/0031-9155/49/9/010.

Li W, Siebers JV, Moore JA (2006) Using fluence separation to account for energy spectra dependence in computing dosimetric a-Si EPID images for IMRT fields. *Med. Phys.* 33: 4468–4480. doi:10.1118/1.2369468.

Liaparinos PF, Kandarakis IS, Cavouras DA, Delis HB, Panayiotakis GS (2006) Modeling granular phosphor screens by Monte Carlo methods. *Med. Phys.* 33: 4502–4514. doi:10.1118/1.2372217.

McCurdy BM, Luchka K, Pistorius S (2001) Dosimetric investigation and portal dose image prediction using an amorphous silicon electronic portal imaging device. *Med. Phys.* 28: 911–924. doi:10.1118/1.1374244.

McCurdy BM, Pistorius S (2000) A two-step algorithm for predicting portal dose images in arbitrary detectors. *Med. Phys.* 27: 2109–2116. doi:10.1109/IEMBS.2000.897910.

McVicar N, Atwal P, Lobo J, Popescu IA (2015) Adaptive patient dose assessment using daily 3D cone beam CTs and Monte Carlo simulations, World Congress of Medical Physics and Biomedical Engineering, Toronto, Canada.

Michail CM, Fountos GP, Liaparinos PF, Kalyvas NE, Valais I, Kandarakis IS. et al. (2010) Light emission efficiency and imaging performance of Gd2O2S: Eu powder scintillator under x-ray radiography conditions. *Med. Phys.* 37: 3694–3703. doi:10.1118/1.3451113.

Monajemi TT, Fallone BG, Rathee S (2006) Thick, segmented CdWO4-photodiode detector for cone beam megavoltage CT: A Monte Carlo study of system design parameters. *Med. Phys.* 33: 4567–4577. doi:10.1118/1.2370503.

Parent L, Fielding AL, Dance DR, Seco J, Evans PM (2007) Amorphous silicon EPID calibration for dosimetric applications: Comparison of a method based on Monte Carlo prediction of response with existing techniques. *Phys. Med. Biol.* 52: 3351–3368. doi:10.1088/0031-9155/52/12/003.

Parent L, Seco J, Evans PM, Fielding A, Dance DR (2006) Monte Carlo modelling of a-Si EPID response: The effect of spectral variations with field size and position. *Med. Phys.* 33: 4527–4540. doi:10.1118/1.2369465.

Persoon LC, Egelmeer AG, Öllers MC, Nijsten SM, Troost EG and Verhaegen F (2013) First clinical results of adaptive radiotherapy based on 3D portal dosimetry for lung cancer patients with atelectasis treated with volumetric-modulated arc therapy (VMAT), *Acta Oncol.* 52: 1484–1489.

Podesta M, Nijsten SM, Persoon LC, Scheib SG, Christof Baltes C and Frank Verhaegen F (2014) Time dependent pre-treatment EPID dosimetry for standard and FFF VMAT, *Phys. Med. Biol.* 59: 4749–4768.

Popescu IA and Lobo J (2013) Monte Carlo simulations for TrueBeam with jaw tracking, using curved or planar IAEA phase spaces: A Source 20 update, International Conference on the Use of Computers in Radiation Therapy, Melbourne, Australia.

Popescu IA, Atwal P, Lobo J, Lucido J, McCurdy BMC (2015) Patient-specific QA using 4D Monte Carlo phase space predictions and EPID dosimetry, *J. Phys.: Conf. Ser.* 573: 012004.

Radcliffe T, Barnea G, Wowk B, Rajapakshe R, Shalev S (2009) Optimization of metal/phosphor screens at megavoltage energies. *Med. Phys.* 20: 1161–1169. doi.org/10.1118/1.596970.

Rowshanfarzad P, McCurdy BMC, Sabet M, Lee C, O'Connor DJ, Greer PB (2010) Measurement and modeling of the effect of support arm backscatter on dosimetry with a varian EPID. *Med. Phys.* 37: 2269–2278. doi:10.1118/1.3369445.

Sabet M, Menk FW, Greer PB (2010) Evaluation of an a-Si EPID in direct detection configuration as a water-equivalent dosimeter for transit dosimetry. *Med. Phys.* 37: 1459–1467. doi:10.1118/1.3476220.

Sawant A, Antonuk LE, El-Mohri Y, Zhao Q, Wang Y, Li Y. et al. (2006) Segmented crystalline scintillators: Empirical and theoretical investigation of a high quantum efficiency EPID based on an initial engineering prototype CsI(TI) detector. *Med. Phys.* 33: 1053–1066. doi:10.1118/1.2178452.

Schach von Wittenau AE, Logan CM, Aufderheide MB, Slone DM (2002) Blurring artifacts in megavoltage radiography with a flat-panel imaging system: Comparison of Monte Carlo simulations with measurements. *Med. Phys.* 29: 2559–2570. doi:10.1118/1.1513159.

Siebers JV, Keall P, Kawrakow I (2005) Monte Carlo dose calculation for external beam radiotherapy; in *The Modern Technology of Radiation Oncology. A Compendium For Medical Physics and Radiation Oncologists* (J van Dyk Ed.) Vol. 2, p. 514, Madison, WI, Medical Physics Publication.

Siebers JV, Kim JO, Ko L, Keall PJ, and Mohan R (2004) Monte Carlo computation of dosimetric amorphous silicon electronic portal images, *Med. Phys.* 31: 2135–2146.

Spezi E, Lewis DG and Smith CW (2001) Monte Carlo simulation and dosimetric verification of radiotherapy beam modifiers, *Phys. Med. Biol.* 46: 3007–3029.

Spezi E and Lewis DG (2002) Full forward Monte Carlo calculation of portal dose from MLC collimated treatment beams, *Phys. Med. Biol.* 47: 377–390.

Star-Lack J, Sun M, Meyer A, Morf D, Constantin D, Fahrig R. et al. (2014) Rapid Monte Carlo simulation of detector DQE(f). *Med. Phys.* 41: 31916. doi:10.1118/1.4865761.

Su S, Atwal P, Lobo J, Popescu IA (2015) Efficient, all-in-one, Monte Carlo simulations of transit EPID cine-mode dose distributions for patient-specific VMAT quality assurance, World Congress of Medical Physics and Biomedical Engineering, Toronto, Canada.

Sun B, Yaddanapudi S, Goddu SM, Mutic S (2015) A self-sufficient method for calibration of Varian electronic portal imaging device. *J. Phys. Conf. Ser.* 573: 12041. doi:10.1088/1742-6596/573/1/012041.

Teke T, Gill B, Duzenli C, Popescu IA (2011) A Monte Carlo model of the Varian IGRT couch top for RapidArc QA, *Phys. Med. Biol.* 56: N295–N305.

Teymurazyan A, Pang G (2012) Monte Carlo simulation of a novel water-equivalent electronic portal imaging device using plastic scintillating fibers. *Med. Phys.* 39: 1518. doi: 10.1118/1.3687163.

Van EA, Depuydt T, Huyskens DP (2004) The use of an aSi-based EPID for routine absolute dosimetric pre-treatment verification of dynamic IMRT fields. *Radiother. Oncol.* 71: 223–234. doi:10.1016/j.radonc.2004.02.018.

van Elmpt WJ, Nijsten SM, Schiffeleers RF, Dekker AL, Mijnheer BJ, Lambin P and Minken A W (2006) A Monte Carlo based three-dimensional dose reconstruction method derived from portal dose images, *Med. Phys.* 33: 2426–2434.

Vial P, Greer PB, Oliver L, Baldock C (2008) Initial evaluation of a commercial EPID modified to a novel direct-detection configuration for radiotherapy dosimetry. *Med. Phys.* 35: 4362–4374. doi:10.1118/1.2975156.

Wang S, Gardner JK, Gordon JJ, Li W, Clews L, Greer PB. et al. (2009) Monte Carlo-based adaptive EPID dose kernel accounting for different field size responses of imagers. *Med. Phys.* 36: 3582–3595. doi:10.1118/1.3158732.

Wowk B, Radcliffe T, Leszczynski KW, Shalev S, Rajapakshe R (1994) Optimization of metal/phosphor screens for on-line portal imaging. *Med. Phys.* 21: 227–235. doi:10.1118/1.597299.

Yeboah C, Pistorius S (2000) Monte Carlo studies of the exit photon spectra and dose to a metal/phosphor portal imaging screen. *Med. Phys.* 27: 330–339. doi:10.1118/1.598835.

PART 2

CLINICAL APPLICATIONS

4

Radiographic imaging

PHILIP VIAL

4.1 INTRODUCTION

Beam's eye view (BEV) radiographic imaging refers to the acquisition of planar X-ray projection images using the megavoltage linear accelerator (LINAC) treatment beam as the X-ray source and an X-ray imaging receptor to collect and display the image for the purpose of patient localization and treatment verification. As described in Chapter 1, radiographic imaging of one type or another has been used throughout the history of external beam radiation therapy. The importance of BEV imaging, otherwise known as *portal imaging*, for treatment accuracy improvement and error reduction was well known before digital imaging became widely available (Reinstein et al. 1987). The introduction of digital BEV imaging in place of film cassette systems vastly improved the imaging workflow and increased the frequency of imaging during the 1990s and 2000s (Herman et al. 2001b). More recently, the widespread availability of in-room kilovoltage (kV) imaging systems has in many cases replaced or at least reduced the role of BEV imaging. However, BEV imaging retains some unique characteristics that cannot be entirely replaced by alternative imaging systems. BEV imaging is the only direct means of verifying the position of the patient anatomy with respect to the treatment portal. BEV imaging verifies, in a single image, both the shape and location of the treatment portal with respect to the patient anatomy. In addition, there are potential applications for tracking moving targets and the quantitative verification of the delivered dose. Other image guidance methods can only indirectly verify patient location via registration of the imaging system to the LINAC isocenter, and they do not capture the delivered treatment field shape or location. Further, the information captured by BEV imaging exists whether we detect it or not, requiring only the addition of an image receptor and acquisition system to make use of this information that is otherwise lost. In this chapter, we describe the components and processes involved in BEV imaging systems as they are routinely used in the modern radiation therapy setting.

4.2 SYSTEM COMPONENTS

4.2.1 IMAGING SYSTEM

The BEV imaging system consists of (1) an X-ray source, the megavoltage (MV) LINAC; (2) an object, typically the patient lying on the treatment table, (3) an image receptor positioned behind the object, typically an amorphous silicon (aSi) digital flat-panel imager, commonly referred to as the electronic portal imaging device (EPID), (4) image acquisition electronics to govern the acquisition parameters, data readout, and data processing to form a digital image matrix; and (5) a display monitor and software for user interface. In addition, off-line review and image archives are managed within an oncology information system (OIS) and an image archive system (Shakeshaft et al. 2014).

Imaging detectors and their principles of operation were described in detail in Chapter 2. We shall describe some of the fundamental characteristics of BEV imaging by comparison to conventional diagnostic radiography as follows:

1. The radiation source: The LINAC operates in pulsed mode. Electron pulses pass through the accelerating waveguide with a maximum pulse repetition frequency

(PRF) of about 400 Hz, impinging on the target to create the clinical X-ray beam. Each pulse is a few microseconds long and delivers about 0.2 mGy (Beierholm et al. 2011). The instantaneous dose rates vary dramatically from very high during a beam pulse (around 70 Gy/s for a typically flattened beam) to almost zero in-between pulses. Individual portal image frames are read out row-by-row at a rate of about 3 to 10 frames/s. To avoid beam pulse artifacts, BEV image acquisition systems usually operate in a synchronized mode by connection to a LINAC pulse signal to pause readout during each beam pulse (Berger et al. 2006, Podesta et al. 2012, Mooslechner et al. 2013). Diagnostic X-ray sources are operated under simple continuous current conditions involving much smaller instantaneous intensities.

2. Energy spectrum: At MV energies, Compton interactions are the predominant process for creating contrast in anatomical structures seen in a BEV radiograph. Attenuation due to Compton interactions has a relatively weak relationship with different anatomical tissues compared to photoelectric processes that are the dominant process in diagnostic radiography (Dance et al. 2014). Therefore, the MV radiation beam exiting the patient contains poorer subject contrast than is the case in a diagnostic X-ray beam. Working in favor of BEV imaging is the fact that a LINAC beam is far more intense and the primary transmission through the patient is much higher than for a diagnostic beam, so the number of primary photons reaching the detector is much higher and the fraction of the total signal from patient scatter is less (Boyer et al. 1992, Jaffray et al. 1994). Unlike diagnostic X-ray systems, the LINAC beam spectrum is typically fixed to the available treatment beamlines and there is no flexibility for optimizing spectrums to specific imaging applications by energy selection or filtration.
3. Image receptor: The MV X-rays incident on the X-ray receptor are much less likely to be detected and contribute to image signal compared to diagnostic X-rays due to the poor quantum absorption efficiency (Jaffray et al. 1995). The detective quantum efficiency (DQE), which is a measure of how efficiently the detector transfers the signal-to-noise properties of the incident radiation beam to the output image relative to an ideal detector, is about 1% for a standard EPID, and about 40%–80% for standard diagnostic detectors (Granfors and Aufrichtig 2000, Herman et al. 2001b). Detector readout circuitry is sensitive to radiation damage, and MV radiation is more difficult to shield. To minimize risk of radiation damage the electronic components in EPIDs are positioned out of the direct beam and are protected by additional shielding. Whereas in diagnostic systems, the electronics can be adequately shielded when positioned under the detector within the exposed area. Antiscatter grids widely used in diagnostic imaging to minimize patient scatter are not effective at MV energies. There is potential for improving MV imaging performance with new detector designs, refer to Chapter 12 for more information on advanced image receptors for BEV imaging.

In summary, the poorer contrast properties of the incident MV beam and the reduced detection efficiency at high energies limits the image quality and requires higher doses to the patient for a useful anatomical image compared to a diagnostic radiography system. Figure 4.1 illustrates how some of these factors impact clinical image quality.

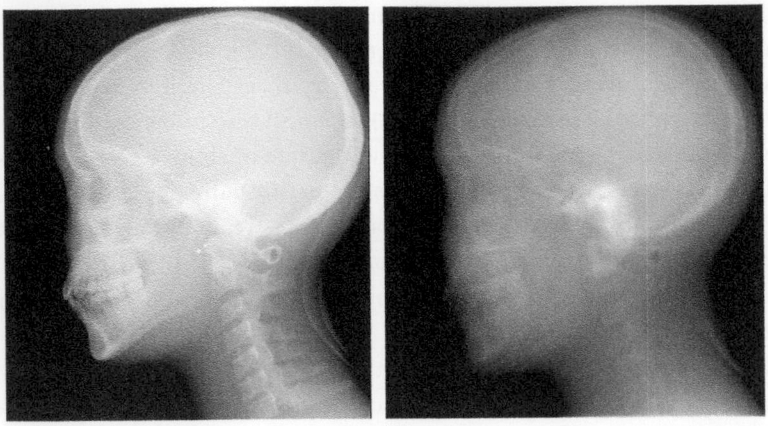

(a) (b)

Figure 4.1 Unfiltered radiographic images of the same head phantom acquired with a kV imaging system (a) and MV imaging system (b). The kV radiograph was acquired at 100 kV, 0.5 mAs. The MV radiograph was acquired at 6 MV with 3 monitor units. In this example, the MV radiograph delivers approximately 1.5–2 orders of magnitude of more doses to the patient (head phantom). Both images have been windowed to the equivalent range of gray-level information contained in each image.

4.2.2 Reference images

Image-guided radiation therapy (IGRT) requires the registration of an image acquired at treatment with a reference image. In modern CT-based radiation therapy, the reference image for BEV imaging is the digitally reconstructed radiograph (DRR), which is a computer-generated virtual radiograph. The DRR simulates a radiographic image by simulating the X-ray transport through the patient (CT) and incident onto the detector plane. DRRs may also contain additional treatment field information such as the location of the collimator central axis and field (portal) shape as defined by collimating jaws or multileaf collimators (MLCs). 3D contours of targets or anatomical structures can also be projected onto the 2D DRR. The DRR must be scaled to match the BEV image dimensions that is dependent on the imaging system geometry.

Several different models for generating DRRs have been developed (Siddon 1985, Sherouse et al. 1990, Galvin et al. 1995). They are all based on ray tracing from a point source through a CT dataset, voxel-by-voxel, and onto a 2D image plane behind the object. The intensity at each pixel of the DRR is modulated according to the attenuation properties of each voxel integrated along a rayline from the source to the pixel through the CT dataset. In the setting of BEV imaging, the aim of the DRR algorithm is to provide a clear and geometrically accurate reference image of those high-contrast anatomical features visible on the BEV image that are important for image registration. The relationship between CT number and attenuation can be manipulated to simulate different beam energies or enhance different contrast details on the DRR. The appearance of the DRR can also be manipulated by postprocessing filters and window level settings similar to any other radiograph. Filters are mainly useful for improving contrast in the measured BEV image. One common image filter is called adaptive histogram equalization (AHE) (Rosenman et al. 1993, Sherrier and Johnson 1987). An example of window leveling and AHE is shown in Figure 4.2. The AHE filter spreads out the information of interest across the histogram to make better use of the gray levels and thereby enhance contrast, at the cost of adding some artifact in sharp high-contrast regions and also increased noise. A variation of AHE called contrast limited AHE (CLAHE) was developed to reduce the enhancement of noise in regions of low contrast (Rosenman et al. 1993).

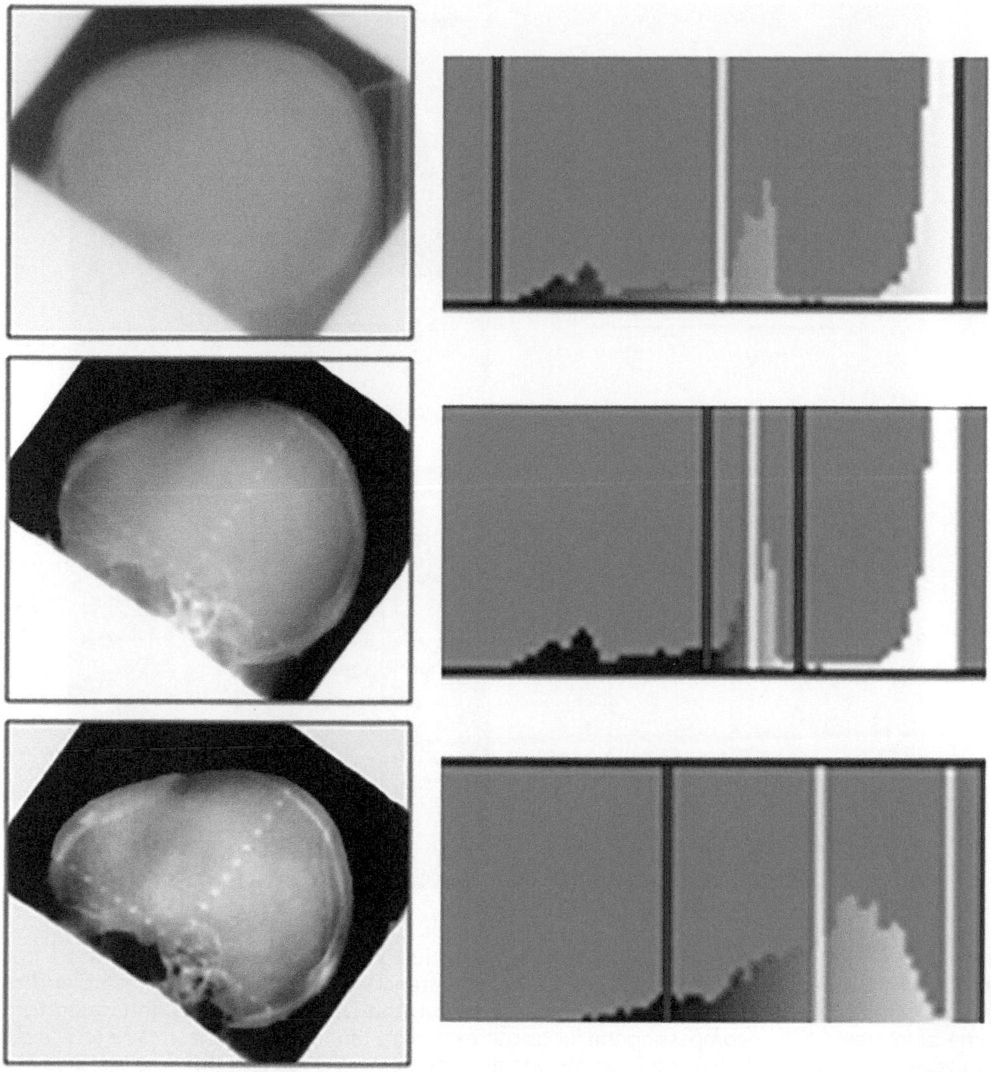

Figure 4.2 BEV images of a whole brain treatment with different windowing and filters applied. The top unfiltered image uses the full window range (red vertical lines on the gray level histogram). The middle unfiltered image has window levels narrowed to the useful gray level. The bottom image has been AHE filtered, as can be seen by the altered distribution of the gray-level histogram.

The optimal DRR settings may vary for different clinical applications, and settings should be developed for each application considering exactly how the DRR is to be used. For example, when using ribs and vertebra as the reference for patient setup, one may choose DRR settings that highlight these bony structures. If it were important to visualize the carina or a soft tissue tumor within the lung, then different DRR settings should be used to bring out the soft tissue contrast (Figure 4.3). For prostate treatments, one may prefer a DRR that highlights implanted gold seed fiducials within the prostate gland. For breast tangents, it may be the position of the lung and chest wall interface relative to the posterior field edge (demonstrating the amount of lung within the treatment portal). There are also DRR tools that can assist where there are metal prosthetics obscuring anatomy within the treatment portal (Lovelock et al. 2005).

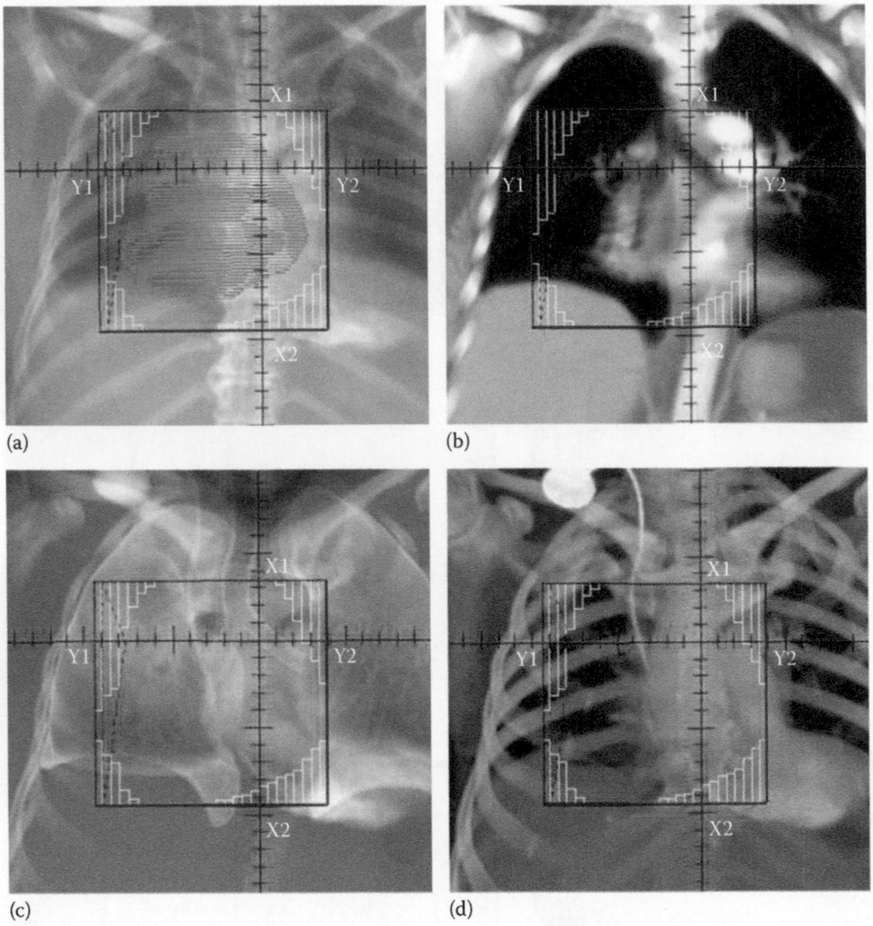

Figure 4.3 DRRs of an anterior beam treating right lung and mediastinum generated from a commercial treatment planning system: Image (a) is a standard DRR using lung presets and ramp filter. The target volume contours are also projected onto the DRR. (b) DRR was generated using a volume of interest (VOI) encompassing the lungs but excluding the ribs and spine. Image (c) is a digital composite radiograph (DCR) using lung presets and the same VOI as (b) DRR. Image (d) is a DCR with ribs preset.

All modern radiotherapy treatment planning systems (RTPS) contain tools for generating and manipulating DRRs. Some RTPS's include tools to define volumes of interest for the generation of DRRs that exclude parts of the anatomy that are not of interest or obscure features of interest.

Various annotations are included or can be added to DRRs in the RTPS. Central axis crosshairs and MLC or collimating jaw outlines are often projected onto the DRR to show the prescribed portal image position and shape. Annotated scales can be useful for confirming the image scaling. Tools to draw lines or points onto the DRR are useful as additional guides for image registration.

The DRR must be in a digital form compatible with other software in the radiotherapy chain. Typically, DRRs and all the information added to them can be exported from the RTPS as DICOM-RT files and can be imported to any OIS or IGRT software applications. The correct generation and transfer of DRRs across different computer systems in the radiotherapy workflow must be verified when commissioning or testing software components within that workflow (IAEA 2004).

4.2.3 Image registration

Image registration is the determination of a common coordinate system in which images can be compared (Van Herk 2000). A key link in the BEV image guidance chain is the registration of the portal image to the DRR. Errors in this process may be propagated directly to treatment, and importantly any system errors in the registration process may be propagated through all treatments of all patients. It is therefore important to understand the principles of operation and to consider the various factors that contribute to image registration accuracy.

There are many methods for registering a portal image to a DRR. Different commercial software systems have different configurable settings available and users need to investigate the details of their own systems or combination of systems. The position of the BEV image with respect to the isocenter may be determined by a physical graticule in the beam or a software-based calibration such as a fixed pixel of the detector (typically the centre pixel). Figure 4.4 shows an example of a BEV-to-DRR image registration. In this example, the isocenter was manually assigned to each image according to the physical graticule in the BEV image and the digital graticule in the DRR. Registration was performed manually to bones aided by contours and the system reports offsets between the reference (DRR) and measured (BEV image) isocenter points.

The registration process may be manual or automatic, using a number of available methods. Manual or automatic registration can be based on contours, points, image gray levels, isocenter, or treatment field edges (Balter et al. 1992, Bijhold et al. 1992, Moseley and Munro 1994, Fritsch et al. 1995, Leszczynski et al. 1995, Hristov and Fallone 1996, Krueder et al. 1998, Bastida-Jumilla et al. 2011). The registration process can be configured for individual images or a linked pair of stereoscopic images, such as an anterior–posterior (AP) and lateral orthogonal image pair. Most modern IGRT systems allow the kV and MV imaging systems to combine such that a stereoscopic pair of images can be generated from any combination of kV and MV images (kV/kV, MV/MV, and kV/MV). Since the kV and MV imaging systems are fixed at orthogonal angles to each other (Figure 4.5), there is no gantry rotation required to create an image pair, thus saving the time it takes to rotate the gantry between images (Mutanga et al. 2008).

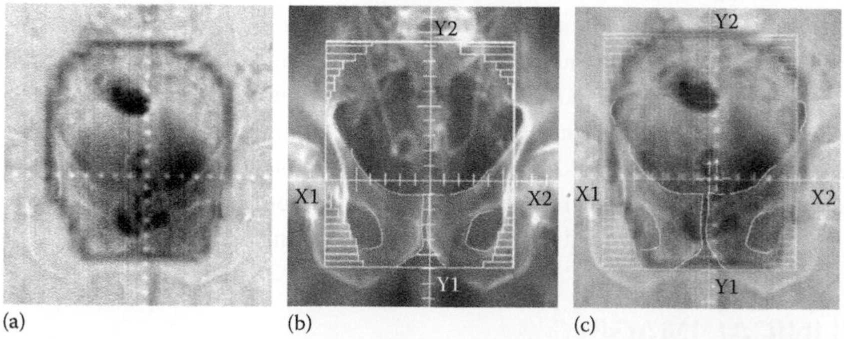

Figure 4.4 The AP pelvis image (a) is a filtered double exposure BEV image with a physical graticule used to assign isocenter in the image registration software. Image (b) is a DRR from the treatment planning system with a digital graticule; MLC shielding is displayed and bony anatomy curves drawn on. Image (c) is an overlay of both images after manual registration to bony anatomy. The two isocenters are indicated by small crosses. In this example, the registration software reported 0.5 and 0.1 cm offsets to the right and superior, respectively.

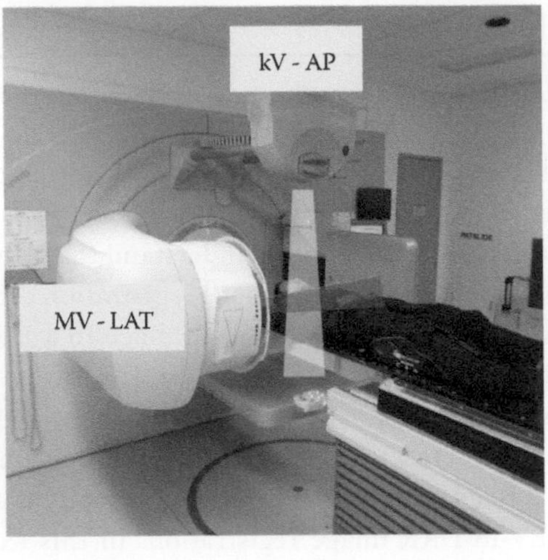

Figure 4.5 Stereoscopic image pairs may be acquired using MV and kV imaging systems without gantry rotation between images.

Any misalignment of the physical or digital graticule with the true X-ray isocenter will propagate through to treatment setup errors. Such system calibration errors are not recognized by the image registration software and are therefore not necessarily visible to the LINAC operator or any subsequent off-line image review. The potentially serious consequences of a systematic misalignment in treatment have prompted recommendations for daily quality assurance (QA) of the isocenter alignment of IGRT systems (Bissonnette et al. 2012).

The result of image registration is a set of treatment table shifts for the transverse, longitudinal, and vertical directions. If a six-degrees-of-freedom table is used then registration may also be correct for rotations and the table offsets include pitch, roll, and yaw values. Table shifts are limited to rigid motions, and therefore only rigid registration methods are used in IGRT. Poor patient setup can cause deformations due to different angles between or within bony structures such as the spinal vertebra, pelvis, skull, and mandible (van Kranen et al. 2009). These errors cannot be completely corrected for with rigid offsets and re-setup of the patient may be required at the discretion of the treating team. Ideally, the table shifts are transferred and executed remotely via computer network to the LINAC control system without the need for manual data entry or to go into the treatment bunker to manually shift the table. Remote table shift and the time it saves in patient setup has been a significant factor in the rapid uptake of daily online IGRT. As with any new technology, remote robotic table controls and automated IGRT procedures bring new risks to patient safety that need to be carefully considered and mitigated where possible (ICRP 2009). For example, the potential for serious collisions with the patient and/or equipment should be reconsidered with new robotic table systems. Regular revision of safety and QA programs are required to capture the evolving clinical IGRT processes.

4.3 CLINICAL IMAGING

There have been several excellent reports on the clinical use of BEV imaging in radiotherapy (Munro 1999, Hurkmans et al. 2001, Herman et al. 2001a, 2005, Radiologists TRCo 2008). As mentioned in this chapter's introduction, IGRT technology has changed dramatically since even these relatively recent reviews were published. In many clinical applications, the role of BEV imaging

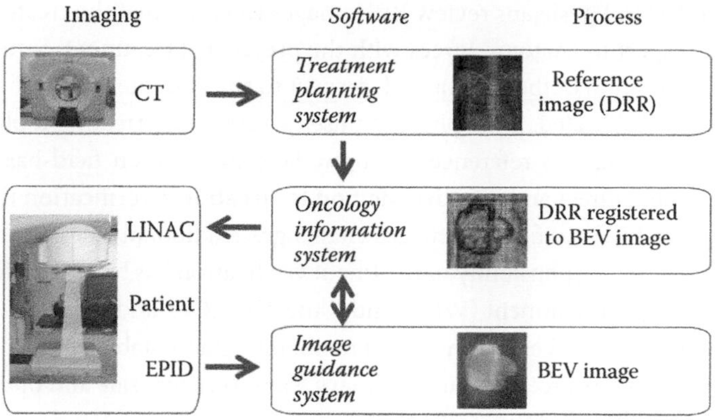

Figure 4.6 Simple schematic diagram of imaging hardware, software, and processes involved in BEV imaging in a typical radiotherapy workflow. Arrows indicate the data flow. Systems may be configured such that the registration of DRR and portal images can take place in the oncology information system or in in the image guidance system. Methods for assigning isocenter, registering images, and reporting offsets may vary between software systems and are configurable within software systems. Careful system verification is required to confirm consistent transfer of data and that processes are robust to all clinical scenarios, such as different patient orientations.

has been replaced, in part if not in full, by in-room kV imaging systems (Fang-Fang et al. 2009, Potters et al. 2010, Korreman et al. 2010). This is partly due to the superior image quality that can be achieved with fewer doses using kV IGRT systems, and partly due to the fact that treatment portals themselves have been largely replaced by dynamically time-variant intensity-modulated beams that greatly reduce the utility of BEV imaging (Figure 4.6). Nevertheless, recent surveys of IGRT practices across different continents indicate that BEV imaging continues to play an important clinical role (Mayles 2010, Simpson et al. 2010, Bridge et al. 2015).

BEV imaging has its origins in the era of field-based treatment planning. As treatments became more conformal, the treatment fields became smaller and more irregular in shape, making them generally less useful for portal imaging. One way to address this difficulty is to use a double exposure method where the portal image is superimposed onto a larger image that captures the surrounding anatomy, as shown in Figure 4.4 (Hatherly et al. 2001). The additional imaging dose to surrounding anatomy from the open-field portion limits the frequency of double exposure imaging. Another limitation is that the choice of treatment beam angles may not be suited for stereoscopic imaging. This problem can be avoided by the use of orthogonal imaging fields, typically an AP and lateral field pair using fields large enough to visualize the bony anatomy around the treatment area. These image-only fields (sometimes referred to as setup fields) can be used with MV imaging but they became more routine with the introduction of kV in-room imaging that adds significantly less dose than equivalent MV setup fields (Walter et al. 2007, Ding and Munro 2013). Image-only field pairs were also somewhat forced onto radiotherapy by the inability to acquire BEV images with intensity-modulated radiation therapy (IMRT) and volumetric modulated arc therapy (VMAT). Today it is a routine clinical practice to image daily with kV field pairs or cone-beam computed tomography (CBCT) prior to treatment.

Although kV imaging and modulated delivery techniques bring obvious benefits to radiotherapy practice, in some clinical scenarios it could be argued that BEV imaging is more appropriate. Imaging fields, whether kV planar, MV planar, or CBCT, can be used for pretreatment setup corrections as described earlier, but only BEV imaging provides a verification image of the actual delivered beam. Traditionally, the physician's review of this verification image was a key aspect of radiotherapy

QA (Reinstein et al. 1987). Physicians review BEV images to verify that the treated area and associated shielding with respect to anatomy agrees with the prescribed treatment. Setup fields or CBCTs can only confirm the position of the patient with respect to the isocenter position, and that assumes the system is accurately calibrated. If the physician uses a field-based treatment planning technique, particularly where shielding was referenced to bony landmarks, then field-based imaging (BEV imaging) provides a more direct and intuitive method for treatment verification than setup fields or CBCTs. The absence of BEV verification images challenges traditional roles and responsibilities of radiotherapy staff, since the responsibility of treatment verification has been transferred to the radiation therapists at the time of treatment (White and Kane 2007, Cox and Jimenez 2009). It also relies critically on the medical physicist establishing and maintaining acceptable performance of IGRT systems and an appropriately resourced multidisciplinary team to ensure safe and optimal procedures is established and followed (Fontenot et al. 2014).

As a general rule, the treatments that are best suited to BEV imaging are those from BEV (field)-based treatment plans. Emerging roles for advanced forms of BEV imaging are discussed in Part 3 (Chapters 8–12) of this book. The following sections describe some of the conventional BEV imaging procedures that are still widely used.

4.3.1 Breast and chest wall radiotherapy

Breast and chest wall cancer patients comprise a large group of radiotherapy patients who still often receive conventional field-based treatments (or modern variants thereof). Recent surveys of breast radiotherapy practice report that forward planned tangential beams are still the predominant technique (Dundas et al. 2015). Although more advanced imaging techniques may provide more information and superior image quality, they will also add dose to organs-at-risk (OAR) in this relatively young patient cohort and have not been demonstrated to provide a net gain in benefit versus risk in this particular scenario.

BEV imaging has a long-standing role in breast radiotherapy (van Tienhoven et al. 1991, Hurkmans et al. 2001, Alía et al. 2005). Figure 4.7 shows a BEV image used for chest wall radiotherapy. Apart from conventional localization and interfraction setup error studies, BEV imaging has also been used for verifying heart in field in left-sided breast radiotherapy (Magee et al. 1997), and CINE-BEV imaging has been used to evaluate intrafraction breast motion (Kron et al. 2004). BEV imaging has also been useful for verification of the increasingly utilized deep inspiration breath hold (DIBH) technique used to reduce heart dose in left-sided breast radiotherapy (Borst et al. 2010, Jensen et al. 2014, Lutz et al. 2015).

4.3.2 Simple 3-Dimensional Conformal Radiation Therapy (3DCRT) pelvis radiotherapy

The pelvic bony anatomy is relatively well visualized in BEV imaging. For some simple treatment techniques, such as three- or four-field bladder or rectum treatments, BEV imaging provides adequate image guidance (Figure 4.8).

Other subgroups of pelvis patients where MV BEV imaging may be of benefit include the following: (1) where the presence of metal prosthetic hips obscure anatomy in kV imaging, (2) where the thickness of an obese patient reduces penetration of kV signal to the extent that imaging requires excessive mAs, or (3) where kV-MV paired imaging is implemented.

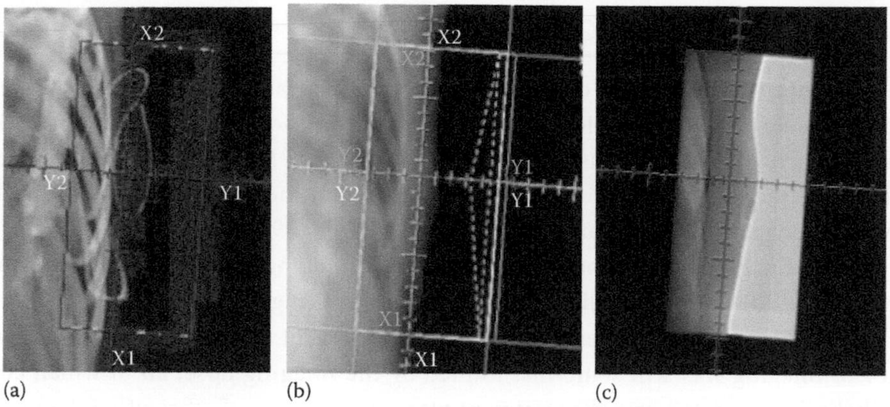

Figure 4.7 Chest wall medial tangent field: (a) Reference images (DCR), (b) DRR, and (c) EPID BEV image. All images include a digital graticule indicating collimator central axis. The DCR highlights high-density structures such as ribs and wire markers on skin. The DRR shows more soft tissue detail including the amount of lung in the field. The BEV image provides verification of the amount of lung in field and the superior/inferior field borders. Reference images also show the jaw orientations and the presence and direction of a wedge (triangle dashed line on anterior field edge), which can be checked against the machine settings at the time of treatment.

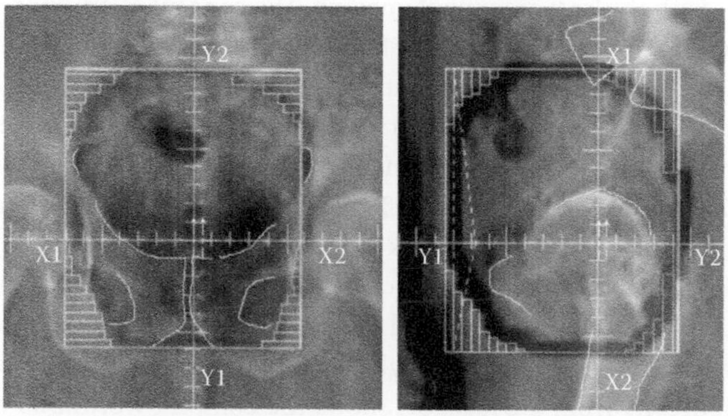

Figure 4.8 Stereoscopic image pair registration of orthogonal EPIDs for a palliative bladder treatment. DRR and double exposure BEV image overlays are viewed with even weightings. The DRR contains MLC shielding and manually drawn contours of pelvic bony anatomy used as the reference for registration.

4.3.3 Field-based palliative radiotherapy

Palliative intent radiotherapy is a significant component of radiotherapy workload, with optimal utilization rates estimated to be 14% of all newly diagnosed cancers (Jacob et al. 2010). The relatively low doses indicated for effective palliation do not generally require the same level of treatment conformality as curative regimens, and this is reflected in IGRT protocols with less frequent imaging (Rybovic et al. 2008, Simpson et al. 2010). Palliative radiotherapy of vertebra, extremities, and brain, for example, often uses simple field-based treatment planning techniques with well-defined bony anatomy landmarks, making them suited to BEV imaging for localization and verification.

4.3.4 Off-line protocols

Off-line imaging protocols are a correction strategy aimed at reducing systematic errors in patient setup using prior information to estimate corrections for future treatment fractions (de Boer and Heijmen 2001, Hurkmans et al. 2001). In the days of film, the procedure for imaging, processing, and measuring setup errors was prohibitively time-consuming and typically limited to weekly imaging with off-line review (Kutcher et al. 1994). The advent of EPIDs greatly streamlined the imaging process making more frequent imaging feasible (Hurkmans et al. 2001, Herman et al. 2001b). The availability of more frequent imaging and the increasingly conformal treatment techniques increased demand for improved patient setup correction strategies. The frequency of imaging was still somewhat limited by the imaging dose, all of which could not necessarily be incorporated into the treatment plan, and the lack of efficiently integrated IGRT tools such as remote robotic table shifts. The widespread availability of both in-room kV imaging and integrated IGRT solutions has made daily pretreatment correction strategies feasible. Such *online* protocols have proven to be more effective and efficient, to the degree that off-line imaging protocols have become increasingly difficult to justify (Valicenti et al. 1994; Kupelian et al. 2008, van der Vight et al. 2009).

The principles of off-line imaging protocols are instructive for understanding the nature of setup uncertainties in radiation therapy. The aim of an off-line protocol is usually to quantify systematic setup errors with the least amount of imaging. The concept of *off-line* refers to the fact that corrections are determined between treatment fractions using retrospective image data, and often applied on subsequent treatment fractions with no imaging. Typically, there is a procedure for detecting any change in the systematic setup error throughout the course of treatment. The literature is rich in descriptions and evaluations of many off-line correction protocols (Amer et al. 2001, Bortfeld et al. 2002, de Boer and Heijmen 2002, van Lin et al. 2003, Herman 2005, Litzenberg et al. 2005, van der Heide et al. 2007, van Herk 2007, Greer et al. 2008, Radiologists TRCo 2008). Limitations of off-line correction strategies include the relatively large amount of missing information, no correction for random errors, and the somewhat cumbersome implementation. The importance of these factors became increasingly apparent when large studies with daily imaging became feasible. Figure 4.9 from a UK report shows the complex workflows of a typical off-line imaging protocol.

4.3.5 Online protocols

As discussed earlier, online protocols have become routine practice in recent years. In principle, the online protocol is much less complicated than off-line protocols because it typically involves the same (or very similar) imaging process being conducted on every treatment fraction regardless of site or technique. For each treatment fraction, the patient is setup on the treatment table and is imaged, registration of X-ray image and reference image performed to obtain offsets, and the treatment table is shifted accordingly, prior to commencing treatment. Online corrections minimize both the random and systematic components of patient setup error each day. Online protocols may stipulate that no shift is necessary if the offset is below a specified *no-action threshold*, or it may be decided to shift on any nonzero offset (zero action threshold). Routine daily online corrections may not be feasible if IGRT is based on BEV imaging alone because BEV imaging typically requires some additional imaging dose. The trade-off between imaging dose and setup accuracy needs to be considered on a site-by-site basis.

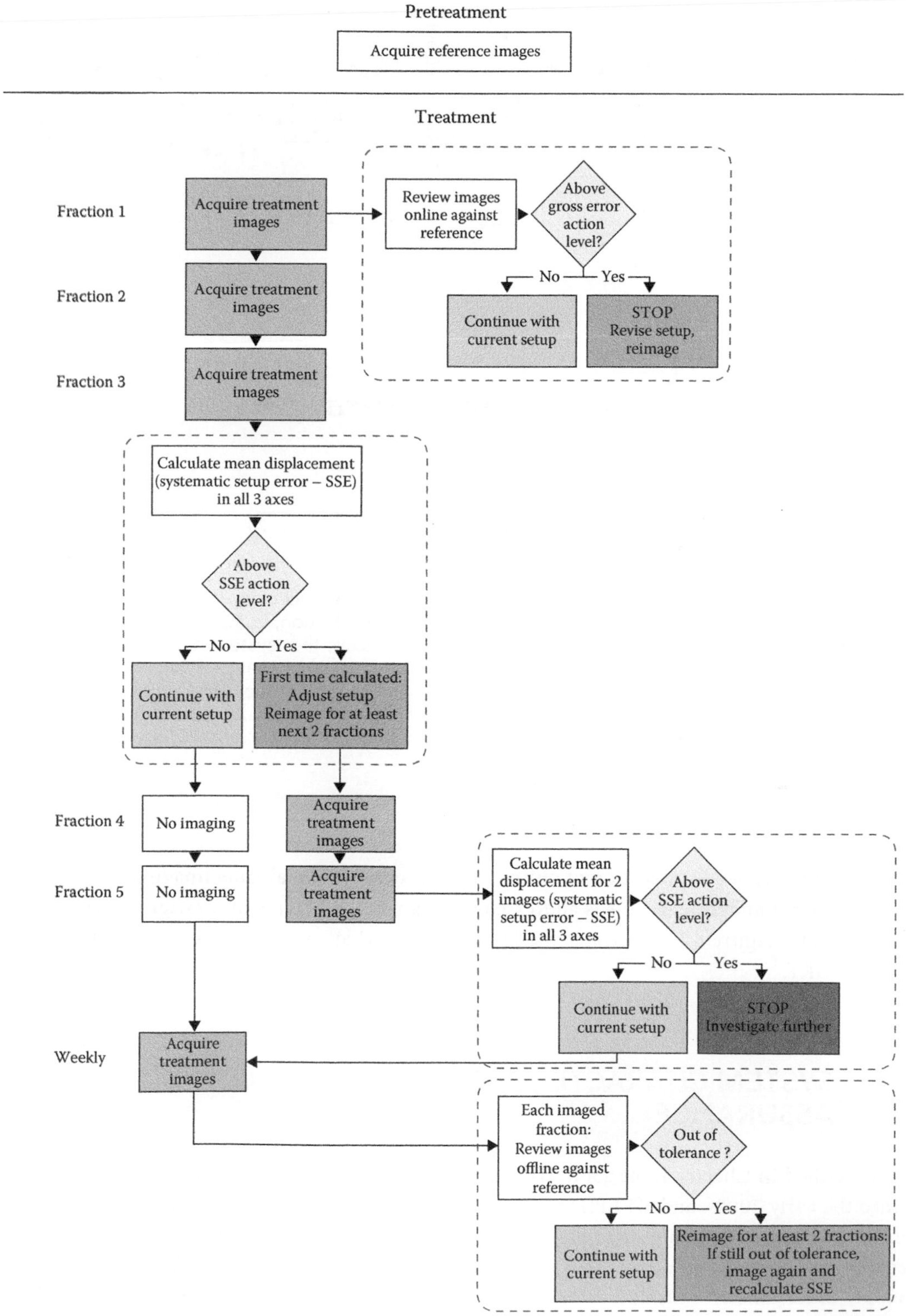

Figure 4.9 An example of an implementation of an off-line correction protocol. (Reproduced with permission from The Royal College of Radiologists. *On Target: Ensuring Geometric Accuracy in Radiotherapy*. London: The Royal College of Radiologists, 2008.)

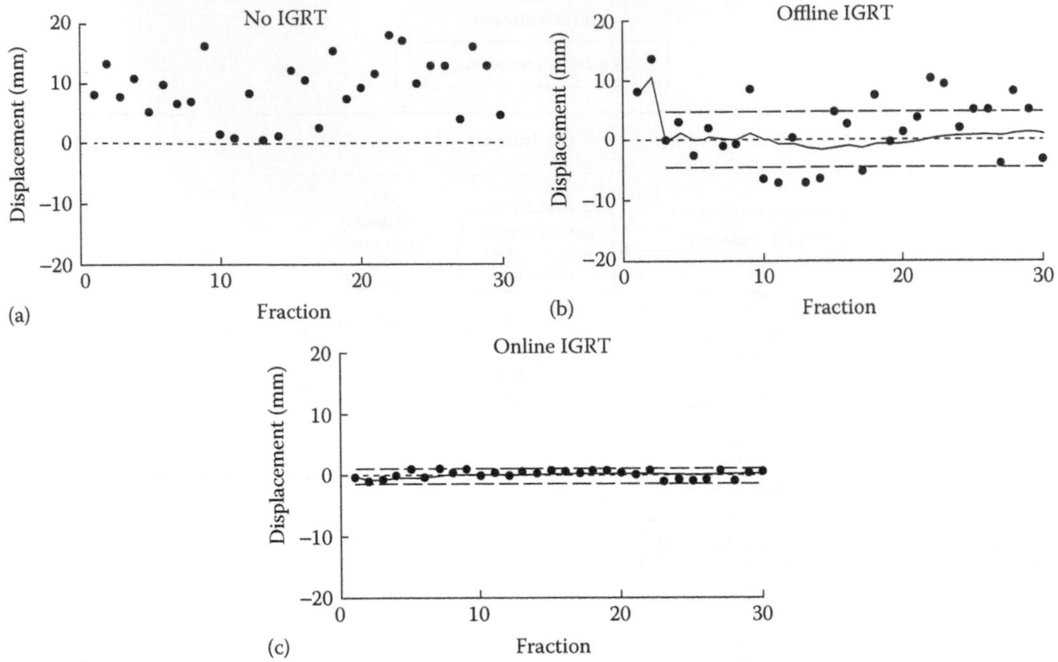

Figure 4.10 Plots of patient setup errors each day of treatment (circles). Image (a) is uncorrected setup data showing large systematic and random error distribution, Image (b) shows setup errors following an off-line correction strategy. The solid line shows that systematic error was reduced from day 3 after correction but random error distribution remain unchanged, and Image (c) shows setup errors following an online correction strategy. The systematic and random errors were reduced from day 1. (Reprinted from *Seminars in Radiation Oncology*, 22, Bujold A. et al., Image-guided radiotherapy: Has it influenced patient outcomes, 50–61., Copyright (2012), with permission from Elsevier.)

A more advanced application of online image protocols is real-time imaging for intrafraction motion management. This is addressed in Chapter 9. In terms of managing interfraction setup errors, Figure 4.10 illustrates the improvements that can be achieved with off-line and online IGRT.

4.4 SYSTEM CHARACTERIZATION AND QUALITY ASSURANCE

As described in Chapter 1, the aSi EPID has been a standard accessory on conventional LINACs since the early 2000s, and aSi EPID technology remains the ubiquitous gold standard in clinical EPIDs (Kirkby and Glendinning 2006). There are two dominant providers of aSi EPID panels in the current radiotherapy market. Varian supplies EPID panels for Varian LINACS and Perkin–Elmer supply EPID panels for Elekta and Siemens LINACs. The specifications of current EPID panels are shown in Table 4.1. Retrofitted camera-based imaging systems are available and in limited use but are not included in this chapter (Odero and Shimm 2009).

Table 4.1 Detector specifications for commonly used aSi EPIDs

Vendor	Varian	Varian	Varian	Perkin Elmer	Perkin Elmer
Detector Model	as500	as1000	as1200	XRD 1640 AL/AG	XRD 1642 AP
Detector type	aSi AMFPI[a], Copper + $Gd_2O_2S:Tb$ scintillator	aSi AMFPI, Copper + $Gd_2O_2S:Tb$ scintillator	aSi AMFPI, Copper + $Gd_2O_2S:Tb$ scintillator	aSi AMFPI, Copper + $Gd_2O_2S:Tb$ scintillator	aSi AMFPI, Copper + $Gd_2O_2S:Tb$ scintillator
Detector dimensions	40 × 30 cm	40 × 30 cm	43 × 43 cm	41 × 41 cm	41 × 41 cm
Pixel dimensions	512 × 384 @ 0.78 × 0.78 mm	1024 × 768 @ 0.392 × 0.392 mm	1280 × 1280 @ 0.336 × 0.336 mm	1024 × 1024@ 0.4 × 0.4 mm	1024 × 1024@ 0.4 × 0.4 mm
Digitization	14 bit	14 bit	16 bit	16 bit	16 bit
Source-to-detector distance	Variable: 120–180 cm	Variable: 120–180 cm	Variable: 120–180 cm	160 cm	160 cm

[a] Active Matrix Flat-Panel Imager

4.4.1 Residual errors

The overall accuracy of treatment depends on many clinical, physical, and process-related components. Important components of treatment accuracy not addressed by BEV imaging systems include target delineation and the choice of target surrogates used for image registration (Bujold et al. 2012). As a physical IGRT system, the end-to-end accuracy of BEV imaging is in-principle the same as a modern in-room kV imaging system. BEV imaging may be slightly more accurate because it is not reliant on the coregistration of the kV imaging and treatment coordinate systems. However, the poorer MV image quality is expected to increase variability in registration accuracy compared with kV systems (Pisani et al. 2000). The overall accuracy of kV (2D or 3D) versus MV imaging systems is reliant on many variables including the contrast of the reference anatomy, characteristics of 3D deformations, clinical IGRT protocols, and other clinical site and patient-specific considerations (Mageras and Mechalakos 2007, Moseley et al. 2007, Topolnjak et al. 2008, White et al. 2009, Hawkins et al. 2011, Gill et al. 2012).

4.4.1.1 Geometric accuracy

The importance of geometric accuracy for BEV imaging depends on the clinical application and the software implementation for IGRT. Most modern applications of BEV imaging will be highly dependent on an accurately calibrated position of the flat panel in the plane perpendicular to the beam axis, and on its distance from the source. As discussed in Section 4.2.3, the method of image registration is also an important consideration and can be divided broadly into two categories:

1. Pixel-based isocenter placement: This method is is commonly used in modern IGRT systems. A pixel location on the flat panel is the surrogate isocenter. Typically, this is the central pixel (e.g., pixel 512,512 of a 1024 × 1024 pixel detector). Corrections for gravitational effects on image panel versus isocenter alignment with gantry rotation may be added to this procedure by built-in look-up tables. Any error in the calibration procedure or any mechanical variations over time may transfer through the IGRT system as a systematic localization error.

2. Measurement-based positioning: A physical reference for isocenter position is included in the BEV image. This may be an in-air graticule (Figure 4.4) for placement of the isocenter, or simply using the field edge. These methods could involve manual or automatic registration of the physical reference from the BEV image. This method is independent of the calibration of the flat-panel position in the plane perpendicular to the beam central axis, but is dependent on the accurate alignment of the physical reference (i.e., in-air graticule or collimators defining field edges) and the ability of the software or user to locate that reference.

The source-detector distance (SDD) also requires accurate calibration. The SDD accuracy will affect the scaling of the measured image, the importance of which increases with distance from the central axis with divergence from a point source.

Modern IGRT software systems often contain multiple user-defined preferences that determine the registration implementation under clinical conditions. Users must understand their local implementation and design their QA procedures to capture local clinical practice. End-to-end IGRT tests that follow a specific clinical process are recommended to ensure system accuracy.

Recommended tolerance values for the mechanical accuracy of EPIDs have been quoted as ≤2 mm for nonstereotactic and ≤1 mm for stereotactic applications (Klein et al. 2009). Coincidence of kV and MV isocenters for modern IGRT systems should be within 1 mm (Bissonnette et al. 2012). Maintaining tolerances within 2 mm at arbitrary gantry angles requires additional corrections given EPIDs typically sag 1–2 mm over the range of gantry angles (Grattan and McGarry 2010, Rowshanfarzad et al. 2012, 2015). The EPID's mechanical accuracy is only one component of residual errors in IGRT and does not include, for example, the uncertainties in image registration and the mechanics of treatment table shifts.

4.4.2 Image Quality

It is important to maximize image quality in order to minimize the uncertainty in image registration while avoiding unnecessary imaging dose. The image quality of each new imaging panel is tested against vendor specifications during acceptance testing. Baselines for image quality are established during commissioning and are used to monitor performance over time as part of a QA program, typically conducted with one of the commercially available phantoms designed for MV imaging (Rajapakshe et al. 1996, Herman et al. 2001b, Kirkby and Glendinning 2006, Klein et al. 2009, Das et al. 2011, Pesznyák et al. 2011, Stanley et al. 2015).

4.4.2.1 Spatial Resolution

Current portal imaging detectors have pixel sizes of approximately 0.4 mm, which when projected back to the patient equates to a still smaller effective resolution depending on SDD (e.g., about 0.25 mm for 160 cm SDD). Such a high spatial resolution is unnecessary for most radiotherapy imaging applications; in fact, the Elekta kV imaging system downsamples kV planar images to 512 × 512 without a noticeable compromise in quality. Reported spatial resolutions from experimental studies using the modulation transfer function (MTF) are in the range of f50 = 0.21–0.6 lp/mm, noting the sensitivity of MTF to measurement technique. Measuring

MTF from line or edge spread functions in the MV setting is complicated and requires specialized equipment to sufficiently shield the beam (Cremers et al. 2004, Sawant et al. 2007, Son et al. 2014, Deshpande et al. 2015). For clinical purposes, a simpler approach using a commercial image quality phantom is sufficient.

4.4.2.2 CONTRAST-TO-NOISE RATIO

Contrast-to-noise ratio (CNR) provides a measure of an imaging system's low-contrast resolution, and hence the system's ability to distinguish soft tissue features on a radiograph. Relatively poor CNR is a limitation of BEV imaging due to the reasons described in Section 4.2.1, and is a key advantage of kV in-room imaging systems. BEV radiographic imaging is used primarily for registering bony anatomy, implanted fiducial markers, or lung and bronchial boundaries where contrast is sufficiently high to overcome the CNR limitations. Commercial image quality phantoms provide convenient tools for measuring CNR for QA and detector comparisons (Menon and Sloboda 2004, Das et al. 2011, Blake et al. 2013).

4.4.2.3 IMAGE ARTIFACTS

Artifacts in digital radiography are relatively straightforward. Problems with faulty electronics are generally obvious in the image. Figure 4.11 gives some examples of clinical images with faulty electronics. Sometimes these types of problems are intermittent and sometimes they can be removed with panel calibration (Huber et al. 2013), but they usually indicate that a component of the panel is failing. Unfortunately, the panels are not designed to be readily repaired by replacing components and typically, the entire panel must be replaced.

Pixel sensitivities may change over the lifespan of the detector due to radiation effects on the detector array or associated electronics (Winkler and Georg 2006, Huber et al. 2013). The imager-calibration process typically requires (1) an offset (dark field) correction that removes the temperature-dependent background image signal and (2) a gain (flood field) correction that

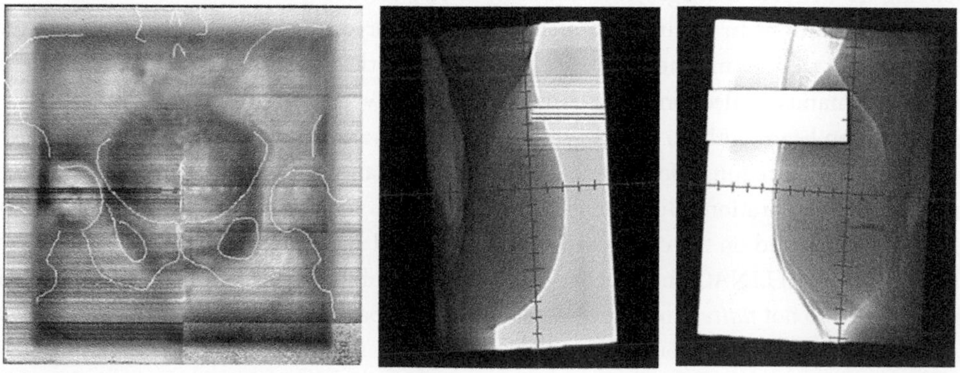

(a) (b) (c)

Figure 4.11 Artifacts indicating faulty imaging panels. Images (a) and (b) show streaking artifacts corresponding to readout lines. Image (c) shows missing data from one readout group of the panel.

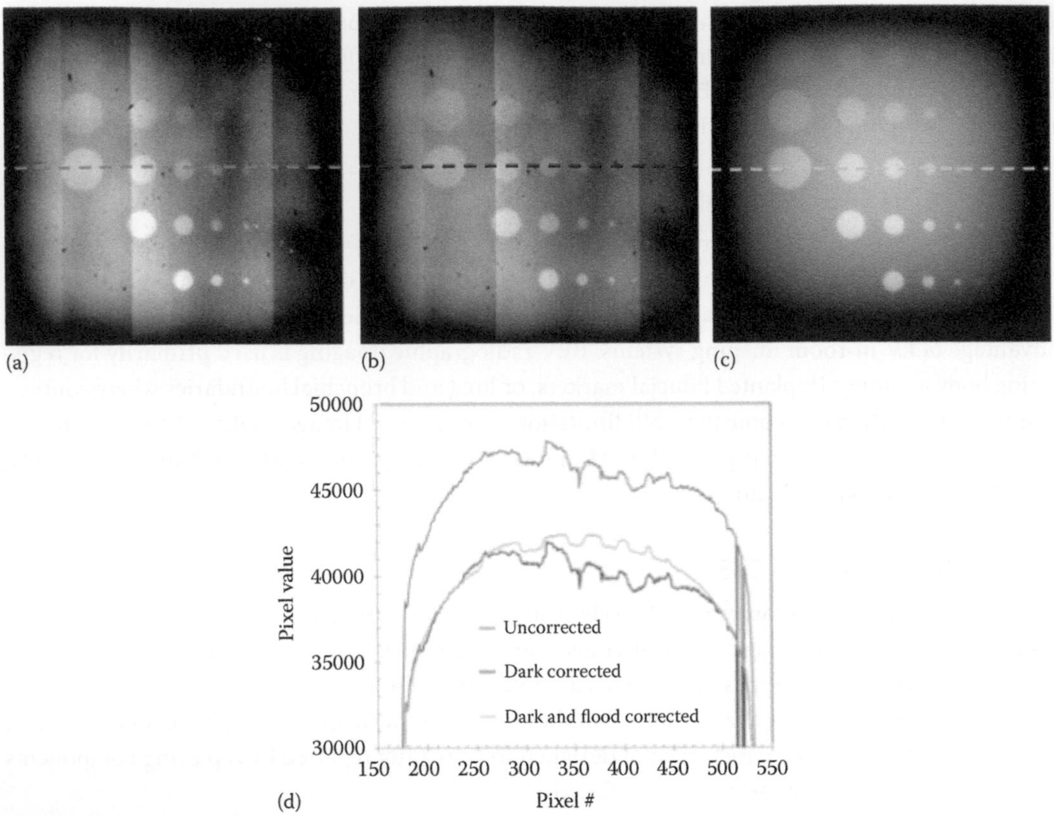

Figure 4.12 EPID images of a Las Vegas phantom: (a) raw image, (b) offset corrected image, (c) offset and gain corrected image, and (d) line profiles across the phantom showing the various stages of correction. Images a, b, and d show the discontinuities in pixel sensitivity corresponding to different readout groups associated with their separate readout electronics.

removes the variations in pixel sensitivities. A third correction to remove damaged pixels is also applied. Figure 4.12 shows images in various corrected states.

There are variations in the implementation of these correction strategies for different vendor panels. Offset corrections may be updated automatically by the system at a configurable interval, or it may require manual collection. Gain corrections are derived from a flood exposure of the entire panel to an open beam with no object in the beam. Calibration files are required for each frame rate, and may also be required for different energies, dose rates, and different panel distances with a multilevel gain calibration (Huber et al. 2013). Since gain file corrections are typically acquired with the panel centered on the collimator central axis, and since pixel response is sensitive to the spectral variations in LINAC photon beams as a function of distance from the central axis (Greer 2007), gain files do not *flatten* images optimally when the panel is moved from its calibrated position, especially for large crossplane or in-plane shifts.

Under certain conditions, LINAC pulses introduce synchronization artifacts in BEV images. This is not normally a problem for conventional BEV imaging where multiple frames are averaged, but tends to become more prominent with fewer frames and less dose per image. Synchronization effects become problematic in imaging applications involving individual frames, such as free-running CINE imaging, particularly at fast frame rates or low PRF (Wertz et al. 2010, Mooslechner et al. 2013). Clinical BEV imaging systems usually operate with an external trigger signal at each LINAC gun pulse to synchronize readout schemes minimizing interference between beam pulses and row readout.

4.4.3 Image dose

It is well established that MV imaging requires one-to-two orders of magnitude of more dose than equivalent kV imaging (Murphy et al. 2007, Walter et al. 2007, Ding and Munro 2013). If BEV imaging is strictly defined as imaging with the treatment beam, then BEV imaging adds no *imaging dose*. In practice, however, some common clinical applications of BEV imaging do add imaging dose, such as the commonly used double exposure method. The double exposure method involves irradiating anatomy outside the treated volume by opening the collimator jaws by some predefined distance. The *open* portion of the double exposure image typically requires between 1 and 3 monitor units (MU), equal to a few cGy, to a relatively large volume outside the treated volume (Kudchadker et al. 2004). Modulated treatments require non-BEV imaging techniques such as setup fields or CBCT. Since kV imaging provides significantly better image quality at significantly less imaging dose, where kV imaging is available it should be used in preference to MV imaging for setup fields. Regardless of modality, all radiographic imaging protocols should be optimized and justified based on the benefit versus risk to the patient in the radiotherapy context. Particular care is required for children and young patients (Olch et al. 2007).

Another characteristic of MV imaging is that, unlike kV imaging, the dose can be included in the treatment plan on any commercial treatment planning system. This makes it possible to incorporate imaging fields into the treatment plan and therefore into the prescribed treatment dose. In practice, this may still be limited by unpredictable variables in the frequency of imaging, especially for off-line imaging protocols. In principle at least, an accurate record of imaging dose could be determined retrospectively in the treatment planning system (TPS).

4.4.4 Quality assurance

All of the mechanical and image quality parameters described earlier should be regularly monitored as part of a routine QA program. Specific tests, frequencies, and tolerances can be based on published guidance documents (Herman et al. 2001b, Klein et al. 2009). Additional tests not described earlier include the correct function of collision interlocks and dry-run system tests that capture the end-to-end clinical process. The consistency of image data (and its transfer between systems) under realistic clinical workflows needs to be captured in a QA program, including verification of different patient orientations, beam directions, collimator, and treatment table rotations.

4.4.5 Field-size limitations

A common practical consideration with BEV imaging is that some treatment portals do not fit within the boundaries of the EPID's detection area. Sometimes the field size will fit but the asymmetric shielding requires the panel to be offset laterally or longitudinally. For example, monoisocentric breast treatments may require a longitudinal panel shift to capture the entire half-beam blocked tangent or superclavicular fields, especially if a double exposure method is used. Technique-specific imaging protocols should include guidance for such panel shifts where required. Irradiating outside the detection area is to be avoided, as it will result in incomplete image capture and risk reducing the lifespan of the detector due to irradiation of the surrounding electronics. For example, the

Elekta EPID is fixed at 160 cm, and the detection area is about 40 cm^2, which means that any jaw positioned greater than 12.5 cm from central axis will overshoot the detection area unless the panel is shifted from its centered location.

4.5 FUTURE

Several factors have impacted the clinical use of BEV imaging over the past decade, including (1) the proliferation of alternative in-room IGRT systems, particularly kV radiographic systems; and (2) the increasing utilization of modulated treatments. On the other hand, there has been renewed interest in BEV imaging in the research setting driven primarily by the search for optimal solutions to real-time intrafraction motion management. The role of BEV imaging is likely to remain complimentary to other in-room imaging modalities that provide improved localization capabilities but lack the fundamental ability to *image* the treatment beam. The unique characteristics of BEV imaging described in this chapter, and the role of BEV imaging in emerging clinical applications outlined in subsequent chapters, make a strong case that BEV imaging has an ongoing role to play in radiographic imaging for radiation therapy.

REFERENCES

Alía, A., Mar, J. and Pastor-Barriuso, R. 2005. Reliability of portal control procedure in irradiation of breast cancer: A Bayesian analysis. *Radiotherapy and Oncology*, 75, 28–33.

Amer, A. M., Mackay, R. I., Roberts, S. A., Hendry, J. H. and Williams, P. C. 2001. The required number of treatment imaging days for an effective off-line correction of systematic errors in conformal radiotherapy of prostate cancer—A radiobiological analysis. *Radiotherapy and Oncology*, 61, 143–150.

Balter, J. M., Pelizzari, C. A. and Chen, G. T. Y. 1992. Correlation of projection radiographs in radiation therapy using open curve segments and points. *Medical Physics*, 19, 329–334.

Bastida-Jumilla, M., Larrey-Ruiz, J., Verdú-Monedero, R., Morales-Sánchez, J. & Sancho-Gómez, J.-L. 2011. DRR and portal image registration for automatic patient positioning in radiotherapy treatment. *Journal of Digital Imaging*, 24, 999–1009.

Beierholm, A., Ottosson, R. O., Lindvold, L. R., Behrens, C. F. and Andersen, C. E. 2011. Characterizing a pulse-resolved dosimetry system for complex radiotherapy beams using organic scintillators. *Physics in Medicine and Biology*, 56, 3033–3045.

Berger, L., Francois, P., Gaboriaud, G. and Rosenwald, J.-C. 2006. Performance optimization of the Varian aS500 EPID system. *Journal of Applied Clinical Medical Physics*, 7, 105–114.

Bijhold, J., Gilhuijs, K. G. A. and Van Herk, M. 1992. Automatic verification of radiation field shape using digital portal images. *Medical Physics*, 19, 1007–1014.

Bissonnette, J.-P., Balter, P. A., Dong, L., Langen, K. M., Lovelock, D. M., Miften, M., Moseley, D. J., Pouliot, J., Sonke, J.-J. and Yoo, S. 2012. Quality assurance for image-guided radiation therapy utilizing CT-based technologies: A report of the AAPM TG-179. *Medical Physics*, 39, 1946–1963.

Blake, S. J.,McNamara A. L., Deshpande, S., Holloway, L. C., Greer, P. B., Kuncic, Z. and Vial, P. 2013. Characterization of a novel EPID designed for simultaneous imaging and dose verification in radiotherapy. *Medical Physics*, 40, 091902.

Borst, G. R., Sonke, J.-J., Den Hollander, S., Betgen, A., Remeijer, P., Van Giersbergen, A., Russell, N. S., Elkhuizen, P. H. M., Bartelink, H. and Van Vliet-Vroegindeweij, C. 2010. Clinical results of image-guided deep inspiration breath hold breast irradiation. *International Journal of Radiation Oncology*Biology*Physics*, 78, 1345–1351.

Bortfeld, T., Van Herk, M. and Jiang, S. B. 2002. When should systematic patient positioning errors in radiotherapy be corrected? *Physics in Medicine and Biology,* 47, N297.

Boyer, A., Antonuk, L. E., Fenster, A., Van Herk, M., Meertens, H., Munro, P., Reinstein, L. E. and Wong, J. 1992. A review of electronic portal imaging devices (EPIDs). *Medical Physics,* 19, 1–16.

Bridge, P., Dempsey, S., Giles, E., Maresse, S., Mccorkell, G., Opie, C., Wright, C. and Carmichael, M.-A. 2015. Practice patterns of radiation therapy technology in Australia: Results of a national audit. *Journal of Medical Radiation Sciences,* 62, 253–260.

Bujold, A., Craig, T., Jaffray, D. and Dawson, L. A. 2012. Image-guided radiotherapy: Has it influenced patient outcomes? *Seminars in Radiation Oncology,* 22, 50–61.

Cox, J. and Jimenez, Y. 2009. The radiation therapist's role in real-time EPI interpretation and decision-making. *European Journal of Radiography,* 1, 139–146.

Cremers, F., Frenzel, T., Kausch, C., Albers, D., Schönborn, T. and Schmidt, R. 2004. Performance of electronic portal imaging devices (EPIDs) used in radiotherapy: Image quality and dose measurements. *Medical Physics,* 31, 985–996.

Dance, D. R., Christofides, S., Maidment, A. D. A., Mclean, I. D. and Ng, K. H. 2014. *Diagnostic Radiology Physics: A Handbook for Teachers and Students,* Vienna, Austria, IAEA.

Das, I. J., Cao, M., Cheng, C.-W., Misic, V., Scheuring, K., Schule, E. and Johnstone, P. A. S. 2011. A quality assurance phantom for electronic portal imaging devices. *Journal of Applied Clinical Medical Physics,* 12, 391–403.

De Boer, H. C. J. and Heijmen, B. J. M. 2001. A protocol for the reduction of systematic patient setup errors with minimal portal imaging workload. *International Journal of Radiation Oncology*Biology*Physics,* 50, 1350–1365.

De Boer, J. C. J. and Heijmen, B. J. M. 2002. A new approach to off-line setup corrections: Combining safety with minimum workload. *Medical Physics,* 29, 1998–2012.

Deshpande, S., Mcnamara, A. L., Holloway, L., Metcalfe, P. and Vial, P. 2015. Feasibility study of a dual detector configuration concept for simultaneous megavoltage imaging and dose verification in radiotherapy. *Medical Physics,* 42, 1753–1764.

Ding, G. X. and Munro, P. 2013. Radiation exposure to patients from image guidance procedures and techniques to reduce the imaging dose. *Radiotherapy and Oncology,* 108, 91–98.

Dundas, K. L., Pogson, E. M., Batumalai, V., Boxer, M. M., Yap, M. L., Delaney, G. P., Metcalfe, P. and Holloway, L. 2015. Australian survey on current practices for breast radiotherapy. *Journal of Medical Imaging and Radiation Oncology,* 59, 736–742.

Fang-Fang, Y., Wong, J., Balter, J. M., Benedict, S., Bissonnette, J.-P., Craig, T., Dong, L. et al. 2009. The role of in-room kV X-ray imaging for patient setup and target localzation. AAPM Report No. 104. AAPM.

Fontenot, J., Alkhatib, H., Garrett, J., Jensen, A., Mccullough, S., Olch, A. J., Parker, B., Yang, C.-C. J. and Fairobent, L. 2014. AAPM medical physics practice guideline 2.a: Commissioning and quality assurance of X-ray-based image-guided radiotherapy systems. *Journal of Applied Clinical Medical Physics,* 15, 3–13.

Fritsch, D. S., Chaney, E. L., Boxwala, A., Mcauliffe, M. J., Raghavan, S., Thall, A. and Earnhart, J. R. D. 1995. Core-based portal image registration for automatic radiotherapy treatment verification. *International Journal of Radiation Oncology*Biology*Physics,* 33, 1287–1300.

Galvin, J. M., Sims, C., Dominiak, G. and Cooper, J. S. 1995. The use of digitally reconstructed radiographs for three-dimensional treatment planning and CT-simulation. *International Journal of Radiation Oncology, Biology, Physics,* 31, 935–942.

Gill, S., Thomas, J., Fox, C., Kron, T., Thompson, A., Chander, S., Williams, S., Tai, K. H., Duchesne, G. and Foroudi, F. 2012. Electronic portal imaging vs kilovoltage imaging in fiducial marker image-guided radiotherapy for prostate cancer: An analysis of set-up uncertainties. *The British Journal of Radiology,* 85, 176–182.

Granfors, P. R. and Aufrichtig, R. 2000. Performance of a 41 × 41-cm2 amorphous silicon flat panel x-ray detector for radiographic imaging applications. *Medical Physics,* 27, 1324–1331.

Grattan, M. W. D. and Mcgarry, C. K. 2010. Mechanical characterization of the Varian Exact-arm and R-arm support systems for eight aS500 electronic portal imaging devices. *Medical Physics,* 37, 1707–1713.

Greer, P. B. 2007. Off-axis dose response characteristics of an amorphous silicon eletronic portal imaging device. *Medical Physics,* 34, 3815–3824.

Greer, P. B., Dahl, K., Ebert, M. A., Wratten, C., White, M. and Denham, J. W. 2008. Comparison of prostate set-up accuracy and margins with off-line bony anatomy corrections and online implanted fiducial-based corrections. *Journal of Medical Imaging and Radiation Oncology,* 52, 511–516.

Hatherly, K. E., Smylie, J. C., Rodger, A., Dally, M. J., Davis, S. R. and Millar, J. L. 2001. A double exposed portal image comparison between electronic portal imaging hard copies and port films in radiation therapy treatment setup confirmation to determine its clinical application in a radiotherapy center. *International Journal of Radiation Oncology*Biology*Physics,* 49, 191–198.

Hawkins, M. A., Aitken, A., Hansen, V. N., Mcnair, H. A. and Tait, D. M. 2011. Set-up errors in radiotherapy for oesophageal cancers—Is electronic portal imaging or conebeam more accurate? *Radiotherapy and Oncology,* 98, 249–254.

Herman, M., Balter, J., Jaffray, D., Mcgee, K., Munro, P., Shalev, S., Van Herk, M. & Wong, J. 2001a. Clinical use of electronic portal imaging: Report of AAPM radiation therapy committee task group 58. *Medical Physics,* 28, 712.

Herman, M. G. 2005. Clinical use of electronic portal imaging. *Seminars in Radiation Oncology,* 15, 157–167.

Herman, M. G., Balter, J. M., Jaffray, D. A., Mcgee, K. P., Munro, P., Shalev, S., Van Herk, M. and Wong, J. W. 2001b. Clinical use of electronic portal imaging: Report of AAPM radiation therapy committee task group 58. *Medical Physics,* 28, 712–737.

Hristov, D. H. and Fallone, B. G. 1996. A grey-level image alignment algorithm for registration of portal images and digitally reconstructed radiographs. *Medical Physics,* 23, 75–84.

Huber, S., Mooslechner, M., Mitterlechner, B., Weichenberger, H., Serpa, M., Sedlmayer, F. and Deutschmann, H. 2013. Image quality improvements of electronic portal imaging devices by multi-level gain calibration and temperature correction. *Physics in Medicine and Biology,* 58, 6429.

Hurkmans, C. W., Remeijer, P., Lebesque, J. V. and Mijnheer, B. J. 2001. Set-up verification using portal imaging; review of current clinical practice. *Radiotherapy and Oncology,* 58, 105–120.

IAEA 2004. Technical Report Series No. 430. Commissioning and quality assurance of computerized planning systems for radiation treatment of cancer. In: IAEA (Ed.) *Technical Report Series.* Vienna, Austria: IAEA.

ICRP 2009. Preventing accidental exposures from new external beam radiation therapy technologies. In: ICRP (Ed.) *ICRP Publication 112.*

Jacob, S., Wong, K., Delaney, G. P., Adams, P. and Barton, M. B. 2010. Estimation of an optimal utilisation rate for palliative radiotherapy in newly diagnosed cancer patients. *Clinical Oncology,* 22, 56–64.

Jaffray, D. A., Battista, J. J., Fenster, A. and Munro, P. 1994. X-ray scatter in megavoltage transmission radiography: Physical characteristics and influence on image quality. *Medical Physics,* 21, 45–60.

Jaffray, D. A., Battista, J. J., Fenster, A. and Munro, P. 1995. Monte Carlo studies of x-ray energy absorption and quantum noise in megavoltage transmission radiography. *Medical Physics,* 22, 1077–1088.

Jensen, C., Urribarri, J., Cail, D., Rottmann, J., Mishra, P., Lingos, T., Niedermayr, T. and Berbeco, R. 2014. Cine EPID evaluation of two non-commercial techniques for DIBH. *Medical Physics,* 41, 021730.

Kirkby, M. C. and Glendinning, A. G. 2006. Developments in electronic portal imaging systems. *British Journal of Radiology,* 79, S50–S65.

Klein, E. E., Hanley, J., Bayouth, J., Yin, F.-F., Simon, W., Dresser, S., Serago, C. et al. 2009. Task group 142 report: Quality assurance of medical accelerators. *Medical Physics,* 36, 4197–4212.

Korreman, S., Rasch, C., Mcnair, H., Verellen, D., Oelfke, U., Maingon, P., Mijnheer, B. and Khoo, V. 2010. The European Society of Therapeutic Radiology and Oncology–European Institute of Radiotherapy (ESTRO–EIR) report on 3D CT-based in-room image guidance systems: A practical and technical review and guide. *Radiotherapy and Oncology,* 94, 129–144.

Kron, T., Lee, C., Perera, F. and Yu, E. 2004. Evaluation of intra- and inter-fraction motion in breast radiotherapy using electronic portal cine imaging. *Technology in Cancer Research & Treatment,* 3, 443–449.

Krueder, F., Schreiber, B., Kausch, C. and Dossel, O. 1998. A structure-based method for on-line matching of portal images for an optimal patient set-up in radiotherapy. *Philip Journal of Research,* 51, 317–337.

Kudchadker, R. J., Chang, E. L., Bryan, F., Maor, M. H. and Famiglietti, R. 2004. An evaluation of radiation exposure from portal films taken during definitive course of pediatric radiotherapy. *International Journal of Radiation Oncology*Biology*Physics,* 59, 1229–1235.

Kupelian, P. A., Lee, C., Langen, K. M., Zeidan, O. A., Mañon, R. R., Willoughby, T. R. and Meeks, S. L. 2008. Evaluation of image-guidance strategies in the treatment of localized prostate cancer. *International Journal of Radiation Oncology*Biology*Physics,* 70, 1151–1157.

Kutcher, G. J., Coia, L., Gillin, M., Hanson, W. F., Leibel, S., Morton, R. J., Palta, J. R. et al. 1994. Comprehensive QA for radiation oncology: Report of AAPM radiation therapy committee task group 40. *Medical Physics,* 21, 581–618.

Leszczynski, L., Loose, S. and Dunscombe, P. 1995. Segmented chamfer matching for the registration of field borders in radiotherapy images. *Physics in Medicine and Biology,* 40, 83.

Litzenberg, D. W., Balter, J. M., Lam, K. L., Sandler, H. M. and Ten Haken, R. K. 2005. Retrospective analysis of prostate cancer patients with implanted gold markers using off-line and adaptive therapy protocols. *International Journal of Radiation Oncology*Biology*Physics,* 63, 123–133.

Lovelock, D. M., Hua, C., Wang, P., Hunt, M., Fournier-Bidoz, N., Yenice, K., Toner, S. et al. 2005. Accurate setup of paraspinal patients using a noninvasive patient immobilization cradle and portal imaging. *Medical Physics,* 32, 2606–2614.

Lutz, C. M., Poulsen, P. R., Fledelius, W., Offersen, B. V. and Thomsen, M. S. 2015. Setup error and motion during deep inspiration breath-hold breast radiotherapy measured with continuous portal imaging. *Acta Oncologica,* 55, 193–200.

Magee, B., Coyle, C., Kirby, M. C., Kane, B. and Williams, P. C. 1997. Use of electronic portal imaging to assess cardiac irradiation in breast radiotherapy. *Clinical Oncology,* 9, 259–261.

Mageras, G. S. and Mechalakos, J. 2007. Planning in the IGRT context: Closing the loop. *Seminars in Radiation Oncology,* 17, 268–277.

Mayles, W. P. M. 2010. Survey of the availability and use of advanced radiotherapy technology in the UK. *Clinical Oncology,* 22, 636–642.

Menon, G. V. and Sloboda, R. S. 2004. Quality assurance measurements of a-si epid performance. *Medical Dosimetry,* 29, 11–17.

Mooslechner, M., Mitterlechner, B., Weichenberger, H., Huber, S., Sedlmayer, F. and Deutschmann, H. 2013. Analysis of a free-running synchronization artifact correction for MV-imaging with aSi:H flat panels. *Medical Physics,* 40, 031906.

Moseley, D. J., White, E. A., Wiltshire, K. L., Rosewall, T., Sharpe, M. B., Siewerdsen, J. H., Bissonnette, J.-P., Gospodarowicz, M., Warde, P., Catton, C. N. and Jaffray, D. A. 2007. Comparison of localization performance with implanted fiducial markers and cone-beam computed tomography for on-line image-guided radiotherapy of the prostate. *International Journal of Radiation Oncology*Biology*Physics,* 67, 942–953.

Moseley, J. and Munro, P. 1994. A semiautomatic method for registration of portal images. *Medical Physics,* 21, 551–558.

Munro, P. 1999. Megavoltage radiography for treatment verification. In: Van Dyk, J. (Ed.) *The Modern Technology of Radiation Oncology.* Madison, WI: Medical Physics Publishing.

Murphy, M. J., Balter, J. M., Balter, S., Bencomo, J. A., Das, I. J., Jiang, S. B., Ma, C. M. et al. 2007. The management of imaging dose during image-guided radiotherapy: Report of the AAPM Task Group 75. *Medical Physics,* 34, 4041–4063.

Mutanga, T. F., De Boer, H. C. J., Van Der Wielen, G. J., Wentzler, D., Barnhoorn, J., Incrocci, L. and Heijmen, B. J. M. 2008. Stereographic targeting in prostate radiotherapy: Speed and precision by daily automatic positioning corrections using kilovoltage/egavoltage image pairs. *International Journal of Radiation Oncology*Biology*Physics,* 71, 1074–1083.

Odero, D. O. and Shimm, D. S. 2009. Third party EPID with IGRT capability retrofitted onto an existing medical linear accelerator. *Biomedical Imaging and Intervention Journal,* 5, e25.

Olch, A. J., Geurts, M., Thomadsen, B., Famiglietti, R. and Chang, E. L. 2007. Portal imaging practice patterns of children's oncology group institutions: Dosimetric assessment and recommendations for minimizing unnecessary exposure. *International Journal of Radiation Oncology*Biology*Physics,* 67, 594–600.

Pesznyák, C., Polgár, I., Weisz, C., Király, R. and Zaránd, P. 2011. Verification of quality parameters for portal images in radiotherapy. *Radiology and Oncology,* 45, 68–74.

Pisani, L., Lockman, D., Jaffray, D., Yan, D., Martinez, A. and Wong, J. 2000. Setup error in radiotherapy: On-line correction using electronic kilovoltage and megavoltage radiographs. *International Journal of Radiation Oncology*Biology*Physics,* 47, 825–839.

Podesta, M., Nijsten, S. M. J. J. G., Snaith, J., Olrandini, M., Lustberg, T., Emans, D., Aland, T. and Verhaegen, F. 2012. Measured vs simulated portal images for low MU fields on three accelerator types: Possible consequences for 2D portal dosimetry. *Medical Physics,* 39, 7470–7479.

Potters, L., Gaspar, L. E., Kavanagh, B., Galvin, J. M., Hartford, A. C., Hevezi, J. M., Kupelian, P. A. et al. 2010. American Society for Therapeutic Radiology and Oncology (ASTRO) and American College of Radiology (ACR) Practice Guidelines for Image-Guided Radiation Therapy (IGRT). *International Journal of Radiation Oncology*Biology*Physics,* 76, 319–325.

Radiologists TRCo. 2008. *On Target: Ensuring Geometric Accuracy in Radiotherapy.* London: Royal College of Radiologists, Society and College of Radiographers, Institute of Physics and Engineering in Medicine.

Rajapakshe, R., Luchka, K. and Shalev, S. 1996. A quality control test for electronic portal imaging devices. *Medical Physics,* 23, 1237–1244.

Reinstein, L., Amols, H., Biggs, P., Droege, R., Filimonov, A., Lutz, W. and Shalev, S. 1987. AAPM Report No. 24, Radiotherapy Portal Imaging Quality. In: AAPM (Ed.) *Report of AAPM Task Group No. 28.* New York: American Institute of Physics.

Rosenman, J., Roe, C. A., Cromartie, R., Muller, K. E. and Pizer, S. M. 1993. Portal film enhancement: Technique and clinical utility. *International Journal of Radiation Oncology*Biology*Physics,* 25, 333–338.

Rowshanfarzad, P., Riis, H. L., Zimmermann, S. J. and Ebert, M. A. 2015. A comprehensive study of the mechanical performance of gantry, EPID and the MLC assembly in Elekta linacs during gantry rotation. *The British Journal of Radiology,* 88, 20140581.

Rowshanfarzad, P., Sabet, M., O'connor, D. J., Mccowan, P. M., Mccurdy, B. M. C. and Greer, P. B. 2012. Detection and correction for EPID and gantry sag during arc delivery using cine EPID imaging. *Medical Physics,* 39, 623–635.

Rybovic, M., Banati, R. B. and Cox, J. 2008. Radiation therapy treatment verification imaging in Australia and New Zealand. *Journal of Medical Imaging and Radiation Oncology,* 52, 183–190.

Sawant, A., Antonuk, L. and El-Mohri, Y. 2007. Slit design for efficient and accurate MTF measurement at megavoltage x-ray energies. *Medical Physics,* 34, 1535–1545.

Shakeshaft, J., Perez, M., Tremethick, L., Ceylan, A. and Bailey, M. 2014. ACPSEM ROSG Oncology-PACS and OIS working group recommendations for quality assurance. *Australasian Physical & Engineering Sciences in Medicine,* 37, 3–13.

Sherouse, G. W., Novins, K. and Chaney, E. 1990. Computation of digitally reconstructed radiographs for use in radiotherapy treatment design. *International Journal of Radiation Oncology*Biology*Physics,* 18, 651–658.

Sherrier, R. H. and Johnson, G. A. 1987. Regionally adaptive histogram equalization of the chest. *Medical Imaging, IEEE Transactions on,* 6, 1–7.

Siddon, R. L. 1985. Fast calculation of the exact radiological path for a three-dimensional CT array. *Medical Physics,* 12, 252–255.

Simpson, D. R., Lawson, J. D., Nath, S. K., Rose, B. S., Mundt, A. J. and Mell, L. K. 2010. A survey of image-guided radiation therapy use in the United States. *Cancer,* 116, 3953–3960.

Son, S.-Y., Choe, B.-Y., Lee, J.-W., Kim, J.-M., Jeong, H.-W., Kim, H.-G., Kim, W.-S., Lyu, K.-Y., Min, J.-W. and Kim, K.-W. 2014. Evaluation of an edge method for computed radiography and an electronic portal imaging device in radiotherapy: Image quality measurements. *Journal of the Korean Physical Society,* 65, 1976–1984.

Stanley, D. N., Papanikolaou, N. and Gutierrez, A. N. 2015. An evaluation of the stability of image-quality parameters of Varian on-board imaging (OBI) and EPID imaging systems. *Journal of Applied Clinical Medical Physics,* 16, 87–98.

Topolnjak, R., Sonke, J.-J., Nijkamp, J., Rasch, C., Minkema, D., Remeijer, P. and Van vliet-vroegindeweij, C. 2008. Breast patient setup error assessment: Comparison of electronic portal image devices and cone-beam computed tomography matching results. *International Journal of Radiation Oncology*Biology*Physics,* 78, 1235–1243.

Valicenti, R. K., Michalski, J. M., Bosch, W. R., Gerber, R., Graham, M. V., Cheng, A., Purdy, J. A. and Perez, C. A. 1994. Is weekly port filming adequate for verifying patient position in modern radiation therapy? *International Journal of Radiation Oncology*Biology*Physics,* 30, 431–438.

Van Der Heide, U. A., Kotte, A. N. T. J., Dehnad, H., Hofman, P., Lagenijk, J. J. W. and Van Vulpen, M. 2007. Analysis of fiducial marker-based position verification in the external beam radiotherapy of patients with prostate cancer. *Radiotherapy and Oncology,* 82, 38–45.

Van der Vight, L. P., Van Lin, E. N. J. T., Spitters-Post, I., Visser, A. G. & Louwe, R. J. W. 2009. Off-line setup corrections only marginally reduce the number of on-line corrections for prostate radiotherapy using implanted gold markers. *Radiotherapy and Oncology,* 90, 359–366.

Van Herk, M. 2000. Image registration using chamfer matching. In: Bankman, I., N. (ed.) *Handbook of Medical Imaging Processing and Analysis.* San Diego, CA: Academic Press.

Van Herk, M. 2007. Different styles of image-guided radiotherapy. *Seminars in Radiation Oncology,* 17, 258–267.

Van Kranen, S., Van Beek, S., Rasch, C., Van Herk, M. and Sonke, J.-J. 2009. Setup uncertainties of anatomical sub-regions in head-and-neck cancer patients after offline CBCT guidance. *International Journal of Radiation Oncology*Biology*Physics,* 73, 1566–1573.

Van Lin, E. N. J. T., Van Der Vight, L., Huizenga, H., Kaanders, J. H. A. M. and Visser, A. G. 2003. Set-up improvement in head and neck radiotherapy using a 3D off-line EPID-based correction protocol and a customised head and neck support. *Radiotherapy and Oncology,* 68, 137–148.

Van Tienhoven, G., Lanson, J. H., Crabeels, D., Heukelom, S. and Mijnheer, B. J. 1991. Accuracy in tangential breast treatment set-up: a portal imaging study. *Radiotherapy and Oncology,* 22, 317–322.

Walter, C., Boda-Heggemann, J., Wertz, H., Loeb, I., Rahn, A., Lohr, F. and Wenz, F. 2007. Phantom and in-vivo measurements of dose exposure by image-guided radiotherapy (IGRT): MV portal images vs. kV portal images vs. cone-beam CT. *Radiotherapy and Oncology,* 85, 418–423.

Wertz, H., Dzmitry, S., Manuel, B., Michael, R., Chris, K., Kevin, B., Uwe, G. et al. 2010. Fast kilovoltage/megavoltage (kVMV) breathhold cone-beam CT for image-guided radiotherapy of lung cancer. *Physics in Medicine and Biology,* 55, 4203.
White, E. and Kane, G. 2007. Radiation medicine practice in the image-guided radiation therapy era: New roles and new opportunities. *Seminars in Radiation Oncology,* 17, 298–305.
White, E. A., Brock, K. K., Jaffray, D. A. and Catton, C. N. 2009. Inter-observer variability of prostate delineation on cone beam computerised tomography images. *Clinical Oncology,* 21, 32–38.
Winkler, P. and Georg, D. 2006. An intercomparison of 11 amorphous silicon EPIDs of the same type: Implications for portal dosimetry. *Physics in Medicine and Biology,* 51, 4189–4200.

5

Megavoltage cone-beam computed tomography

OLIVIER GAYOU

5.1 INTRODUCTION AND HISTORY

Two-dimensional (2D) imaging, using either film or electronic portal imaging (EPI), has long been the gold standard for initial patient setup and weekly verification of patient positioning. Weaknesses inherent to 2D imaging for patient setup include the lack of information of tumor position relative to bony anatomy and out of plane rotation. The need for a more precise patient positioning has led to the development of three-dimensional (3D) imaging systems such as cone-beam computed tomography (CBCT), which yield an image in a format similar to that of the image acquired during simulation and allows for a direct comparison of patient positioning and tumor localization.

It was recognized early in the development of portal imagers that a rotating 4 or 6 MV treatment beam could be used in conjunction with the imager to produce a tomographic image. Early systems developed in the 1980s used a 4 MV beam collimated to produce a 1 cm thick fan beam and a linear array of plastic scintillators (Simpson et al. 1982), later improved with a scintillation crystal-photodiode detector (Morton et al. 1991). The second half of the 1990s saw the development of 2D X-ray detectors (Mosleh-Shirazi et al. 1998), enabling to take full advantage of the divergence of the beam. One of the first such systems (Midgley et al. 1998) used a liquid-filled matrix ionization chamber located in a cassette mounted on a retractable arm of a Varian linear accelerator (LINAC) (Varian Medical Systems, Palo Alto, CA). All these early systems, ultimately, could not be used clinically because of the high imaging dose necessary to obtain an image of sufficient quality, but they contributed greatly to the development and widespread adoption of CBCT in general, both with kilovoltage (kV) and megavoltage (MV) systems.

In collaboration with the University of California, San Francisco (UCSF), Siemens Medical Solutions (Concord, CA) developed a low-dose MV-CBCT, that was demonstrated to be useful clinically in a landmark paper in 2005 (Pouliot et al. 2005). This work was the basis of the MVision™ product that is available on Siemens LINACS, and is currently the only commercially available MV-CBCT system. This chapter will discuss this system in detail, with an emphasis on clinical and practical aspects.

5.2 TECHNOLOGY OF MV-CBCT IMAGING

5.2.1 MV-CBCT ACQUISITION

A CBCT image is reconstructed from a number of 2D projections acquired on a flat-panel detector in small angular steps as the source-detector system rotates around the object. In MV-CBCT, the X-ray source is the treatment LINAC itself, as illustrated in Figure 5.1, set to its lowest available photon energy, typically 6 MV. On the Siemens clinical MVision™ system, the detector consists of 1024 × 1024 amorphous silicon (aSi) photodiodes connected to thin film transistors. The diodes are spaced every 400 μm in both directions, for a total active area of 409.6 × 409.6 mm^2. The panel is located at a source-to-imager distance of 145 cm, so that the maximum imaging field size at the isocenter plane is 27.4 × 27.4 cm^2. The bit depth of each pixel is 16 bits and the maximum readout time for the entire detector is 285 ms.

There are several different acquisition protocols, in which parameters can be varied. The parameter having the largest impact on image quality is the total exposure. As the treatment beam is used, exposure to the patient can be directly expressed as absorbed dose using the usual formalism for dose calculation with MV therapy beams. The beam is calibrated to deliver 1 monitor unit (MU) per cGy to water at the depth of maximum dose at a source-to-axis distance (SAD) of 100 cm, using a field size of 10 cm×10 cm. The total exposure is then selected by choosing the number of MU that will be delivered throughout the arc. In general, more MU corresponds to better signal-to-noise. Clinically useful protocols range from 3 to 20 MU. It is technically challenging to use less than 3 MU because the gantry speed calibration is out of range for such a fast acquisition. Protocols in excess of 20 MU are technically feasible but not used clinically due to the prohibitive dose associated with imaging.

Next, the number of projections and the sampling rate, that is, the angular travel between two projections, can be selected during acquisition. These two parameters combine to form either a full arc of 360° or a partial arc of 200°. The number of projections can range from 100 to 600, and the sampling rate from 0.5 to 2.0 per projection. When partial arcs are selected, the start angle can be chosen freely between 180° (beam posterior to the patient in supine position) and 340°. A recent study showed that none of these acquisition parameters had a significant effect on image quality

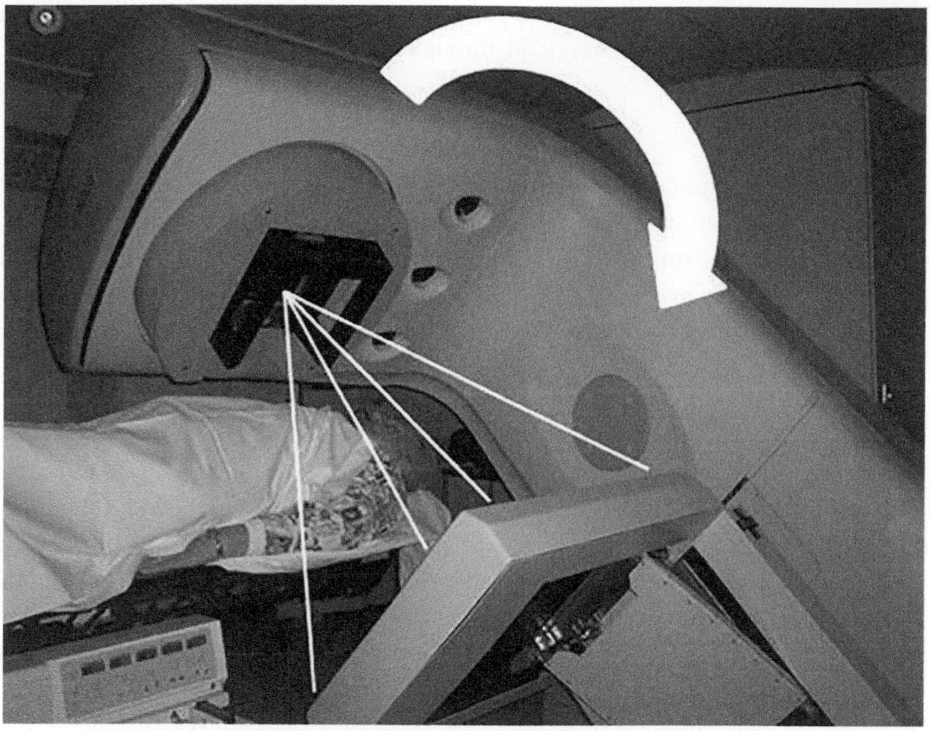

Figure 5.1 MV-CBCT system during acquisition. The megavoltage treatment beam is used in arc mode with a field size covering the amorphous silicon portal imager. Beam pulse and detector acquisition trigger are synchronized to acquire an image at a few degree intervals, typically 1°, over an arc of 200° or 360°. The result is a reconstructed volume of maximum dimensions 27.4 × 27.4 × 27.4 cm^3.

within their clinically available range (Gayou 2012). However, considerations other than image quality could be taken into account to select the acquisition parameters. For example, the use of a higher number of projections will lead to a longer acquisition time, as the gantry will have to move slowly to allow time for the detector to readout the data before another projection is acquired, with acquisition time variations on the order of 15 s. Similarly, a higher number of projections will require a longer reconstruction time. So, choosing a coarse sampling could lead to a more efficient clinical workflow without a cost in image quality. In addition, if analyzing the unreconstructed projections themselves is of interest, a lower number of projections are desirable, since for the same total exposure a higher exposure per projection is realized, therefore leading to better contrast on the projections themselves.

The image quality of the reconstructed MV-CBCT image is directly affected by the image quality of the individual projections. When the imager acquires data while the beam is continuously on, the beam pulse structure and the detector trigger combine to create undesirable artifacts on the projection. In order to avoid such artifacts, the readout trigger is synchronized with the beam pulse when the gantry reaches the next angle, and the beam is held while the detector is in readout mode, effectively removing the beam pulse artifact.

5.2.2 MV-CBCT RECONSTRUCTION

The principle of 3D reconstruction of tomographic images from a set of 2D projections is well established (Feldkamp et al. 1984) and has been described in the context of MV-CBCT (Pouliot et al. 2005, Morin et al. 2009). At any gantry angle ϑ, a voxel V identified by coordinates (x, y, z)

in the reconstruction volume is related to its projection pixel p on the detector plane identified by transverse and longitudinal coordinates (u, v), through a projection matrix $P(\vartheta)$ such that

$$p(u, v, \vartheta) = P(\vartheta)V\,(x, y, z) \tag{5.1}$$

The projection image first undergoes a series of steps designed to improve image quality prior to performing the reconstruction. These steps first include offset, gain, and dead pixel corrections, and transmission filtering, using either an average or a diffusion filter to reduce noise. The diffusion filter uses different levels of averaging depending on the local gradient. A logarithmic conversion and normalization is then applied to the corrected pixel value $I(u, v)$:

$$N(u, v, \vartheta) = ln\left(\frac{I_0(u, v, \vartheta)}{I(u, v, \vartheta)}\right) \tag{5.2}$$

where $I_0(u, v)$ is the intensity of the pixel without attenuation obtained in the MV-CBCT gain calibration procedure without any object between the source and the imager. A backprojection filter is applied in the frequency domain, either edge-enhancing, edge-preserving, or smoothing. All these steps are described in detail in the literature (Morin 2007).

The voxel at position V in the reconstruction volume is obtained by summing the contributions of the corresponding projection pixel at each gantry angle ϑ, identified through Equation 5.1. The projection matrices P are obtained via a calibration procedure that involves a geometric phantom with 108 embedded tungsten beads with known location on the phantom. The phantom, shown in Figure 5.2, is placed at the LINAC isocenter, and a series of 200 projections are acquired. On each projection, corresponding to a different gantry angle, the position of each bead on the imaging plane is automatically detected and mapped to the actual bead position in the reconstruction volume on the phantom. This gives an overdetermined sample of projection equations. Even though gantry-mounted imaging systems do not have a rigid isocentric geometry and source-to-image plane relationship, the projection matrices calculated in this calibration procedure take into account detector movements and possible sag and mispositioning.

The reconstruction process results in a $27.4 \times 27.4 \times L\ cm^3$, where L is the Y-jaw opening in the cranio-caudal direction, which can be adjusted from 5.0 to 27.4 cm. Restricting the jaw opening reduces scatter in the imaging field resulting in better image quality, but also hides potentially useful anatomical information for image registration. Therefore, the selection of this parameter will depend on the anatomical site and user preference. Users can also choose between three different reconstructed resolutions: 0.5, 1.0, or 2.0 mm voxel size in all three dimensions.

The last step of the reconstruction process is the application of a uniformity correction to eliminate cupping artifacts, which increases noise in the center of the image due to beam scatter and hardening. This correction uses a geometric model whose parameters are empirically determined using water phantoms of different sizes. In the Siemens systems, the parameters themselves are not available to the clinical user, who can only make a selection between three anatomical sites (pelvis, thorax, and head-and-neck) and three sizes (small, medium, and large).

The MV-CBCT acquisition takes 45–60 s depending on the selected MU protocol, and reconstruction starts on a dedicated workstation immediately on arc completion. The reconstruction time is approximately 60–110 s, depending on the selected image resolution. Table 5.1 summarizes the acquisition and reconstruction parameters available to the user of the Siemens MVision™ system.

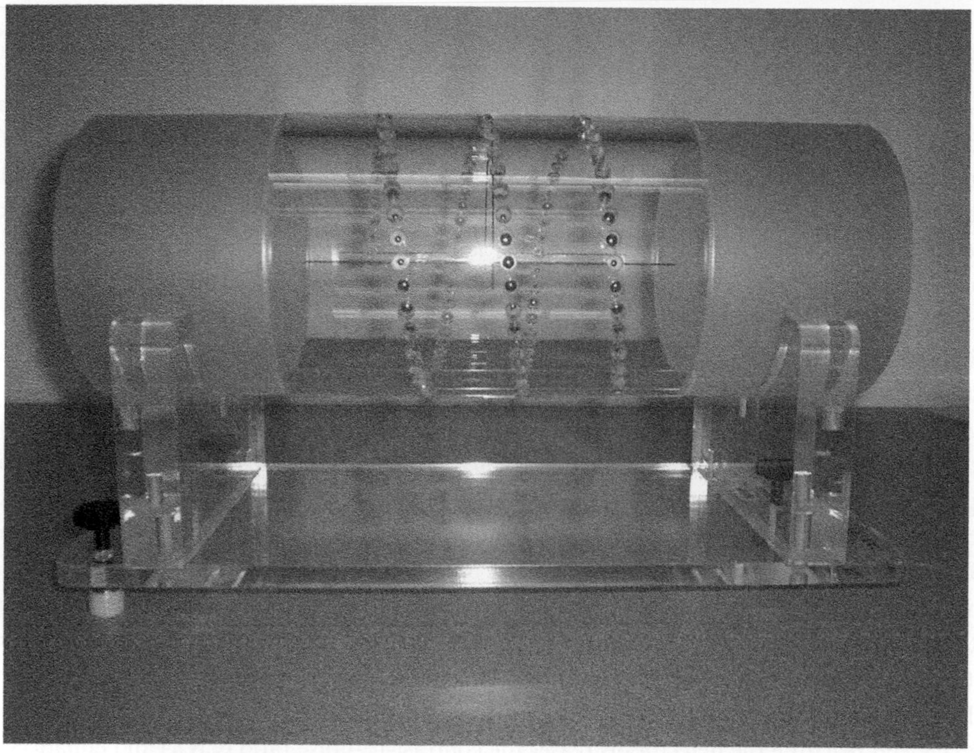

Figure 5.2 Photograph of the manufacturer-provided geometry calibration phantom. One hundred and eight tungsten beads of known position are placed in a helical fashion around the cylindrical phantom. The relationship between the beads' absolute position in the reconstruction volume and their extracted position on the projections acquired in the calibration process is used to build the projection matrices for each gantry angle.

Table 5.1 Summary of acquisition and reconstruction parameters available to the user of the Siemens MVision™ system, and their possible values

MV-CBCT parameter	Possible values
Source-to-image distance	145 cm
Source-to-axis distance	100 cm
Typical clinical exposure protocols	3–20 MU
Beam output	1 cGy/Mu at d_{max}
Detector pixel size and number	400 μm, 1024 × 1024
Detector size	409.6 × 409.6 mm^2
Arc length	200° or 360°
Number of projections	100, 180, 200, 360, 400, 450, 600
Sampling rate	0.5°, 0.6°, 0.7°, 0.8°, 0.9°, 1.0°, 2.0°
Transmission filter	Average or diffusion
Backprojection filter	Edge-enhancing, edge-preserving, or smoothing
Uniformity correction	Pelvis, thorax, or head-and-neck
Y-jaw opening (L)	5.0–27.4 cm
Reconstructed volume size	27.4 × 27.4×L cm^3
Voxel size	0.5, 1.0, 2.0 mm

5.3 IMAGE QUALITY AND IMAGING DOSE

The quality of the reconstructed MV-CBCT image is of paramount importance to the image-guidance process, where it is registered to a planning dataset acquired using a conventional fan-beam kV CT. The image should be of sufficient quality to distinguish different anatomical objects of different size and providing different contrast with the surrounding anatomy, which is usually case dependent. For example, if the object to register is the prostate, then the MV-CBCT image must provide enough contrast with the surrounding soft tissue, like bladder and rectum, so that it can be clearly identified and aligned to the planning CT. The level of contrast needed for the prostate is higher than the one needed to register bones, which are more easily visible on the image because of their high density.

A comprehensive study of the effects on image quality of the reconstruction parameters presented in Table 5.1 was performed (Morin et al. 2009). It showed that most of these parameters have a significant impact on image quality. However, it also showed that for each of these parameters, there was an optimal value that was not case dependent. The authors recommend using a 0.5 mm voxel size, whereas displaying a slice thickness of 5 mm; using a diffusion transmission filter; and using the *smoothing* backprojection filter. Furthermore, it was mentioned earlier in this chapter that acquisition parameters such as number of projections and sampling rate do not have a significant impact within the range that is available to the user.

In all photon-based imaging systems, increasing the number of photons traversing the object to image and reaching the detector decreases the contrast-to-noise ratio (CNR) and enhances visibility. However, when using high-energy photons, increasing the number of photons increases the absorbed dose to the patient, to levels that can reach a clinically significant percentage of the dose used for therapy. It therefore becomes critical in MV-CBCT imaging to use the minimum dose that can produce the visibility required for each particular case, in a way similar to the familiar as low as reasonably achievable (ALARA) principle of radiation safety. From a practical standpoint, the user really has only two parameters to select when creating an imaging protocol to setup the patient: the total exposure, expressed with the total number of MU to be delivered, and the uniformity correction model, which depends on anatomical site and patient size.

5.3.1 QUANTIFICATION OF IMAGE QUALITY

5.3.1.1 UNIFORMITY

Image quality is quantified by three main characteristics: uniformity, CNR, and spatial resolution. A more uniform reconstructed image makes the process of image viewing and registration easier by allowing for the use of general preset level and window settings. Uniformity can be evaluated in a phantom of uniform density, comparing the mean pixel value around the periphery of the phantom to that around the center. The use of the proper uniformity correction model, based on anatomical site and patient size, yields a uniformity reaching 98% in both the axial and cranio-caudal direction (Morin et al. 2009), using the following formula, normalized to the difference between water and air mean pixel values:

$$U = \left[1 - \frac{\text{mean(periphery)} - \text{mean(center)}}{\text{mean(water)} - \text{mean(air)}}\right] \times 100 \tag{5.3}$$

5.3.1.2 SPATIAL RESOLUTION

Spatial resolution represents the ability of a system to distinguish a small object of high contrast from its surrounding. A system with high spatial resolution is able to resolve smaller objects. There are several ways to estimate spatial resolution, each having their own limitations. One way is to image a phantom containing different high-contrast line pairs of different sizes and separation, and evaluate which is the smallest line pair visible. Using this method, it was found that the MVision™ system was able to resolve 0.3 line pair per mm, which is lower than the 0.5 line pair/mm obtained with a conventional kV-CT with the same phantom (Gayou and Miften 2007a). Another method consists of extracting the point spread function (PSF) of the system, which represents the system's response to a point stimulus, with the use of a small object of known size. This method involves deconvolving the PSF from the object size. The study of spatial resolution of the MVision™ system using the second method found that the full width half maximum of the PSF was around 1.9 mm (Morin et al. 2009).

5.3.1.3 CONTRAST-TO-NOISE RATIO

CNR is the most important image quality characteristic in image-guided clinical applications. It characterizes the ability to distinguish a structure from its surrounding background. A system with a high CNR is able to resolve a structure even if the density difference with the surrounding anatomy is low. The major disadvantage of the MV-CBCT system compared to its kV counterparts lies in its ability to detect contrast. In kV imaging, the photons interacting with the anatomy are more likely to interact via the photoelectric effect rather than the Compton effect. As the photoelectric effect is highly sensitive to the atomic number Z of the material, contrast will be greater for the same level of noise, compared to an MV imaging system, leading to a higher CNR. The MV system also derives contrast from photoelectric effect, but in a much smaller proportion. This is especially true for bony anatomy with high Z components, but is also reflected in lower resolution of soft tissue.

Any given system is characterized by its CNR as a function of the difference in electron density between an object and its background, as the electron density is the factor driving contrast. For clinical applications purposes, the CNR is measured using a water phantom containing inserts of known electron density in the range typically found in the human body, and extracting the mean pixel values within the insert and the background and is given by the following formula:

$$\text{CNR} = \frac{\text{mean(insert)} - \text{mean(background)}}{\text{mean}\left[\text{standard deviation(insert)}, \text{standard deviation(background)}\right]}$$

For example, dense bone (electron density relative to water 1.512) or lung in the inhaling respiratory phase (0.190) will provide a higher CNR than adipose tissue (0.952), muscle (1.043), or liver (1.052), against a water-equivalent background. The ability to resolve low-contrast objects ultimately depends on the level of noise, which according to counting statistics, varies as the inverse square root of the number of photons hitting the detector. Therefore, for any density insert, the CNR will increase proportionally by the square root of the exposure, that is, the square root of the absorbed dose by the patient. As a clinical example, with a low MU protocol (4 MU), the CNR for muscle is around 3, which leads to limited visibility. With a high MU protocol (15 MU), the CNR for muscle increases to 5, improving the ability to distinguish it from the background. For dense bone, the CNR is around 21 for a 3 MU protocol. Therefore, when registering bony anatomy for image guidance, a low MU, low-dose protocol is largely sufficient.

5.3.2 Imaging dose

5.3.2.1 IMAGING DOSE AND IMAGE QUALITY

The dose related to MV cone-beam imaging has always been a somewhat controversial topic. Some contend that it is not possible to use MV-CBCT for daily imaging as the associated dose can reach a few percent of the therapeutic dose, and could therefore be clinically significant. Others argue that radiation dose from daily kV CBCT can also be significant, and yet usually ignored in the planning process. A comprehensive study of MV-CBCT dose and its relationship with image quality was performed (Gayou et al. 2007b). Central and peripheral dose was measured using an ion chamber in a cylindrical acrylic phantom, and the dose distribution was characterized inside an anthropomorphic phantom using film and thermoluminescent dosimeters (TLDs). The dose to the center of the pelvic-sized phantom for the 15 MU protocol was around 9 cGy; the dose to the center of the head-sized phantom for the 4 MU protocol was around 2.5 cGy. These values agreed with other studies (Morin et al. 2007a) and followed expectations for the given beam energy, field size, and depth. The dose posterior to the isocenter was 20%–50% lower, depending on the phantom size, whereas the dose anterior to the isocenter was 50% higher, since the arc was acquired anteriorly from the right to the left side of the phantom. Figure 5.3 shows the effects of the choice of MU protocol, that is, the imaging dose, on image quality of a phantom with inserts of different density, for a 200° arc. The CNR clearly increases when the dose increases, so that on the phantom objects that cannot be resolved with 2 cGy imaging dose can easily be seen with 12 cGy imaging. It should be noted that even with 48 cGy, which is excessive for daily clinical use, the image quality does not

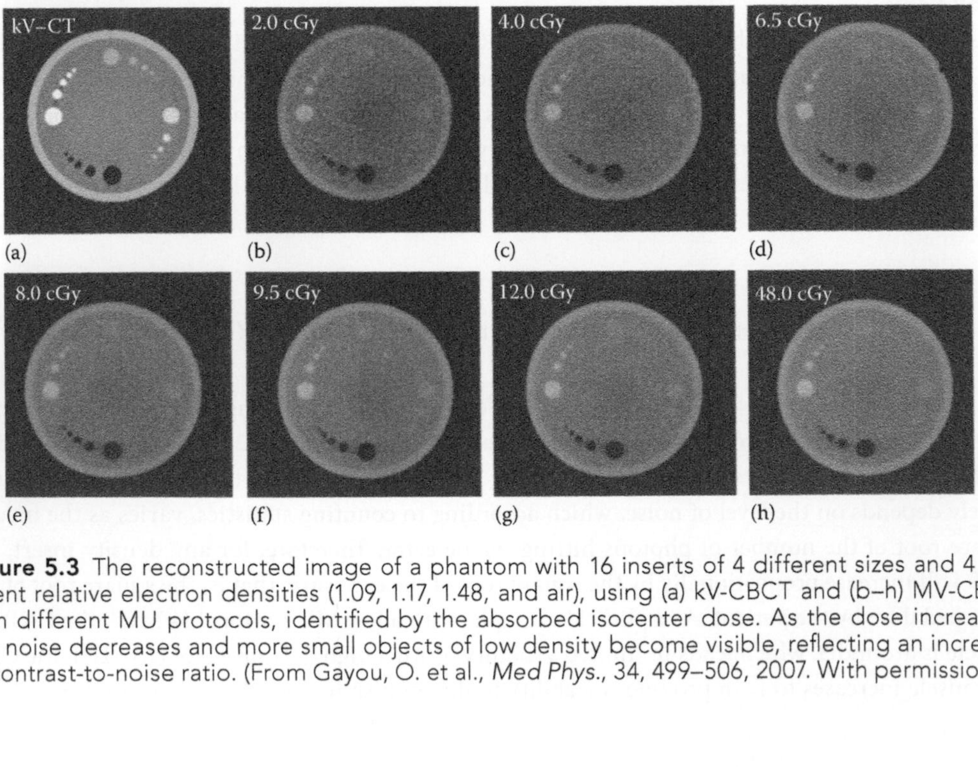

Figure 5.3 The reconstructed image of a phantom with 16 inserts of 4 different sizes and 4 different relative electron densities (1.09, 1.17, 1.48, and air), using (a) kV-CBCT and (b–h) MV-CBCT with different MU protocols, identified by the absorbed isocenter dose. As the dose increases, the noise decreases and more small objects of low density become visible, reflecting an increase in contrast-to-noise ratio. (From Gayou, O. et al., *Med Phys.*, 34, 499–506, 2007. With permission.)

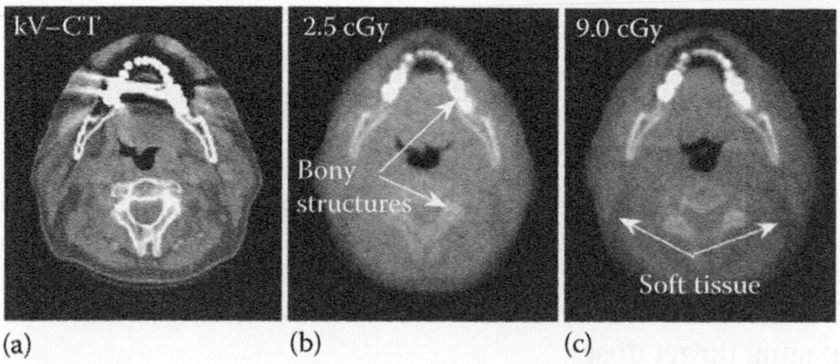

Figure 5.4 Illustration of the increased contrast-to-noise ratio with imaging dose in a head-and-neck patient. With 2.5 cGy (b), only bony anatomy can be used as guide for image registration to the planning CT (a). With 9.0 cGy (c), soft tissue resolution is increased. (From Gayou, O. et al., *Med Phys.*, 34, 499–506, 2007. With permission.)

equal that of kV fan-beam CT. Figure 5.4 illustrates this point further on a head-and-neck patient, where 2.5 cGy imaging dose is sufficient for visualization of bony anatomy, but 9.0 cGy is required to resolve soft tissue in the neck.

From a practical clinical standpoint, the choice of imaging dose should come from a cost/benefit analysis, which on one side considers the imaging dose relative to the therapeutic dose, and on the other side evaluates the clinical benefit of the imaging procedure in terms of tumor control and possible decrease in toxicity to organs-at-risk (OAR). For example, the imaging dose deposited by a 15 MU MV-CBCT scan used for image guidance of a prostate intensity-modulated radiation therapy (IMRT) treatment, is around 9–10 cGy. As the prostate moves daily depending on bladder and rectal content, daily imaging is recommended if margins on the order of 5 mm are used, and a high-dose protocol is necessary to resolve the prostate, bladder, and rectum. In this case, the imaging dose amounts to 5% of the usual fractional dose of 180–200 cGy, which is clinically significant. If bony anatomy is the basis for image guidance, a smaller dose can be used, as the CNR for bone is larger than that for soft tissue. A 2.5 cGy imaging dose represents only 1.2%–1.3% of a conventional fractional dose. For stereotactic cranial or body radiation therapy applications, where a high fractional therapeutic dose is used, a high-dose imaging protocol is not only acceptable as the imaging dose is a smaller fraction of the therapeutic dose, it is also recommended as high quality image guidance is of prime importance in daily setup of high-dose treatments, which typically have smaller margins.

5.3.2.2 IMAGING DOSE MANAGEMENT

Recognizing that delivering a target imaging dose on the order of 4%–5% of the therapeutic dose is excessive, but also realizing that the benefits of volumetric imaging in terms of soft tissue definition or identification of rotation, deformation, or anatomical changes over time are of great value, methods were developed to integrate the imaging dose into the radiation treatment plan. One of the great advantages of MV-CBCT is the fact that it uses the treatment beam, and therefore is already modeled in the treatment planning system (TPS). Including the imaging dose only consists of calculating an arc of given start and stop angles, field size, and total MU. There is no additional quality assurance (QA) other than verifying the constancy of the output of the beam in imaging mode. One method of dose integration first optimizes the therapy plan, then calculates the total imaging dose for the prescribed imaging modality and frequency. The two are then added together with a compensation factor weighing the treatment plan dose, so that the total target dose corresponds to the prescribed dose, according to the following formula (Morin et al. 2007a):

$$T_{Tx} = CF \cdot T_{Tx} + T_{MV-CBCT} \cdot n_f \quad (5.4)$$

where:

T_{Tx} is the planned target dose
$T_{MV\text{-}CBCT}$ is the imaging target dose
CF is the compensating factor, which is usually between 93% and 100%
n_f is the number of imaging sessions

As a result, the target dose is identical to the plan in which the imaging dose would be ignored. OAR surrounding the target receive more dose than if no imaging was performed, but less than if it was performed and not taken into account.

A second method first calculates the imaging dose from the MV-CBCT arc, and then performs an IMRT optimization taking into account the fixed imaging dose (Miften et al. 2007). This method results in similar target coverage as the first method, but it also has the potential to improve the sparing of the OAR that are affected by at least one of the IMRT beams, compared to a plan ignoring the imaging dose. The reason is that since the optimizer already knows what dose such organs are receiving from the imaging process, it can optimize the IMRT beams, so that the total dose does not violate the constraints. Organs that are not directly affected by the IMRT beams but receive imaging dose see a net increase in total dose, compared to a plan ignoring the imaging dose.

The ease of modeling the imaging dose in any TPS actually constitutes an advantage over kV imaging modalities. TPS are usually not equipped to handle low-energy beams, and modeling these fields can be very difficult. However, the total dose from a kV-CBCT may not be negligible, especially dose to bony structures, whose high Z components are sensitive to low-energy photons (Ding et al. 2010). The kV-CBCT imaging dose can even be 2 to 3 times higher for children than for adults due to their smaller size (Deng et al. 2012).

5.4 PERFORMANCE AND QUALITY ASSURANCE

Image quality is an important aspect of MV-CBCT. If the anatomy of interest cannot be resolved on the image, the registration cannot be performed. However, positioning accuracy is of primary importance in the registration process. An image of good quality can be incorrectly registered to the planning CT if an error is introduced in the reconstruction process leading to the introduction of an offset that does not accurately reflect the real patient geometry. Positioning errors introduced during image reconstruction tend to be systematic in nature. Two potential causes are (1) mispositioning of the flat-panel detector, whose deploying mechanism can wear out over time; and (2) bad projection matrices due to a faulty geometry calibration setup. The accuracy and reproducibility of the flat-panel deployment can and should be checked daily with the help of a metallic crosshair inserted in the accessory mount. The stability of the position of the projection of this crosshair can be assessed and corrected, if necessary. A detector shift in the longitudinal direction of 1.5 mm will translate to a longitudinal position error of 1 mm at isocenter. A lateral detector shift may result in a rotational error of the reconstructed image.

A careful calibration of the projection matrices is of primary importance to the positioning accuracy of the system. From the user's perspective, most care must go into setting the calibration phantom accurately, with the help of the room lasers or the gantry crosshair at the cardinal angles. The rest of the process, that is, the acquisition of the calibration arc and the construction of the projection matrices, is performed automatically by the computer. The stability of the calibration process was verified over a 12-month period and found to be excellent, with variations well under 1 mm (Morin et al. 2009).

The positioning accuracy of the system can be verified by taking an image of a phantom with metallic seeds of known absolute position, and assess the reconstructed position of the seed. Such a test should be done daily with a simple cube phantom aligned to room lasers. In the case of the Siemens MVision™ system, a cylindrical QA phantom is provided, with a total of 12 metallic beads embedded in three axial slices of the phantom. When the phantom is carefully aligned around the LINAC isocenter, the positions of the 12 beads are known, and their reconstructed position should fall within ±2 mm, according to the manufacturer's specifications. In practice, the accuracy of the MVision™ system is closer to 1 mm. It should be noted that another advantage over a gantry-mounted kV volumetric imaging system is that the imaging and treatment system share the same isocenter by design, since they share the same source. As such the QA step of verifying isocenter coincidence is removed.

The phantom described previously can also be used to perform tests of image quality reproducibility. The phantom is divided into several sections, each measuring a different quality characteristic. One section is made of solid water for quantification of uniformity and noise. The section containing line pair patterns assesses spatial resolution. Two sections containing inserts of different electron density relative to water and different sizes measures the stability of the CNR ratio. One of these slices is shown in Figure 5.3. These phantom tests need to be repeated on a monthly basis. A noted degradation of image quality is usually corrected for with a recalibration of the detector gain, obtained by irradiating the flat panel with no object in the path.

Image artifacts can also be introduced by issues with other components of the LINAC, such as the gantry rotation. These artifacts are generally apparent on the image and are noticed immediately by the daily users. Corrective action can then be taken in a short time.

5.5 CLINICAL USE

The primary use of MV-CBCT is for localization, that is, verification and correction of the patient setup on the treatment table according to the setup used during simulation. In the MVision™ system, the process is incorporated in the workflow. The imaging arc is scheduled in the record-and-verify system (R&V) and transferred to the treatment workstation, which provides the interface between the R&V and the LINAC. During acquisition of the arc, the projections are automatically transferred to the workstation, and the image reconstruction process begins immediately. The planning CT image, which was transferred from the R&V system prior to the first treatment, is loaded automatically, and a rigid registration is performed with a maximum mutual information algorithm, using translation in the longitudinal, vertical, and lateral directions. A visual evaluation and manual correction of the registration is then performed, and the corrective couch offsets are calculated. The couch is moved accordingly and the treatment can proceed.

For localization purposes, the reduced CNR compared to kV-CBCT does not constitute a great handicap. MV-CBCT–guided therapy is used routinely for daily imaging of all anatomical sites. Its accuracy has been proven for soft tissues such as prostate, comparing favorably with ultrasound and evenly with stereoscopic markers (Gayou and Miften 2008). Anatomical sites that rely essentially on bones for registration, such as brain or head-and-neck, can use lower MU protocols. In fact MV-CBCT has been used routinely in accuracy studies evaluating head-and-neck setups (Graff et al. 2013, Motegi et al. 2014). The use of MV-CBCT to image fiducial markers to help localizing pancreatic tumors were also reported (Packard et al. 2015). Finally, a study of the effects of tumor motion on MV-CBCT lung imaging found that for motion less than 15 mm and the positioning accuracy of the system was not compromised (Gayou and Colonias 2015).

One of the great advantages of volumetric information obtained from CBCT imaging is the ability to not only correct for mispositioning but also to monitor changes in anatomy over the course of treatment. For example, patients treated for head-and-neck cancer tend to lose a lot of weight as a result of dysphagia and odynophagia caused by treatment-induced irritation. In some cases, weight loss can be so severe that the planning CT no longer represents proper patient geometry and the dose distribution does not conform to the actual disease. Daily volumetric imaging is a great tool for monitoring such changes. Figure 5.5 shows the planning CT with initial contours of target and OAR for an oropharyngeal cancer patient, along with a MV-CBCT image taken after 4 weeks of treatment. The MV-CBCT image shows an approximately 3 cm reduction in neck diameter, leading the planning target volume to lie outside the patient contour. Subsequently, a second planning CT was acquired for that patient, and a new plan was created and used for the remainder of the treatment. In this case, the daily use of MV-CBCT imaging directly led to this simple form of adaptive therapy.

A significant milestone in adaptive therapy is the ability to adapt the treatment plan everyday just prior to treatment, based on information obtained from the image-guided radiation therapy (IGRT) image. The steps involved in such a process include fast and accurate contouring and fast and accurate dose calculation. Accurate contouring relies primarily on good image quality, typically with high CNR. This is an area where MV-CBCT would likely underperform, relative

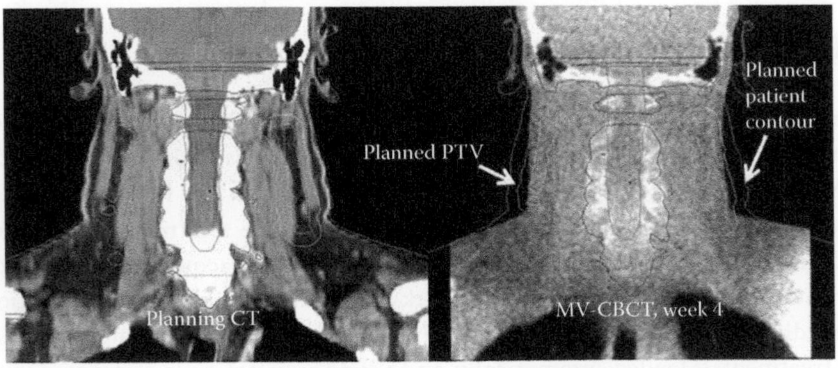

Figure 5.5 Clinical example of adaptive radiotherapy enabled by the use of MV-CBCT in a head-and-neck patient. After 4 weeks of radiation therapy, the patient lost a considerable amount of weight. The MV-CBCT image shows a decrease in neck diameter of 3 cm, with planning contours protruding outside the patient, potentially leading to misadministration. The patient underwent a midcourse resimulation and a new plan was designed for the remainder of the treatments.

to kV-CBCT. The literature is poor on the subject of autocontouring or propagating contours on MV-CBCT images, but it stands to reason that MV imaging is not the modality of choice, at least not without the need for manual contour editing, which adds time to the adaptive process. In terms of dose calculation, two parameters are of primary importance: the correspondence between CT number and electron density (CT-to-ED curve), and the uniformity of the image (independence of the CT-to-ED curve on location and patient size). An early study of dose calculation based on MV-CBCT showed the linearity of the CT-to-ED curve over the clinical range (Morin et al. 2007b). It was also demonstrated that with the proper anatomy-dependent uniformity correction algorithm, as discussed earlier in this chapter, dose-volume histograms from dose calculations on MV-CBCT and conventional CT were in excellent agreement. The same group proposed a *dose-guided radiation therapy* workflow, where the dose-of-the-day is calculated based on the image-guidance dataset, and can be used to determine when a patient requires replanning (Cheung et al. 2009). They demonstrated the applicability of MV-CBCT for that purpose.

The relatively small portion of low-energy photons in the MV-CBCT beam spectrum leading to a small probability of photoelectric interaction is a detriment to image quality in general, because of the loss of contrast. However, it allows for a reduction of streaking artifacts in patients with metallic implants, such as dental fillings, prosthetic hip, or surgical clips. With low-energy beams these high-density, high-atomic number materials cause photon starvation that greatly reduces visibility around them. With MV imaging, these artifacts are reduced significantly, as illustrated in Figure 5.6. In these cases, a MV-CBCT image can be taken at simulation time in conjunction with and registered to the planning CT, to help not only contouring but also assigning electron density in the affected region for dose calculation purposes.

(a)

(b)

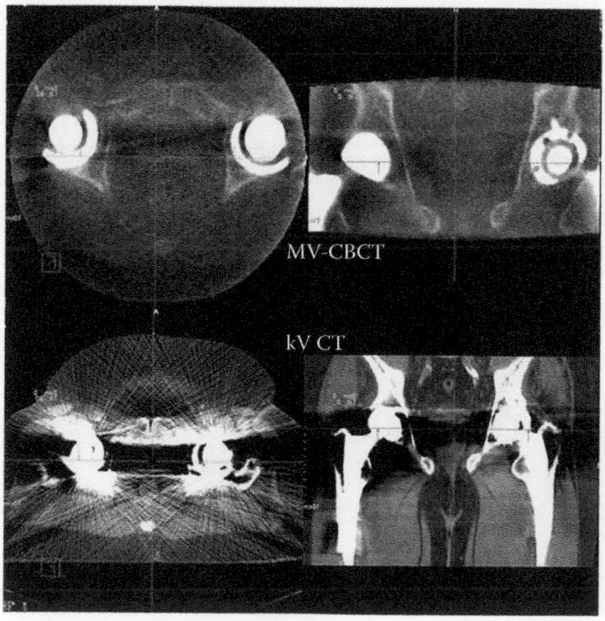

Figure 5.6 Illustration of photon starvation artifacts introduced by high-Z metallic implants in kilovoltage imaging (b). The anatomy of interest is not visible on the planning CT of this prostate patient with bilateral prosthetic hips. By contrast, the MV-CBCT image of the same patient shows a much reduced artifact due to the higher energy spectrum (a). This image can be used in conjunction with the planning CT for contouring purposes.

5.6 FUTURE OF MV-CBCT

The MV-CBCT system described in this chapter is the first-generation system developed and commercialized by Siemens Medical Solutions. There are two main ways to improve the image quality through increased CNR and spatial resolution, while maintaining the simplicity of the MV-CBCT system: (1) development of EPIDs with better detective quantum efficiency (DQE), and (2) increase of the low-energy photon fluence in the beam spectrum. These topics are covered in Chapters 12 and 8, respectively. However, it should be noted that the latter is the focus of the second-generation systems developed by Siemens, commercialized under the InLine kView™ name. The manufacturer created an imaging beamline in which components in the treatment head that contribute to low-energy photon absorption or scatter have been reduced to a minimum (Faddegon et al. 2008). The tungsten target was replaced with a low-Z carbon target. The tapered primary collimator was replaced with an untapered collimator. The flattening filter (FF) and mirror were removed and the ceramic monitor chamber was replaced with a thin-wall ionization chamber. Finally, the energy of the electron beam hitting the target was reduced from 6 to 4.2 MeV to curtail electron leakage from the target. As a result, the energy spectrum of the beam was significantly shifted down, which not only had the effect of increasing the high-contrast component of the beam but also to better match the detector-response window, thereby reducing noise. Measurements showed that the CNR was increased by a factor up to 3 compared to the first generation treatment beamline systems. The resulting effect is illustrated in Figure 5.7, which shows much better resolution of soft tissue even with an imaging dose of 2.5 cGy, similar to the contrast observed with 9 cGy in Figure 5.4.

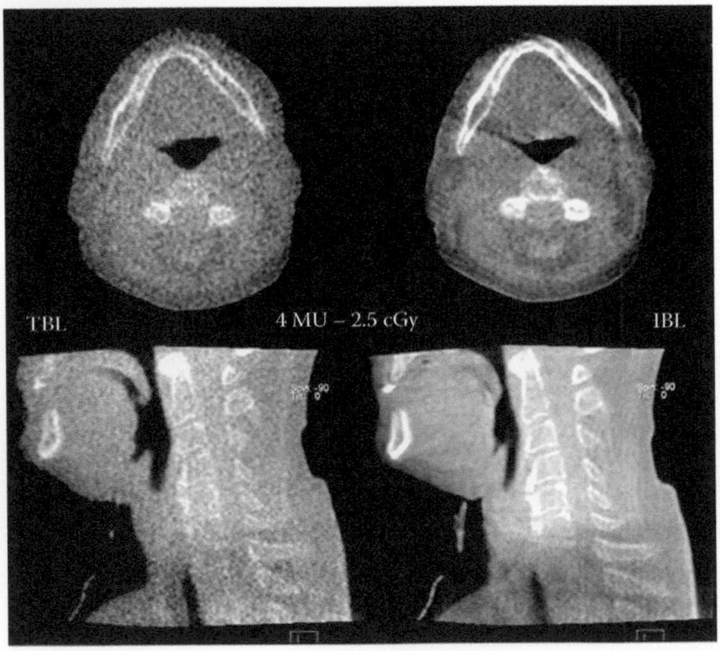

Figure 5.7 Illustration of the increased contrast-to-noise ratio obtained with a new imaging beamline with the same isocenter imaging dose for head-and-neck cancer patient. The new imaging beamline uses a 4.2 MeV electron beam striking a low-Z carbon target without flattening filter, resulting in a relative increase in the low-energy photon fluence in the beam spectrum.

Of the three major LINAC vendors who developed gantry-mounted CBCT systems, Siemens was the only one to pursue the simpler, more cost-effective route of MV energy imaging, at the cost of decreased image quality. Varian and Elekta both developed kV systems, with an X-ray tube and flat-panel detector mounted 90° from the treatment beam. However, in 2012, Siemens decided to pull out of the LINAC market, and stopped manufacturing and selling systems, while still offering services on existing systems until 2022. Therefore, there is effectively no commercially available new MV-CBCT system on the market for purchase at this time. The development work done by Siemens and their collaborators at the University of California, San Francisco (UCSF), however, provides a solid foundation for other manufacturers to design MV-CBCT systems if they found it of interest to add it to their imaging arsenal. The characteristics might then be slightly different than those discussed in this chapter, but the foundation would undoubtedly be the same.

REFERENCES

Cheung J et al. (2009) Dose recalculation and the dose-guided radiation therapy (DGRT) process using megavoltage cone-beam CT. *Int J Radiat Oncol Biol Phys* 74(2):583–592.

Deng J, Chen Z, Roberts KB, Nath R (2012) Kilovoltage imaging doses in the radiotherapy of pediatric cancer patients. *Int J Radiat Oncol Biol Phys.* 82(5):1680–1688.

Ding A, Gu J, Trofimov AV, Xu G (2010) Monte Carlo calculation of imaging doses from diagnostic multidetector CT and kilovoltage cone-beam CT as part of prostate cancer treatment plan. *Med Phys.* 37:6199–6204.

Faddegon B et al. (2008) Low dose megavoltage cone beam computed tomography with an unflattened 4 MV beam from a carbon target. *Med Phys.* 35:5777–5786.

Feldkamp LA, Davis LC, Kress JW (1984) Practical cone-beam algorithm. *J Opt Soc Am A.* 1(6):612–619.

Gayou O, Miften M (2007a) Commissioning and clinical implementation of a megavoltage cone-beam CT system for treatment localization. *Med Phys.* 34(8):3183–3192.

Gayou O, Parda DS, Johnson M, Miften M (2007b) Patient dose and image quality from megavoltage cone-beam computed tomography imaging. *Med Phys.* 34(2):499–506.

Gayou O, Miften M (2008). Comparison of megavoltage cone-beam computed tomography prostate localization with online ultrasound and fiducial markers methods. *Med Phys.* 35:531–538.

Gayou O (2012) Influence of acquisition parameters on MC-CBCT image quality. *J App Clin Med Phys.* 13(1):14–26.

Gayou O, Colonias A (2015) Imaging a moving tumor with megavoltage cone-beam computed tomography. *Med Phys.* 42(5):2347–2353.

Graff P et al. (2013) The residual setup errors of different IGRT alignment procedures for head and neck IMRT and the resulting dosimetric impact. *Int J Radiat Oncol Biol Phys.* 86(1):170–176.

Midgley S, Millar RM, Dudson J (1998) A feasibility study for megavoltage cone beam CT using a commercial EPID. *Phys Med Biol.* 43:155–169.

Miften M, Gayou O, Reitz B, Fuhrer R, Leicher B, Parda DS (2007) IMRT planning and delivery incorporating daily dose for megavoltage cone-beam computer tomography imaging. *Med Phys.* 34(10):3760–3767.

Morin O (2007). The Development and role of megavoltage cone-beam computed tomography in radiation oncology. Joint Thesis. University of California, San Francisco, CA and University of California, Berkeley, CA.

Morin O et al. (2007a) Patient dose considerations for routine megavoltage cone-beam CT imaging. *Med Phys.* 34:1819–1827.

Morin O et al. (2007b) Dose calculation using megavoltage cone-beam CT. *Int J Radiat Oncol Biol Phys.* 67(4):1201–1210.

Morin O et al. (2009). Physical performance and image optimization of megavoltage cone-beam CT. *Med Phys.* 36:1421–1432.

Morton EJ, Swindell W, Lewis DG, Evans PM (1991) A linear array scintillation crystal-photodiode detector for megavoltage imaging. *Med Phys.* 18:681–691.

Mosleh-Shirazi MA, Swindell W, Evans PM (1998) Optimization of the scintillation detector in a combined 3D megavoltage CT scanner and portal imager. *Med Phys.* 25:1880–1890.

Motegi K et al. (2014) Evaluating positional accuracy using megavoltage cone-beam computed tomography for IMRT with head-and-neck cancer. *J Radiat Res.* 55(3):568–574.

Packard M, Gayou O, Gurram K, Weiss B, Thakkar S, Kirichenko A (2015) Use of implanted gold fiducial markers with MV-CBCT image-guided IMRT for pancreatic tumours. *J Med Imaging Radiat Oncol.* 59(4):499–506.

Pouliot J et al. (2005) Low-dose megavoltage cone-beam CT for radiation therapy. *Int J Radiat Oncol Biol Phys.* 61:552–560.

Simpson RG, Chen CT, Brubbs EA, Swindell W (1982) A 4 MV CT scanner for radiation therapy: The prototype system. *Med Phys.* 9:574–579.

6

Pretreatment EPID-based patient-specific QA

PETER GREER

6.1 INTRODUCTION

6.1.1 Pretreatment QA

Widespread clinical implementation of intensity-modulated radiation therapy (IMRT) using conventional linear accelerator (LINAC) equipment increased rapidly in this century following a decade of development of the equipment and techniques. In the early years of clinical adoption, there was much debate in the medical physics and radiation oncology community on the quality assurance (QA) requirements and procedures that were needed for this new technology. Previous QA paradigms relied on the fact that essentially a small library of delivered fluences was used for clinical treatment using either the open field or a set of wedged field fluence profiles. These fluences could, therefore, be verified with routine regular QA techniques that evaluated beam symmetry, beam flatness, and wedge profiles. However, the new IMRT fluence deliveries were unique to every patient plan and involved a much more complex planning and delivery sequence using technology and techniques that were new to most treatment centers. Naturally, a conservative approach was taken and the dominant view became that these treatments should be individually measured to ensure that the fluence or dose delivered for the patient corresponded to the treatment plan known as pretreatment QA. Some centers later scaled back the use of pretreatment QA on an individual patient basis once they felt they had gained sufficient experience with the techniques and confidence in the delivery accuracy. However, the majority of centers continue to perform pretreatment QA for their IMRT patients with some jurisdictions making this a mandatory component of patient treatment. The debate on the requirement to perform pretreatment QA on every patient is still ongoing, with arguments that, for example, rigorous equipment QA combined with independent planning calculations or combinations of these may suffice. It is not the purpose of this chapter to enter this debate, rather to accept that the medium-term future in radiation oncology will see continued widespread pretreatment QA.

There is considerable variety in the methods and equipment used to perform pretreatment QA; however, they mostly follow the same general principle. The patient treatment plan is used to calculate a dose on a phantom that corresponds to the measurement phantom or dosimeter that will be used on the LINAC. This is referred to here as the reference plan. The reference or patient plan is then delivered at the LINAC and the dose delivered is measured with the dosimetric system. The measured dose and the reference dose are then compared using an evaluation metric usually the gamma evaluation criteria [1]. The reference plan and measurements can be performed on an individual beam-by-beam basis or with a combined delivery. Particularly for volumetric modulated arc therapy (VMAT), a combined delivery approach is often used. A variety of commercial dosimetry systems are available and widely used for these measurements. The major drawbacks of these systems are the setup time involved and the generally low resolution or sampling of the delivered dose.

This chapter will discuss the role of electronic portal imaging devices (EPIDs) in this context and outline how they have been used for IMRT and VMAT individual patient pretreatment QA. This will be limited in scope to EPID nontransmission dosimetry measurements performed prior to treatment. Nontransmission dosimetry is defined by van Elmpt et al. [2] as "determination of the dose in the detector, patient or phantom or determination of the incident energy fluence, based on measurements without an attenuating medium between the source and the detector, i.e. phantom or patient." This is distinct from transmission (or transit) dosimetry where the beam has passed through the patient and is used to assess delivered dose either at the plane of the EPID or within a phantom or patient dataset (*in vivo* dosimetry). The latter is discussed in more detail in Chapter 11. Although a considerable body of work exists to perform dosimetry with camera-based or other design EPIDs, these have been superseded by flat-panel amorphous silicon (aSi)–based technology for some time now and therefore will not be discussed. The advantages and disadvantages of aSi-type EPIDs for pretreatment QA will be outlined along with the current methods employed for their use and some potential future developments.

6.1.2 Brief dosimetric properties of EPID

6.1.2.1 IMAGE FORMATION

There is considerable literature on the dosimetric properties of EPIDs and these will only be briefly introduced here. For more information, the reader is referred to review articles by van Elmpt et al. [2] or Greer and Vial [3]. In a typical aSi EPID design, X-rays and electrons produced in the copper conversion layer interact with the phosphor layer (terbium-doped gadolinium oxysufide) to produce visible light photons that then interact in the aSi-based photodiode array. A small percentage of the signal produced in the photodiode array is due to direct detection of X-rays and electrons. This has been experimentally measured by blocking light from the phosphor layer from reaching the photodiodes to be approximately 8% of the total (light + direct) signal [4].

6.1.2.2 EPID SCATTER

Scatter of the incident beam will occur in the conversion layer and the phosphor layer and potentially also in layers distal to the detector. Both forward and backward radiation scatter can contribute to dose deposition in the EPID. Considerable optical scattering also occurs in the phosphor layer [5,6]. The EPID response is also affected to an extent by low-energy head scatter that varies with field size and which is not completely filtered out by the conversion layers of the EPID [7]. This is particularly apparent at high energies such as 18 MV where the field-size response of the EPID varies more than that measured with ion chamber [7]. The EPID energy distribution due to a delta-impulse of photon fluence (the *dose kernel* of the EPID) can be modeled directly with

Monte Carlo (MC) methods, or measured with line-spread function experiments. Comprehensive models of EPID scatter that include optical photon scatter and energy deposition in the photodiodes have been developed by Kirkby et al. [6] and more recently by Blake et al. using the GEANT4 MC system [5]. They concluded that modeling of EPID dosimetry using energy deposition in the phosphor layer without explicit modeling of optical scatter provided an adequate description of EPID response.

A feature of Varian EPIDs that is not present for other vendors is backscattered radiation from the support arm that has components seated directly underneath the active detection layer. Backscattered radiation introduces an additional nonuniform backscatter signal to the image, which is dependent on where the fluence impinges on the detector [7–9]. Introducing a shielding layer with thicknesses of 2–5 mm of lead between the support arm and the detector to intercept the backscattered radiation has been found to be effective [8,10–12]. The newer Varian aS1200 EPID design contains a shielding layer between the support arm structure and the active detector. Experiments with a prototype model showed that the backscatter signal was reduced to less than 0.5% of the total signal by this shielding [13]. Methods to model or correct the backscatter artifact from acquired images have also been developed and are discussed in Section 6.3.1.3.

6.1.2.3 ENERGY-DEPENDENT RESPONSE

EPIDs exhibit a strong energy-dependent response due to the high atomic number conversion and phosphor layers. As the photoelectric effect is much more probable at low energies, low-energy X-rays and electrons are preferentially absorbed in the phosphor layer producing a much greater EPID signal per unit incident energy than for higher energy X-rays and electrons.

This energy-dependence manifests as an increased response to off-axis radiation due to the lower average photon energy [14,15]. At less than 15 cm off-axis, the EPID response is already 10% higher than ion-chamber response relative to central axis. This effect is much reduced in flattening filter free (FFF) beams. This off-axis dependence is corrected by the flood-field division of images that are acquired, producing a flat-EPID profile. However, this also removes beam off-axis dosimetric information. The energy dependence also affects measurement of IMRT beams where a significant proportion of the signal produced in a given pixel can be due to multileaf collimator (MLC) transmission. As this is a considerably harder beam than the open field, the EPID response is reduced. The result is a reduced signal in low-dose regions where the majority of the dose is due to MLC transmitted beam. This effect is of the order of 30% in the relative EPID response to open and MLC transmission components, which has been experimentally measured and MC modeled [15,16].

6.1.2.4 PIXEL SENSITIVITY VARIATION

Each individual pixel in the photodiode matrix can exhibit a unique offset or background signal (no radiation present) and unique gain response to incident radiation. These are corrected using a dark-field image and a flood-field image or gain image. The dark field is acquired with the beam off and is subtracted from the raw images. The flood field is acquired with the field completely irradiating the imager to simulate a uniform incident beam fluence, and raw images are divided by the flood-field image. As discussed earlier, the flood-field image signal increases with off-axis distance due to the energy variation in the incident beam. This has two effects: (1) this off-axis response is removed from corrected EPID images and (2) artifacts are introduced if images are acquired at a different EPID position from the flood-field acquisition position [17]. An alternative approach has been proposed where the pixel sensitivity variation (matrix) is measured without the additive off-axis response and is used to correct raw EPID images [14,17].

6.1.2.5 RESPONSE LINEARITY WITH DOSE

Another dosimetric feature of EPIDs that should be considered for dosimetry is the deviation from linearity that can occur for small monitor unit (MU) deliveries or with particular types of imaging modes. These deviations can be due to incomplete recording of the delivered signal due to missing frames at the start or end of acquisition [18] and image lag and ghosting effects [19–26]. Recent work has shown that the majority of nonlinearity effects at low MU are due to incomplete capture of partial dose frames, which can be largely corrected by vendor acquisition improvements [18,27]. Ghosting is a decrease in pixel sensitivity, and hence EPID responses in an irradiation due to a prior irradiation while lag is the charge carry over to subsequent frames due to incomplete detector readout [19]. The nonlinearity can be measured and correction factors derived as a function of delivered MU.

This effect is also important for cine imaging used for VMAT dosimetry, which is discussed in Section 6.4.1.1.

6.1.3 UNCERTAINTIES IN EPID NONTRANSMISSION DOSIMETRY

The uncertainties in the measurements made of IMRT or VMAT deliveries with EPID can be grouped into three broad categories: (1) uncertainties in the spatial localization of the detector measurement with respect to the beam; (2) uncertainties in the measured image values; and (3) uncertainties in conversion of image to dose or prediction of EPID image, depending on which type of method is being used.

Some form of image registration is required to compare EPID images, which are acquired in the pixel-based coordinate system of the EPID and the reference plan dose, which is in a coordinate system referenced to the beam isocenter. The planning system also assumes a perfect isocenter that does not vary with gantry angle and LINAC uncertainties are ignored. The simplest approach is to determine the beam isocenter location on the EPID panel to translate pixel coordinates to isocenter referenced coordinates. This can be achieved by some form of calibration generally using field centers, average of 180° rotated field centers, graticule-type phantoms, matching of image to the corresponding beam outline from the plan, or more advanced calibration phantoms [28–31]. This calibration can also be extended, so that the isocenter location on the panel is described as a function of gantry angle. Measurements will generally include a combined EPID sag/flex and gantry wobble of the LINAC. A method that separates gantry and EPID sag components has been reported and used to study variations across multiple LINAC platforms [29,32]. Newer EPID systems such as the ones on the Varian Truebeam can also reposition the EPID at any gantry angle based on a pre-measured sag/flex calibration function. Provided careful calibration measurements are performed and the EPID sag/flex is stable over time, the uncertainty due to EPID position can be small and certainly, submillimeter accuracy is achievable.

Some software systems that compare EPID to measured or predicted doses do not use this form of calibration and perform either manual or automatic rigid registration of the EPID data to the reference data. Although this can overcome positioning uncertainties, it could also potentially remove real differences in spatial location of the delivered dose (e.g., MLC position systematic offsets).

Uncertainties in the measured image values can arise from a number of sources. Pixels and electronics can degrade and become defective over time due to radiation damage and age.

This can cause increased noise in the images, unstable, or missing signal in pixels or panel areas, and changes in pixel sensitivity or gain. Careful monitoring of EPID performance as part of a pretreatment QA program is essential to ensure consistent results. These programs can include frequent dark and flood-field (gain) calibrations, if required. Another effective approach is to normalize acquired images with a large image acquired of the entire detector. This can remove variations in the pixel response; however, it will also remove any real changes in beam profile that may have occurred. This can also be a useful method to correct for dosimetry performed at EPID positions different from the flood-field calibration position, as this difference in position introduces artifacts [17]. Other sources of signal uncertainty include those described earlier such as backscatter and nonlinear response at smaller MU settings due to incomplete signal capture and/or ghosting and image lag characteristics. Backscatter artifacts, if not corrected, can degrade results and can be difficult to interpret. In the absence of a correction method, potential backscatter work-arounds can include collimator rotations to minimize the beam incident on the support arm or to place the long axis of the beam orthogonal to the main direction of the flood-field division backscatter component.

Finally, uncertainties in conversion of image to dose or prediction of EPID image are important. All models have limitations and inaccuracies. It is important that the user understands the model that is being used and identifies potential limitations that could lead to discrepancies between EPID results and treatment planning system (TPS) or modeled EPID images. A thorough testing of the model used over a wide range of conditions is essential.

6.2 NONDOSIMETRIC METHODS

6.2.1 Multileaf collimator position verification

With very high-resolution pixels (0.3–0.4 mm) and large detector areas, EPIDs have the potential for the measurement of MLC position during IMRT and VMAT deliveries (Figure 6.1). These could provide measurements to validate trajectory log files or separately as an independent measure of leaf position, that can be compared to the planned leaf trajectories or used to reconstruct delivered dose [33]. EPID measurements must be at a high sampling rate to avoid MLC position blurring on the images and to provide sufficient leaf position localization accuracy. Sampling rates of EPIDs were typically in the 2.5–10 Hz range until recently when higher speeds of up to 25 Hz have become available. A prototype complementary metal-oxide-semiconductor (CMOS) detector with a sampling rate of up to 400 Hz has been proposed as an alternative system for these types of measurements [34]. With current clinical systems, it can be difficult to acquire and access the individual cine images or image frames at a sufficient sampling rate required for accurate MLC position analysis. Alternatively, separate frame-grabber systems that use a one-way connection to the clinical system to receive image frames during acquisition are being utilized in the research setting and can provide the required image data as discussed in Section 6.4.1.1.

Another challenge with these measurements is that the EPID operates on a time basis with each image acquired in regular fixed time intervals, whereas the treatment planning MLC positions are specified at control points as a function of delivered MU for IMRT or gantry angle for VMAT. Therefore, the common approach is to determine the delivered MU (gantry angle) as a function of time during the delivery that allows the EPID measured MLC positions to be synchronized to the treatment plan. Several early approaches used the MU signal from the LINAC with a CCD-based EPID system using custom-built circuits to trigger acquisition at specified dose points or by applying percentage MU to measured frames [35–38]. Determining the gantry angle as a function of time during VMAT delivery can be challenging and is discussed later in Section 6.4.1.3.

Analysis methods for detecting and comparing MLC positions have varied with many reports qualitatively overlaying detected MLC positions and planned positions for real-time analysis during delivery with more quantitative off-line analysis [39–42]. Real-time quantitative analysis of MLC trajectories has also been reported [43]. Methods to verify positioning during MLC tracking deliveries have also been developed using planned apertures adjusted for target motion projected onto the EPID [44]. Considerations must also be made when comparing EPID-derived MLC positions and TPS or MLC delivery file positions as these can specify different leaf position parameters. Leaf offset tables applied at the LINAC also have to be considered. An interesting new development with the previously mentioned frame-grabber systems is that leaf positions are written into image information rows enabling real-time comparison using only acquired image data.

6.2.2 Dose recalculation or reconstruction

Measured leaf positions can also be used to recalculate the dose delivered to a patient model [33,45]. Potentially these or similar methods could be applied to pretreatment QA. These methods are particularly useful for MLC tracking deliveries where the planned apertures are no longer actually delivered because the MLC is constantly adjusting position in response to a target motion signal.

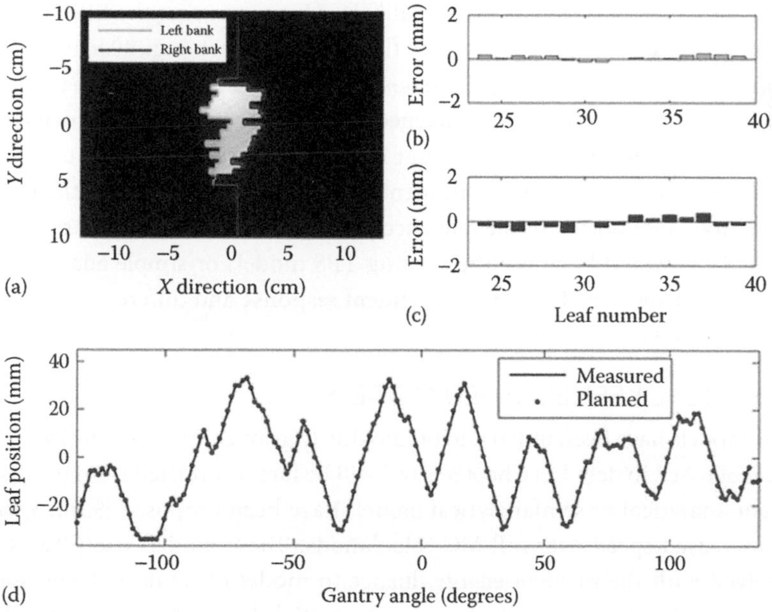

Figure 6.1 A measured beam aperture from an EPID image frame. (a) The planned MLC positions at the corresponding gantry angle have been superimposed on the grayscale EPID image; the difference between the planned and measured MLC positions at this measured EPID image for (b) the left bank, and (c) the right bank; and (d) a measured single leaf trajectory compared to the planned leaf trajectory. The solid line represents the EPID measured trajectory and the points show the MLC position specified at each control point. (From Zwan, B.J. et al. An EPID-based system for gantry-resolved quality assurance for VMAT, Journal of Applied Clinical Medical Physics, 2016, 17(5): 348–356.)

The recorded delivered apertures are used to construct *delivered* DICOM plan files that can be imported to the TPS for dose calculation. These methods are discussed in detail in Chapter 11.

6.3 TECHNIQUES FOR PRETREATMENT DOSIMETRIC IMRT VERIFICATION WITH EPID

Various commercial softwares are available to perform pretreatment verification using EPID acquired images. These include, for example, the Varian Portal Dosimetry, epiQA, EPIdose, Dosimetry Check, and Adaptivo. These systems use principles and methods that are discussed in the following sections.

6.3.1 SIMULATION OF EPID GRAYSCALE VALUES

There have been two major approaches to dosimetry with EPID: (1) models that convert EPID signal to water-equivalent dose, and (2) models that predict the EPID grayscale signal using information from the treatment plan, referred to here as simulation of grayscale values. This latter approach remains to date the most widely used in clinical EPID pretreatment QA. These types of models independently calculate the EPID signal expected from delivery of each beam of the patient treatment plan. The two major components of these prediction models are to (1) model the fluence in-air from the accelerator head due to the prescribed MLC motions, MU delivery, and jaw positions, and (2) model the interaction of this fluence in the EPID layers to produce the EPID signal from the beam. The absolute grayscale can be calibrated by comparing the EPID signal for a reference delivery to the model prediction value with the same geometry (e.g., 10×10 cm^2, 100 MU, and EPID at a specified position). As the measured dose is not directly compared to the TPS dose, these types of models can confirm that expected fluence calculated from the plan is delivered accurately by the LINAC. They do not necessarily account for errors in modeling by the TPS, for example, MLC effects. If the model uses the same incident fluence model as the TPS, then only scattering and dose-deposition components in the reference phantom dose calculation are not included. An attractive proposition would be to use the existing TPS models or simple adaptations of these to calculate EPID dose. However, the energy-dependent response and difference in scatter kernels of the EPID and water make this challenging.

6.3.1.1 ANALYTICAL KERNEL-BASED MODELS

MC prediction models have been utilized to predict EPID grayscale images for IMRT beams; however, these are discussed in detail in Chapter 3 and will be largely omitted from the discussion here. Several different analytical or semianalytical models have been proposed [9,16,23,46,47] with the advantage of increased speed over full MC calculations. These models use EPID scatter kernels that are convolved with the incident-energy fluence to model EPID dose deposition. The kernel represents the normalized point spread function or spatial distribution of energy deposition per unit incident energy for a delta-impulse of photon fluence. It is reasonable to assume that the dose-deposition kernel is independent of the particular location on the EPID as the EPID layers are uniform in structure. This assumption breaks down to an extent with support arm backscatter in the Varian system; however, this can be accounted for in an additional backscatter calculation model that is additive to the basic uniform EPID dose-deposition model. The EPID properties are highly energy dependent; however, and particularly the primary dose deposition increases considerably with off-axis distance. The incident energy fluence will be comprised of both primary and head-scattered components and will include both photons and electrons. Both analytical and MC methods have been incorporated to derive the incident-energy fluence distribution at the plane of the EPID [16,48,49].

A comprehensive analytical mode was reported by Chytyk et al. [49]. Their model includes multiple monoenergetic EPID dose kernels calculated with MC modeling, each describing the EPID dose-deposition pattern in the phosphor layer for a monoenergetic incident pencil beam of photon fluence:

$$D(\mathbf{r}) = \sum_{i=1}^{N} \Psi(\mathbf{r}, E_i) \otimes K(r, E_i)$$

where:

$D(\mathbf{r})$ is the predicted EPID image signal distribution integrated over all incident photon energies
$\Psi(\mathbf{r}, E_i)$ is the incident-energy fluence distribution at the plane of the phosphor layer for energy bin E_i
$K(r, E_i)$ is the normalized EPID dose-deposition (scatter) kernel, assumed to be radially symmetric
$\otimes$ is the convolution operator

Note that the energy dependence of the EPID image formation is accounted for by summing the convolution of incident fluence with monoenergetic EPID dose kernels, over all discrete energy bins in the fluence spectrum [50]. The incident fluence distribution for each energy bin is calculated using a two-source focal and extrafocal model that models both a primary fluence distribution and a scattered fluence distribution. Specific aspects of the MLC and secondary collimators were also modeled including jaw and MLC transmission, MLC rounded leaf tips, tongue and groove effect, and interleaf leakage. This type of model can comprehensively account for the energy-dependent response of the EPID at the expense of increased calculation time. The authors also determined in a separate work that the nonperpendicular incidence of the primary beam off-axis that results in *tilted* dose-deposition kernels could be ignored, and kernels calculated with perpendicular incidence can be used [51].

Simpler versions of the above-mentioned model have been developed that use polyenergetic dose-deposition kernels. These kernels operate on the integrated or total energy fluence distribution (subscript T indicating total energy):

$$D(\mathbf{r}) = O(r)\Psi_T(\mathbf{r}) \otimes K_T(r)$$

Note that a radially dependent off-axis response factor $(O(r))$ is now required to model the energy dependence of the EPID to lower energy components off-axis. In practice, this factor can be ignored when predicting flood-field corrected EPID images, as the flood field acts to cancel the off-axis dependence of the EPID. To predict EPID images that are corrected only for pixel sensitivity, then the off-axis dependence is required.

A two-kernel model was developed by Li et al. based on separation of primary and MLC transmitted energy fluence components [16]. An estimate of the energy fluence distribution at the EPID plane for the open beam and the MLC transmitted beam components is calculated using an MC LINAC model. Separate kernels were derived from an MC model of the EPID for the open beam-energy spectrum and the MLC transmitted energy spectrum. These were also calculated as a function of off-axis distance to model the off-axis energy dependence. The central magnitude of the kernel was ~30% lower for the MLC transmission than for open beam due to the energy-dependent response of the EPID that agrees well with experimental measurements [14,15]. This model was further developed by Wang et al. [52] to include detector-dependent kernels that account for the differences in field-size response for different EPIDs. This was done by varying the thickness of backscatter material used in the MC calculation of the kernels (Figure 6.2).

6.3.1.2 VARIAN PORTAL DOSIMETRY METHOD

A simple polyenergetic kernel model, the widely used van Esch model [23] for the Varian Portal Dosimetry image per MU is expressed as

$$PD(x, y, \mathrm{SDD}) = [(F(x, y, \mathrm{SDD}) \cdot \mathrm{OAR(SDD)} \otimes RF_{PI}] \cdot CSF_{XY} \cdot 1/\mathrm{MUfactor}$$

$F(x, y, \mathrm{SDD})$: It is the incident fluence scaled to the EPID source-detector distance (SDD). This fluence is generated by the Eclipse TPS and for each point is the fraction of the MU that receives open beam through the MLC. This is sometimes referred to as an open-density matrix. This fluence does not contain any off-axis dependence of fluence; however, it does consider MLC transmission and the offset of the 50% field-edge intensity from the rounded leaf-edge tip of the Varian MLC.

$OAR(SDD)$: It is an off-axis factor matrix that is entered by the user during commissioning and is usually a dose profile measured in water at depth = $d_{\max}$.

RF_{PI}: The kernel used in the convolution is a triple Gaussian function. The Gaussian parameters are optimized using comparison of model prediction to a measured EPID image for a pyramid-like test pattern made with static MLC shapes.

$$RF_{PI} = \sum_{i=1}^{3} exp^{-(r/k_i)^2} \text{ with } \sum_{i=1}^{3} w_i = 1$$

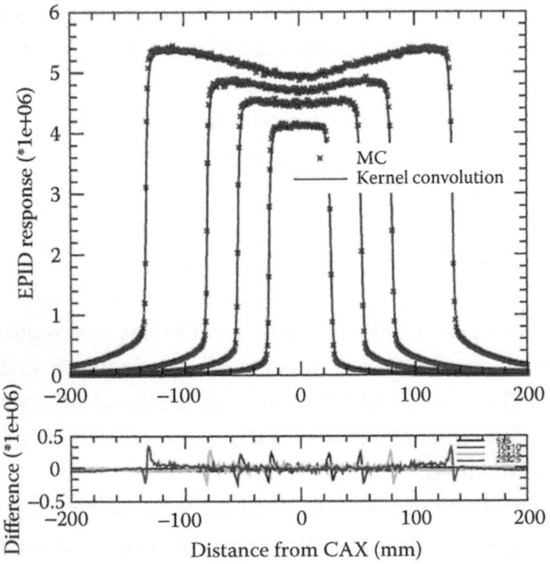

Figure 6.2 Agreement between an analytical model and a Monte Carlo model predicting EPID grayscale response for open fields of 5 × 5, 10 × 10, 15 × 15, and 25 × 25 cm² fields. (From Wang S. et al., *Med. Phys.*, 36(8), 3582–3595, 2009.)

CSF_{XY}: As the fluence model does not consider an output factor due to the collimator opening, a separate field-size dependent scalar factor is included as a look-up table. This is based on the ratio of measured to predicted images at central axis for various symmetric and asymmetric fields. This factor is normalized to the model prediction per MU under calibration conditions of an open 10×10 cm^2 field at 100 cm source-EPID distance. This normalization is expressed in CU units, a custom image unit used by Varian. This normalization factor is usually set to 0.01 CU per MU, so that in the above calibration conditions a planned 1 MU will result in a 0.01 CU prediction model value. The measured EPID image must similarly be calibrated to CU units by recording the image under the calibration conditions.

$1/MUfactor$: To obtain a prediction per MU, the prediction is divided by the MUfactor. To determine the final Varian Portal Dosimetry image, the model result per MU must be multiplied by the MU set for the IMRT beam.

In addition to the CU image calibration, an off-axis factor matrix is applied to the measured image to correspond to the OAR(SDD) added to the prediction model. This is because the incident fluence profile is removed from the EPID image by the flood-field calibration procedure. Usually the same dose in water profile at d_{max} is used. A limitation of the Varian portal dosimetry model is that there is no explicit fitting of the off-axis model predictions and the measured EPID data. To improve the agreement of off-axis predictions and EPID measurements, Bailey et al. [53,54] modified the off-axis factor matrix based on fitting to measured EPID images.

Correction factors to account for the EPID response to the MLC transmitted component at off-axis distances have also been empirically derived [55]. This work identified that off-axis correction factors were required to give good agreement to EPID measurements even for open beams in agreement with Bailey et al. [53,54]. Howell et al. reported a gamma analysis of 1152 treatment fields from 152 treatment plans using the Varian Portal Dosimetry product [56].

6.3.1.3 BACKSCATTER MODELING

For the Varian aS500/1000 systems, a variety of correction methods have been proposed. The simplest and most widely used is to correct the backscatter artifact in the flood-field image. The flood-field image itself has a large backscatter signal component. Therefore, the inverse of this component is introduced to flood-field corrected images by division of the measured image by the flood-field image. By modeling or measuring the backscatter signal component in the flood-field and removing this from subsequent flood-field corrected images, dosimetry for smaller fields can be improved [57,58]. However, larger fields will exhibit greater backscatter artifact, so this method can be considered to shift the backscatter artifact from small fields to larger fields. An implementation of this method has been included in the Varian Portal Dosimetry system using the beam profile correction matrix that is applied to the measured flood-field corrected images [58,59].

Some more sophisticated methods that consider the field-size dependence of the backscatter have been developed based on explicit modeling of the backscatter [60,61] or empirical corrections based on measurements with varying field sizes [62]. As the backscatter introduces asymmetry largely into the in-plane direction of the EPID signal, a method has been derived that uses reflection of pixel values about the crossplane center of the EPID for open fields to determine a series of field-size dependent correction matrices [62]. MC modeling approaches have also been developed where the support arm is modeled as a set of slabs of nonuniform thickness [63]. The model of Wang with kernels developed for variable backscatter thickness can model the effect of backscatter on central axis field-size dependence but not the off-axis asymmetry. The thickness of a uniform water slab varied from 1 to 1.6 cm for different imagers with a 2.1 cm air gap to the detector layer [52]. Rowshanfarzad et al. [60] derived a backscatter kernel from empirical measurements using narrow beams incident on the EPID with the support arm present and with the EPID removed from the support arm. A curve fit to the subtraction of EPID response measured off arm from that measured on arm was used to determine the backscatter kernel. The backscatter prediction model is as follows:

$$S(\mathbf{r}) = WF \cdot \Psi_T(\mathbf{r}) M(\mathbf{r}) \otimes B(\mathrm{r})$$

where:

$S(\mathbf{r})$ is the backscatter component of the EPID image

WF is the weighting factor to model the additional component of EPID central axis signal due to backscatter

$M(\mathbf{r})$ is a binary mask that isolates the incident fluence component that will interact with the backscattering material

$B(\mathrm{r})$ is the backscatter kernel that is modeled as a broad Gaussian function representing the fact that the backscattered radiation is widely scattered

This model was added to an analytical EPID prediction model and an improvement in EPID image prediction was demonstrated for open and IMRT fields. Removal of the backscatter component of the measured image is also feasible. An adaptation of the above model by King et al. [61] has been derived to iteratively estimate and remove the backscatter component of an acquired image. More recently a similar algorithm was developed by Podesta et al. [64].

6.3.2 Conversion of grayscale values to dose in water

To simulate water equivalent dosimetry with an EPID, models that convert EPID signal to water-equivalent dose have been developed. Using EPID data acquired in-air, the dose in a phantom or patient model is estimated. The following models in many cases derive from earlier work performed with video-camera–based EPIDs and scanning liquid ion-chamber–based EPIDs.

6.3.2.1 Deconvolution–convolution methods

A dose plane or planes in a water equivalent phantom can be estimated from the EPID image using kernel methods similar to the analytical prediction models discussed earlier. The most common method is deconvolution of the EPID image with an EPID scatter kernel to obtain incident fluence followed by a model that calculates dose in a water phantom from the incident fluence. As these models begin with the energy integrated EPID image, the energy dependence of the EPID response is difficult to incorporate and they generally use polyenergetic kernels to derive the model. Deconvolution to obtain incident fluence can be expressed as follows:

$$\Psi(\mathbf{r}) = C(r) D(\mathbf{r}) \otimes^{-1} K(r)$$

where $C(r)$ is an off-axis correction factor. For the case of a flood-field corrected EPID image, the correction factor can restore the radially dependent profile of the incident fluence. For an EPID image corrected only with a pixel sensitivity matrix, the correction factor will remove the off-axis energy dependence of EPID response.

One of the early methods used with modern flat-panel imagers derived an EPID scatter kernel using MC modeling of the dose deposition in the phosphor layer from a delta fluence beam combined with an empirically derived optical scatter kernel [65]. This was used to derive primary fluence by deconvolution with the image. The fluence was then convolved with an MC-generated dose in water kernel to calculate dose at a defined depth. The fluence was then compared to 2D dose distributions measured with film in a solid water phantom for open fields and three IMRT fields.

Further work calculated the EPID scatter kernel using MC modeling that included an explicit MC model of the optical light scattering in the gadolinium phosphor layer [6]. The MC kernel was also compared to measured line-spread function data for validation. EPID images deconvolved with the kernel were compared to in-air fluence measured with a diamond detector. The inclusion of the optical kernel improved agreement with the measured fluence. An analytical three exponential function was also fitted to the MC kernel results.

More recently, a similar approach was developed using the functional kernel form derived by Kirkby et al. to obtain incident fluence using the above formalism [13]. This was combined with an analytical dose in water calculation at four different depths using depth-dependent kernels combined with two off-axis factor functions. An implementation of this model by Varian Medical Systems has been presented with dose calculated at 5 cm depth in water for a Truebeam accelerator with an aS1200 backscatter shielded EPID. The model results were compared to MatriXX ion-chamber array results for various test and IMRT fields (Figure 6.3). A correction for the difference in EPID response to primary and MLC transmitted beam components have also been derived that is applied to the EPID image before dose calculation [66]. A similar model formalism was recently reported by Podesta et al. [64] based on the earlier model developed by Nijsten et al. for a Siemens LINAC and Perkin–Elmer EPID [67]. Dose in water is determined at 5 cm depth from EPID images using exponential function kernel parameters obtained by fitting directly to dose-in-water data.

6.3.2.2 DIRECT CONVOLUTION METHODS

The dose in water phantom can also be estimated without the deconvolution to fluence, with a direct convolution of the EPID image with a kernel to estimate dose-in-water plane.

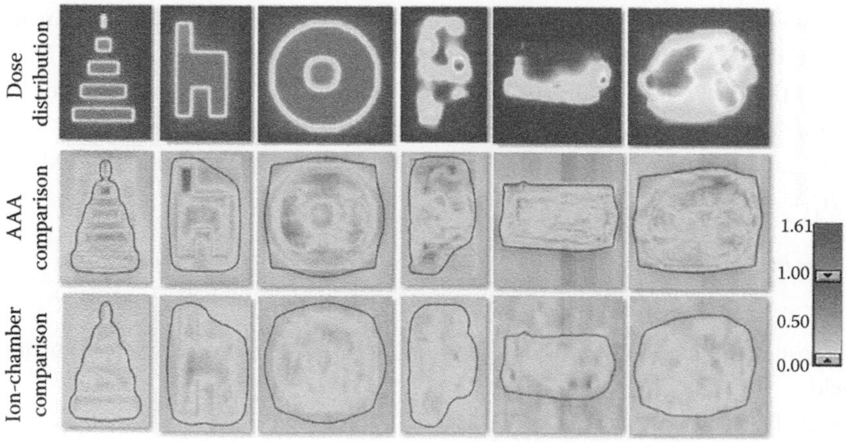

Figure 6.3 EPID images converted to dose-in-water with a deconvolution–convolution approach compared to TPS calculations and measurements for 6 MV IMRT fields. First row: dose distributions. Second row: Gamma comparison to AAA prediction. Third row: Gamma comparison to 2D ionization chamber array. Gamma comparisons for 3%/3 mm. Dark line: 5% isodose line. (From Keller P, EPI2K14.)

$$D_{w,d}(\mathbf{r}) = D(\mathbf{r}) \otimes K_{\text{epid}\to w}(r)$$

where $K_{\text{epid}\to w}(r)$ is a kernel designed to estimate dose-in-water from the EPID signal directly. This limits the depths in water that the dose can be estimated to depths greater than the depth where scattering in the EPID and water are similar, which is around 5–6 cm or greater. This method was used with kernel parameters derived with optimization to calculate dose at 10 cm depth in water in a virtual flat phantom [68]. A similar approach was developed by Nelms et al. [69] and used in the EPIDose commercial software. They also incorporated a correction factor to the EPID response based on each segment size, similar to the calibration method of Nicolini et al. [70] described in the following section.

6.3.2.3 CALIBRATION METHODS

Nonconvolution calibration-based methods to estimate a dose-in-water from the EPID image have also been developed. The simplest method of EPID pixel conversion to dose-in-water was developed by Lee et al. [71]. This method determines the depth in water that most closely matches the scattering properties of the EPID by comparing field-size response and beam penumbra shape for various depths in a water phantom to EPID data. After determining the depth, EPID pixel values are converted to dose-in-water plane at that depth simply by multiplying the EPID signal by a calibration factor, which is the ratio of the dose at the central axis at that depth-in-water for a 10 × 10 cm^2 field to the EPID signal for the same field size given the same number of MUs.

$$D_{w,d_{opt}}(\mathbf{r}) = CF \cdot D(\mathbf{r})$$

where CF is the calibration factor and d_{opt} is the depth-in-water. This method was found to yield gamma pass rates when compared to TPS calculations averaging 97% for 14 IMRT fields with 3%/3 mm, >10% of maximum dose threshold criteria using a depth of 5 cm in water for a 6 MV beam.

Methods based on linear calibration curves of EPID pixel responses to dose-in-water have been investigated. Chang et al. [72] used a calibration method for aSi EPIDs similar to that used previously for the scanning liquid ion-chamber EPID to derive dose-in-water. Nicolini et al. [70] measured the linear calibration curves of EPID pixel value to dose-in-water for varying jaw defined field sizes and the same fields but with the aperture fully blocked by the MLC. This yielded a library of linear curves as a function of field size for both primary beam (*pr*) and MLC transmitted beam (*tr*).

$$D_{pr} = m_{pr}(\text{EwwF}) \cdot P_{pr} + q_{pr}(\text{EwwF}), \quad D_{tr} = m_{tr}(\text{EwwF}) \cdot P_{tr} + q_{tr}(\text{EwwF})$$

where:

- D is the dose in Gy measured in water phantom with ion-chamber at the center of the square field
- EwwF is the field size
- P is the average pixel value for a region of interest at the center of the field
- m and q are the fitted linear parameters with $q \sim 0$

Using the MLC delivery file each IMRT beam was divided into N separate segments s, each with a shape and normalized weight w_s. For each pixel of the image, signal due to primary beam and to MLC transmitted beam is considered. If the pixel with total signal $P(i, j)$ falls within the s_{th} segment aperture the fractional primary signal contribution from that segment $P_{s,pr}(i, j)$ is assumed to be

$$P_{s,pr}(i, j) = w_s \cdot P(i, j)$$

An equivalent field size (EwwF_s) is then calculated for the s_{th} segment and $P_{s,pr}(i, j)$ is used to calculate the dose-in-water from the linear calibration function for pixel to dose for that field size (with field-size interpolation between the library of linear calibration functions)

$$D_{s,pr}(i, j) = m_{pr}(\text{EwwF}_s)P_{s,pr}(i, j) + q_{pr}(\text{EwwF}_s)$$

After repeating this conversion for all segments and summing the results, the dose-in-water due to the primary beam is derived from the EPID image. The fractional pixel value component due to MLC transmitted beam is then estimated from the estimated total primary signal in each pixel subtracted from the actual measured pixel signal. This transmitted component of EPID signal is then calibrated to dose-in-water using the linear equations as given earlier but now measured for the MLC transmitted beam. Finally the dose due to the primary beam and the dose due to MLC transmitted beam is added together to give the dose-in-water calibration for the pixel.

$$D_w(i, j) = D_{tr}(i, j) + \sum_{s=1}^{N} D_{s,pr}(i, j)$$

As the method is based on the calibration of the central region of square fields, the model accuracy will decrease for pixels in segments exhibiting different scattering conditions. The method was termed GLAsS and has been developed into a commercial software package (EpiQA).

6.3.2.4 CALCULATION OF DOSE IN MORE COMPLEX PHANTOMS OR PATIENT MODEL

A further set of EPID pixel to dose conversion methods can be described that estimate dose-in-complex phantoms or a patient model based on nontransmission EPID measurements. Ansbacher derived the dose at 10 cm depth in a flat phantom and then converted this dose to the midplane dose of a virtual cylindrical phantom (radius 10 cm), using an off-axis correction matrix and extension to three-dimensional dose in the phantom using exponential depth dose modeling [68]. This method allows the dose from multiple beams to be combined to determine the overall composite delivered dose that can then be compared to TPS calculations in the phantom. Steciw et al. [73] derived incident fluence by deconvolution of an EPID scatter kernel. This fluence was then used as an input into a commercial TPS for dose calculation. Doses and dose-volume histogram comparisons with the planned dose were then performed. A commercially available system (Dosimetry Check) similarly obtains incident fluence, which is then used with a separate dose calculation engine to estimate dose distributions within the patient CT model for comparison to the planned dose [74]. EPID kernel parameters were optimized to minimize the variance between measured dose profiles in a water tank and EPID derived dose. Isodose overlays and gamma comparisons were demonstrated for two clinical IMRT cases: a prostate and a head and neck case.

6.4 VMAT VERIFICATION WITH EPID

6.4.1 TIME AND GANTRY-ANGLE RESOLVED IMAGING

The rapid introduction of VMAT into widespread clinical use has brought considerable challenges for pretreatment verification methods using EPID imaging. To fully validate VMAT delivery it is essential to determine that *the correct dose is delivered at the correct gantry angle*. Therefore, images should be acquired during the delivery as a function of time or gantry angle and the dose assessed in some manner as a function of angle, a more difficult proposition than using a single integrated image. This can take the form of an assessment of dose delivered over small subarcs of the delivery or reconstruction of a three-dimensional dose distribution from the measured images using projections at specific gantry angles.

6.4.1.1 CINE IMAGING

To acquire these images requires a cine-mode type acquisition where images or image frames are acquired continuously during the VMAT delivery and are available to the user for postdelivery analysis. Cine mode has received much less attention for dosimetry than integrated mode imaging [75,76]. The availability of these types of images from the accelerator vendors is currently variable. The Varian C-Series IAS3 EPID systems have a cine-mode acquisition with a frame rate of up to 10 Hz and potentially higher for half-resolution mode (spatial resolution is reduced to 512 × 384 from 1024 × 768). The system allows the user to specify the number of frames that are averaged for each cine image. This mode suffers from some limitations in that two partial frames at the start and two partial frames at the end of acquisition are discarded and buffer overflow issues can occur when too many frames are acquired. The former means that the signal/MU in the clinical images reduces markedly for small MU and this must be accounted for particularly when calibrating signal to dose. Operating the EPID in half-resolution mode can reduce the latter problem but the increased signal can potentially lead to image saturation with EPID positions close to the source.

The Varian Truebeam system and the Elekta iView system offer movie mode images; however, these are normalized for storage and therefore dosimetric information is lost. Truebeam has an image processing service option that can be used to obtain individual frame images where each frame is the cumulative image to that point. The Perkin–Elmer service software (XIS) used with the Elekta EPID can be utilized to acquire each frame of the delivery in a single image file; however, no gantry-angle information is available for the stored image frames. Mans et al. modified in-house image acquisition software to acquire image frames with the iView at ~2.5 Hz for VMAT verification with their *in vivo* EPID dosimetry software [77]. A new version of the Elekta iView system is currently being released that may address the issue of imaging during VMAT.

6.4.1.2 IMAGE SCANNING ARTIFACTS

Another problem that has been identified with cine imaging is image artifacts that occur from the interplay of image readout scanning and the discrete beam pulsing or dose rate of the accelerator [78,79]. As images are read out sequentially, for example, row-by-row, each row has a shifted time interval over which dose is integrated between two readouts. Therefore, if the dose pulse rate varies over comparable time intervals, different image rows can integrate different doses leading to large signal differences within a single frame. This can hamper dosimetry as the TPS or image prediction currently does not include these variations. Some methods to alleviate this have been reported with algorithms to predict and hence remove the arfifact [78] and frame-rate modifications can provide some canceling of the artifact [79]. This is typically not a problem with integrated images, since if the artifacts are in different locations on the individual frames they are then markedly reduced after averaging.

6.4.1.3 GANTRY-ANGLE DETERMINATION

It is critical that an accurate gantry angle can be assigned to each image. Images are acquired at fixed time intervals that are not necessarily easily related to gantry angle, since gantry speed can vary during VMAT delivery. A variety of different methods have been proposed for this purpose. The Varian C-Series IAS3 system records a gantry angle in the header of each cine image; however, this has been shown to have errors up to 3° [80–82]. An iCom connection to the treatment machine has been used with the Elekta system to record gantry angles and assign to each acquired frame with a measured lag of ~0.4 s [77]. Separate inclinometers can be placed on the LINAC and the angle reading time base synchronized to the image readout time base to assign the angle. Trajectory log files can be used in a similar way. Radiographic inclinometers, where a phantom is imaged on each frame to determine the gantry angle [81], have been utilized to compare with other methods and also to determine angle during delivery by retracting some MLC leaves and jaw positions [83].

6.4.2 Planar dose methods

6.4.2.1 INTEGRATED IMAGE-BASED METHODS

Preexisting methods such as the Varian Portal Dosimetry system, the GLAaS system and the EPIDose system have been applied to VMAT verification using EPID images integrated over relatively large subarcs or a single integrated image of the entire delivery [84–86]. The Portal Dosimetry system splits the delivery into a few separate arc deliveries each covering a fixed angle range with a single integrated image acquired for each. This is then compared to the predicted image for that angle range. Note that each is a separate arc delivery with beam on and off occurring for each. These arc deliveries typically encompass large angle ranges to reduce workload with 60° a typical value. Some limitations of this approach are that the actual patient delivery is not used for the verification and that low dose avoidance regions can give significant differences. High levels of gamma evaluation agreement have been reported with the integrated image approaches for large numbers of patients.

6.4.2.2 TIME OR GANTRY-ANGLE RESOLVED METHODS

Although more difficult to perform, methods that use gantry-resolved dosimetry provide a comprehensive verification of the delivery. These methods acquire separate images over small subarcs of the delivery (a few degrees) and compare these to either a predicted image for that subarc or a TPS calculation of dose to a flat-water phantom for the subarc. A difficulty with these methods is the alignment of the measured and reference doses over these small ranges of angles. Errors in alignment can result in large differences in the comparisons. Several approaches to overcome this problem have been investigated. One approach is to determine the angular range of the measured data and to then calculate the Varian Portal Dosimetry prediction for this angular range [87]. Another method involved measurement of the EPID images over small ~3° subarcs of the delivery and interpolation of an in-house predicted image set before comparison [88]. The interpolation was based on matching the cumulative image signal in both datasets. A time-dependent comparison has been made where the expected dose to water for particular fixed time intervals is calculated based on a measured gantry angle versus time relationship and compared to the dose to water modeled from images acquired in these time intervals [64]. An additional time-dependent gamma parameter was also employed for comparisons.

An interesting issue arises with these methods for how to display results for the angle-dependent comparisons. A *pseudo-3D* method was used by Liu et al. [87] where the predicted image and measured sets as a function of gantry angle are stacked to form 3D *cubes* with each image in the *x*–*y* plane and the *z* direction corresponding to gantry angle. The data are then projected through the stack onto three planes representing leaf motion axis versus gantry angle, crossleaf axis versus gantry angle and leaf motion axis versus crossleaf axis. Gamma evaluation calculations were then performed on these 2D projected data planes. Woodruff et al. [88] displayed gamma pass rate results versus gantry angle as line charts along with the angular differences between the aligned angle using cumulative signal and the target gantry position. Podesta et al. [64] used radial plots to display the gamma evaluation results as a function of gantry angle after these were translated back from time to gantry angle.

6.4.3 3D METHODS

6.4.3.1 3D DOSE ESTIMATION BASED ON TIME RESOLVED IMAGES

The aforementioned problems encountered with aligning small angular ranges of doses can be overcome by using the measured cine images separately calculating a 3D dose in phantom for each image using the gantry angle associated with each image. Uncertainties in gantry angle for the acquired images have a more limited effect particularly if these are random in nature. Several of the methods used for IMRT that involve these types of calculations can be utilized for VMAT in a straightforward manner with limited adaptation in the methods required. The Dosimetry Check software can be used in this case provided a gantry angle is available for the acquired images. The method of Ansbacher is suited to the VMAT geometry by using a virtual cylindrical phantom. Some preliminary results with this method have been reported [80] (Figure 6.4).

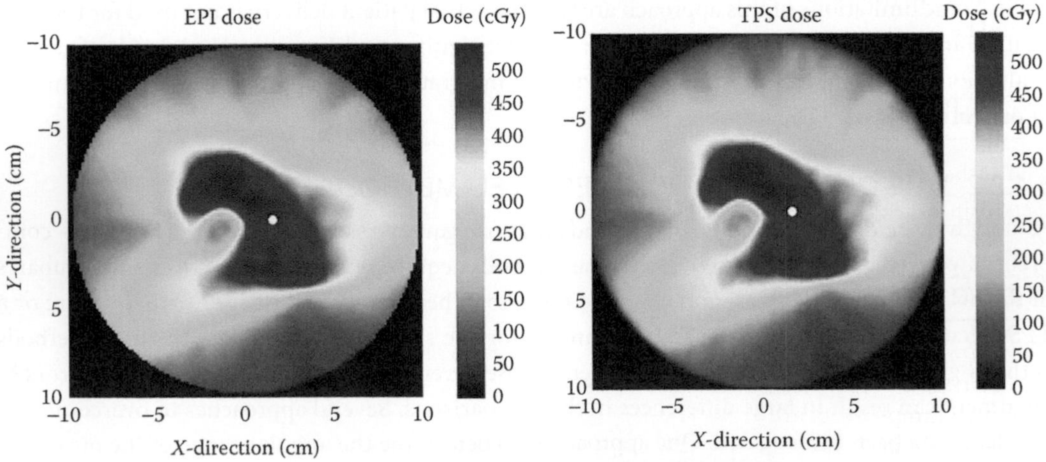

Figure 6.4 Illustration of a 3D dose calculation in a virtual cylindrical phantom using cine-EPID images for a VMAT high dose delivery. The left is from EPID and the right is from the TPS. The dose in the cylindrical phantom is calculated from each image using recorded gantry angles and the doses for all images summed. In this case, 148 images from two arcs were used.

6.5 FUTURE DEVELOPMENTS

6.5.1 Gold standard models

There is a general trend toward producing LINAC that have consistent dosimetric characteristics that can then be used with gold-standard TPS datasets for ease of commissioning. LINAC characteristics are generally very consistent across machine types. This brings the possibility for similar *gold-standard* EPID dosimetry models where the same image acquisition parameters are used across different centers. This will reduce commissioning workload and acceptance of these systems, and enable comparison of data between different centers [59,89].

6.5.2 Clinical trial auditing

Methods for pretreatment verification with EPID are currently being extended to clinical trial auditing. The European Organization for Research and Treatment of Cancer (EORTC) recently reported on a study where centers performed their own in-house pretreatment QA, which could include EPID methods and the results were analyzed centrally [90]. However, one-third of the center's data could not be analyzed and interpretation of results is difficult. A study has recently commenced to use EPID alone for this type of assessment [91]. This aims to standardize the measurement system as much as possible as almost all centers have EPID and these are similar in construct. The conversion of EPID image to dose and comparison to TPS is performed at a central site to standardize the analysis.

6.5.3 Frame-grabber systems

An interesting new development is the use of systems such as separate computers with frame-grabber cards and software to obtain a copy of the individual frames that are acquired in the clinical software. The clinical system can be operated using either integrated or cine-mode imaging in this case. Image lines at the end of each frame can be coded with information on the machine state such as gantry angle, MUs, and even MLC positions. These systems are ideal for cine imaging and other applications including real-time use. Future availability of these systems as clinical tools would be of great benefit to the community.

ACKNOWLEDGMENTS

The author gratefully acknowledges the comments provided on this chapter by Boyd McCurdy and Ben Mijnheer.

REFERENCES

1. Low, D. et al. A technique for the quantitative evaluation of dose distributions. *Medical Physics*, 1998. **25**(5): 656–661.
2. van Elmpt, W. et al. A literature review of electronic portal imaging for radiotherapy dosimetry. *Radiotherapy and Oncology*, 2008. **88**(3): 289–309.
3. Greer, P.B. and P. Vial, Epid Dosimetry, in *Concepts and Trends in Medical Radiation Dosimetry*, Rosenfeld A. B. et al., (Eds.). 2011. 129–144.
4. Vial, P. et al. Initial evaluation of a commercial EPID modified as a novel direct-detection dosimeter for radiotherapy. *Medical Physics*, 2008. **35**(10): 4362–4374.

5. Blake, S.J. et al. Characterization of optical transport effects on EPID dosimetry using Geant4. *Medical Physics*, 2013. **40**(4): 041708.
6. Kirkby, C. and R. Sloboda, Comprehensive Monte Carlo calculation of the point spread function for a commercial a-Si EPID. *Medical Phyics*, 2005. **32**(4): 1115–1127.
7. Gustafsson, H. et al. EPID dosimetry: Effect of different layers of materials on absorbed dose response. *Medical Physics*, 2009. **36**(12): 5665–5674.
8. Ko, L., J.O. Kim, and J.V. Siebers, Investigation of the optimal backscatter for an aSi electronic portal imaging device. *Physics in Medicine and Biology*, 2004. **49**: 1723–1738.
9. Greer, P.B. et al. An energy-fluence convolution model for amorphous silicon EPID dose prediction. *Medical Physics*, 2009. **36**(2): 547–555.
10. Moore, J.A. and J.V. Siebers, Verification of the optimal backscatter for an aSi electronic portal imaging device. *Physics in Medicine and Biology*, 2005. **50**(10): 2341–2350.
11. Rowshanfarzad, P. et al. Reduction of the effect of non-uniform backscatter from an E-type support arm of a Varian a-Si EPID used for dosimetry. *Physics in Medicine and Biology*, 2010. **55**(22): 6617–6632.
12. Rowshanfarzad, P. et al. Improvement of Varian a-Si EPID dosimetry measurements using a lead-shielded support-arm. *Medical Dosimetry*, 2011. **37**(2): 145–151.
13. King, B.W., D. Morf, and P.B. Greer, Development and testing of an improved dosimetry system using a backscatter shielded electronic portal imaging device. *Medical Physics*, 2012. **39**(5): 2839–2847.
14. Parent, L. et al. Monte Carlo modelling of a-Si EPID reponse: The effect of spectral variations with field size and position. *Medical Physics*, 2006. **32**(12): 4527–4540.
15. Greer, P.B. et al. Experimental investigation of the response of an amorphous silicon EPID to intensity modulated radiotherapy beams. *Medical Physics*, 2007. **34**(11): 4389–4398.
16. Li, W., J.V. Siebers, and J.A. Moore, Using fluence separation to account for energy spectra dependence in computing dosimetric a-Si EPID images for IMRT fields. *Medical Physics*, 2006. **33**(12): 4468–4480.
17. Greer, P.B., Correction of pixel sensitivity variation and off-axis response for amorphous silicon EPID dosimetry. *Medical Physics*, 2005. **32**(12): 3558–3568.
18. Greer, P.B., 3D EPID based dosimetry for pre-treatment verification of VMAT–methods and challenges. *Journal of Physics: Conference Series*, 2013. **444**(1): 012010.
19. Zhao, W. et al. Ghosting caused by bulk charge trapping in direct conversion flat-panel detectors using amorphous selenium. *Medical Physics*, 2005. **32**(2): 488–500.
20. Siewerdsen, J.H. and D.A. Jaffray, A ghost story: Spatio-temporal response characteristics of an indirect-detection flat panel imager. *Medical Physics*, 1999. **26**(8): 1624–1641.
21. Partridge, M., B.-M. Hesse, and L. Muller, A performance comparison of direct and indirect-detection flat-panel imagers. *Nuclear Instruments and Methods in Physical Research Section A*, 2002. **484**: 351–363.
22. Greer, P.B. and C.C. Popescu, Dosimetric properties of an amorphous silicon electronic portal imaging device for verification of dynamic intensity modulated radiation therapy. *Medical Physics*, 2003. **30**(7): 1618–1627.
23. Van Esch, A., T. Depuydt, and D.P. Huyskens, The use of an aSi-based EPID for routine absolute dosimetric pre-treatment verification of dynamic IMRT fields. *Radiotherapy and Oncology*, 2004. **71**: 223–234.
24. McDermott, L.N. et al. Dose-response and ghosting effects of an amorphous silicon electronic portal imaging device. *Medical Physics*, 2004. **31**(2): 285–295.
25. Winkler, P., A. Hefner, and D. Georg, Dose-response characteristics of an amorphous silicon EPID. *Medical Physics*, 2005. **32**(10): 3095–3105.
26. Winkler, P. and D. Georg, An intercomparison of 11 amorphous silicon EPIDs of the same type: Implications for portal dosimetry. *Physics in Medicine and Biology*, 2006. **51**(17): 4189–4200.

27. Podesta, M. et al. Measured vs simulated portal images for low MU fields on three accelerator types: Possible consequences for 2D portal dosimetry. *Medical Physics*, 2012. **39**(12): 7470–7479.
28. Gao, S. et al. Evaluation of IsoCal geometric calibration system for Varian linacs equipped with on-board imager and electronic portal imaging device imaging systems. *Journal of Applied Clinical Medical Physics/American College of Medical Physics*, 2014. **15**(3): 4688.
29. Rowshanfarzad, P. et al. An EPID-based method for comprehensive verification of gantry, EPID and the MLC carriage positional accuracy in Varian linacs during arc treatments. *Radiation Oncology*, 2014. **9**(1): 249.
30. Rowshanfarzad, P. et al. A comprehensive study of the mechanical performance of gantry, EPID and the MLC assembly in Elekta linacs during gantry rotation. *British Journal of Radiology*, 2015. **88**(1051): 20140581.
31. Wendling, M. et al. A simple backprojection algorithm for 3D in vivo EPID dosimetry of IMRT treatments. *Medical Physics*, 2009. **36**(7): 3310–3321.
32. Rowshanfarzad, P. et al. Detection and correction for EPID and gantry sag during arc delivery using cine EPID imaging. *Medical Physics*, 2012. **39**(2): 623–635.
33. Lee, L., W. Mao, and X. L, The use of EPID-measured leaf sequence files for IMRT dose reconstruction in adaptive radiation therapy. *Medical Physics*, 2008. **35**(11): 5019–5029.
34. Zin, H.M. et al. Towards real-time VMAT verification using a prototype, high-speed CMOS active pixel sensor. *Physics in Medicine and Biology*, 2013. **58**(10): 3359–3375.
35. James, H.V. et al. Verification of dynamic multileaf collimation using an electronic portal imaging device. *Physics in Medicine and Biology*, 2000. **45**(2): 495–509.
36. Partridge, M. et al. Independent verification using portal imaging of intensity-modulated beam delivery by the dynamic MLC technique. *Medical Physics*, 1998. **25**(10): 1872–1879.
37. Ploeger, L.S. et al. A method for geometrical verification of dynamic intensity modulated radiotherapy using a scanning electronic portal imaging device. *Medical Physics*, 2002. **29**(6): 1071–1079.
38. Sonke, J.J. et al. Leaf trajectory verification during dynamic intensity modulated radiotherapy using an amorphous silicon flat panel imager. *Medical Physics*, 2004. **31**(2): 389–395.
39. Bakhtiari, M. et al. Using an EPID for patient-specific VMAT quality assurance. *Medical Physics*, 2011. **38**(3): 1366–1373.
40. Yang, Y. and L. Xing, Quantitative measurement of MLC leaf displacements using an electronic portal image device. *Physics in Medicine and Biology*, 2004. **49**: 1521–1533.
41. Chang, J. et al. Use of EPID for leaf position accuracy QA of dynamic multi-leaf collimator (DMLC) treatment. *Medical Physics*, 2004. **31**(7): 2091–2096.
42. Samant, S.S. et al. Verification of multileaf collimator leaf positions using an electronic portal imaging device. *Medical Physics*, 2002. **29**(12): 2900–2912.
43. Fuangrod, T. et al. An independent system for real-time dynamic multileaf collimation trajectory verification using EPID. *Physics in Medicine and Biology*, 2014. **59**(1): 61–81.
44. Han-Oh, S. et al. Verification of MLC based real-time tumor tracking using an electronic portal imaging device. *Medical Physics*, 2010. **37**(6): 2435–2440.
45. Poulsen, P.R. et al. A method of dose reconstruction for moving targets compatible with dynamic treatments. *Medical Physics*, 2012. **39**(10): 6237–6246.
46. Pasma, K.L. et al. Portal dose image (PDI) prediction for dosimetric treatment verification in radiotherapy. I. An algorithm for open beams. *Medical Physics*, 1998. **25**(6): 830–840.
47. McCurdy, B.M.C., K. Luchka, and S. Pistorius, Dosimetric investigation and portal dose image prediction using an amorphous silicon electronic portal imaging device. *Medical Physics*, 2001. **28**(6): 911–924.
48. Siebers, J.V. et al. Monte Carlo computation of dosimetric amorphous silicon electronic portal images. *Medical Physics*, 2004. **31**(7): 2135–2146.
49. Chytyk, K. and B.M.C. McCurdy, Comprehensive fluence model for absolute portal dose image prediction. *Medical Physics*, 2009. **36**(4): 1389–1398.

50. Papanikolaou, N. et al. Investigation of the convolution method for polyenergetic spectra. *Medical Physics*, 1993. **20**(5): 1327–1336.
51. Chytyk, K. and B.M.C. McCurdy, Investigation of tilted dose kernels for portal dose prediction in a-Si electronic portal imagers. *Medical Physics*, 2006. **33**(9): 3333–3339.
52. Wang, S. et al. Monte Carlo-based adaptive EPID dose kernel accounting for different field size responses of imagers. *Medical Physics*, 2009. **36**(8): 3582–3595.
53. Bailey, D.W., L. Kumaraswamy, and M.B. Podgorsak, An effective correction algorithm for off-axis portal dosimetry errors. *Medical Physics*, 2009. **36**(9): 4089–4094.
54. Bailey, D.W. et al. A two-dimensional matrix correction for off-axis portal dose prediction errors. *Medical Physics*, 2013. **40**(5): 051704.
55. Vial, P. et al. The impact of MLC transmitted radiation on EPID dosimetry for dynamic MLC beams. *Medical Physics*, 2008. **35**(4): 1267–1277.
56. Howell, R.M., I.P.N. Smith, and C.S. Jarrio, Establishing action levels for EPID-based QA for IMRT. *Journal of Applied Clinical Medical Physics*, 2008. **9**(3): 16–25.
57. Greer, P.B. et al. An energy fluence-convolution model for amorphous silicon EPID dose prediction. *Medical Physics*, 2009. **36**(2): 547–554.
58. Vinall, A.J. et al. Practical guidelines for routine intensity-modulated radiotherapy verification: Pre-treatment verification with portal dosimetry and treatment verification with in vivo dosimetry. *British Journal of Radiology*, 2010. **83**(995): 949–957.
59. Van Esch, A. et al. Optimized Varian aSi portal dosimetry: Development of datasets for collective use. *Journal of Applied Clinical Medical Physics/American College of Medical Physics*, 2013. **14**(6): 4286.
60. Rowshanfarzad, P. et al. Measurement and modelling of the effect of support arm backscatter on dosimetry with a Varian EPID. *Medical Physics*, 2010. **37**(5): 2269–2278.
61. King, B.W. and P.B. Greer, A method for removing arm backscatter from EPID images. *Medical Physics*, 2013. **40**(7): 071703.
62. Berry, S.L., C.S. Polvorosa, and C.S. Wuu, A field size specific backscatter correction algorithm for accurate EPID dosimetry. *Medical Physics*, 2010. **37**(6): 2425–2434.
63. Cufflin, R.S. et al. An investigation of the accuracy of Monte Carlo portal dosimetry for verification of IMRT with extended fields. *Physics in Medicine and Biology*, 2010. **55**(16): 4589–4600.
64. Podesta, M. et al. Time dependent pre-treatment EPID dosimetry for standard and FFF VMAT. *Physics in Medicine and Biology*, 2014. **59**(16): 4749–4768.
65. Warkentin, B. et al. Dosimetric IMRT verification with a flat-panel EPID. *Medical Physics*, 2003. **30**(12): 3143–3155.
66. Zwan, B.J. et al. Dose-to-water conversion for the backscatter-shielded EPID: A frame-based method to correct for EPID energy response to MLC transmitted radiation. *Medical Physics*, 2014. **41**(8): 081716.
67. Nijsten, S.M. et al. A global calibration model for a-Si EPIDs used for transit dosimetry. *Medical Physics*, 2007. **34**(10): 3872–3884.
68. Ansbacher, W., Three-dimensional portal image-based dose reconstruction in a virtual phantom for rapid evaluation of IMRT plans. *Medical Physics*, 2006. **33**(9): 3369–3382.
69. Nelms, B.E., K.H. Rasmussen, and W.A. Tome, Evaluation of a fast method of EPID-based dosimetry for intensity-modulated radiation therapy. *Journal of Applied Clinical Medical Physics*, 2010. **11**(2): 140–157.
70. Nicolini, G. et al. GLAaS: An absolute dose calibration algorithm for an amorphous silicon portal imager. Applications to IMRT verifications. *Medical Physics*, 2006. **33**(8): 2839–2851.
71. Lee, C. et al. A simple approach to using an amorphous silicon EPID to verify IMRT planar dose maps. *Medical Physics*, 2009. **36**(3): 984–992.
72. Chang, J. and C.C. Ling, Using the frame averaging of aS500 EPID for IMRT verification. *Journal of Applied Clinical Medical Physics*, 2003. **4**(4): 287–299.

73. Steciw, S. et al. Three-dimensional IMRT verification with a flat-panel EPID. *Medical Physics*, 2005. **32**(2): 600–612.
74. Renner, W.D., K. Norton, and T. Holmes, A method for deconvolution of integrated electronic portal images to obtain incident fluence for dose reconstruction. *Journal of Applied Clinical Medical Physics*, 2005. **6**(4): 22–39.
75. McCurdy, B.M.C. and P.B. Greer, Dosimetric properties of an amorphous-silicon EPID used in continuous acquisition mode for application to dynamic and arc IMRT. *Medical Physics*, 2009. **36**(7): 3028–3039.
76. Yeo, I.J. et al. Conditions for reliable time-resolved dosimetry of electronic portal imaging devices for fixed-gantry IMRT and VMAT. *Medical Physics*, 2013. **40**(7): 072102.
77. Mans, A. et al. 3D Dosimetric verification of volumetric-modulated arc therapy by portal dosimetry. *Radiotherapy and Oncology*, 2010. **94**(2): 181–187.
78. Mooslechner, M. et al. Analysis of a free-running synchronization artifact correction for MV-imaging with aSi:H flat panels. *Medical Physics*, 2013. **40**(3): 031906.
79. Woodruff, H.C. and P.B. Greer, 3D dose reconstruction: Banding artefacts in cine mode EPID images during VMAT delivery. *Journal of Physics: Conference Series*, 2013. **444**(1).
80. Ansbacher, W., C.L. Swift, and P.B. Greer, An evaluation of cine-mode 3D portal image dosimetry for volumetric modulated arc therapy. *Journal of Physics: Conference Series*, 2010. **250**: 108–111.
81. Rowshanfarzad, P. et al. Gantry angle determination during arc IMRT: evaluation of a simple EPID-based technique and two commercial inclinometers. *Journal of applied clinical medical physics/American College of Medical Physics*, 2012. **13**(6): 3981.
82. McCowan, P.M. et al. Precise gantry angle determination for EPID images during rotational IMRT. *Medical Physics*, 2011. **38**: 3534.
83. Adamson, J. and Q. Wu, Independent verification of gantry angle for pre-treatment VMAT QA using EPID. *Physics in Medicine and Biology*, 2012. **57**: 6587–6600.
84. Iori, M. et al. Dosimetric verification of IMAT delivery with a conventional EPID system and a commercial portal dose image prediction tool. *Medical Physics*, 2010. **37**(1): 377–390.
85. Nicolini, G. et al. The GLAaS algorithm for portal dosimetry and quality assurance of RapidArc, an intensity modulated rotational therapy. *Radiation Oncology*, 2008. **3**(1): 24.
86. Bailey, D.W. et al. EPID dosimetry for pretreatment quality assurance with two commercial systems. *Journal of Applied Clinical Medical Physics*, 2012. **13**(4): 82–99.
87. Liu, B. et al. A novel technique for VMAT QA with EPID in cine mode on a Varian TrueBeam linac. *Physics in Medicine and Biology*, 2013. **58**(19): 6683–6700.
88. Woodruff, H.C. et al. Gantry-angle resolved VMAT pretreatment verification using EPID image prediction. *Medical Physics*, 2013. **40**(8): 081715.
89. Hanson, I.M. et al. Clinical implementation and rapid commissioning of an EPID based in-vivo dosimetry system. *Physics in Medicine and Biology*, 2014. **59**(19): N171–N179.
90. Weber, D.C. et al. IMRT credentialing for prospective trials using institutional virtual phantoms: Results of a joint European Organization for the Research and Treatment of Cancer and Radiological Physics Center project. *Radiation Oncology*, 2014. **9**(1): 123.
91. Miri, N. et al. Virtual EPID Standard Phantom Audit (VESPA) for remote IMRT and VMAT dosimetric credentialling, *Physics in Medicine and Biology*, 2017 **62**: 4293–4299.

7

Beam's eye view imaging for patient safety

ERIC FORD

7.1 INTRODUCTION: PATIENT SAFETY AND THE LINK TO BEAM'S EYE VIEW IMAGING

Patient safety and quality of care is integral to the practice of health care, but is of special concern in radiation oncology due to the complexity of the processes, the many hand-offs between various professionals, the protracted timeline for the preparation and execution of treatment, and the many technical components from multiple vendors (Donaldson 2008). This complexity can be appreciated in the extremely intricate workflow diagrams of the process of care (Ford et al. 2009). Given this complexity, it is not surprising that errors can and do occur. In fact, error is an inevitable part of any process (Reasons 1997). In radiation oncology, the results of such errors can be dramatic as seen in some of the recent tragic accidents in the field: the overdose of a young patient in Scotland, Lisa Norris (Executive 2006), the miscalibration of a stereotactic radiotherapy dose system (Bogdanich and Rebelo 2010; CoxHealth 2010), and the large overdose of a patient in New York due to a failed intensity-modulated radiation therapy (IMRT) delivery (Bogdanich and Ruiz 2010). Reports of these accidents in the U.S. national media have galvanized the field toward improving patient safety (Hendee and Herman 2011).

Catastrophic errors are rare in radiation oncology. They do, however, point to a deeper underlying issue with the safety and quality of care. The quality of radiotherapy plans can be thought of as lying on a spectrum (Dunscombe et al. 2013) with the lowest quality plans being obviously unsafe. In the middle, however, there are plans which, though not catastrophically unsafe, are of lower quality than standard of care or best practices would dictate. Evidence of this quality gap can be found in data from cooperative group trials. A reanalysis of the TROG02.02 trial for head-and-neck cancer, for example, showed that 12% of plans were considered as seriously noncompliant on re-review (Peters et al. 2010). Furthermore, it has been shown that these low-quality plans are linked to poor outcomes and worse overall survival (Peters et al. 2010). Other cooperative group trials show an effect like this as described in several recent meta-analyses (Fairchild et al. 2013; Ohri et al. 2013). The issues of patient safety and quality of care extend well beyond radiation oncology of course. Recent studies have updated the 1999 report from the Institute of Medicine and suggest that iatrogenic harm may be the third leading cause of death in the United States (James 2013).

There is then a clear need to improve and standardize the quality of treatments in radiation therapy. Numerous efforts are underway to address this problem including the use of incident learning (Zeitman et al. 2012; Clark et al. 2013; Hoopes et al. 2015), techniques in quantitative risk assessment and mitigation (Huq et al. 2008, 2016; Thomadsen et al. 2013), and improved use of peer review and quality audits (Marks et al. 2013). A thorough treatment of this topic can be found in the *2013 AAPM Summer School proceedings on quality and safety* (Thomadsen et al. 2013).

In this context, the question becomes: What is the role of beam's eye view (BEV) imaging for improving the safety and quality of care? There are two main aspects of BEV imaging that are directly relevant:

1. *BEV imaging as an "end-to-end test"*: A great deal of attention has been given recently to the *end-to-end* test as a measure of safety and quality. This test is typically performed with a phantom that includes an embedded object and/or films. The phantom serves as a surrogate for the patient and follows the entire workflow up through the point of treatment. The concept here is that unintended deviations in planning or localization can be identified in the irradiated phantom. Such tests are recommended in the *ASTRO Safety White Papers* and elsewhere (Fraass et al. 2011; Moran et al. 2011). BEV imaging can also be thought of as an end-to-end test. Though it is not possible to embed films or measurement devices, one can perform localization measurements and even calculate back-projected dose through the patient (e.g., Chapter 11).
2. *BEV imaging as an automated safety net*: Unlike many quality control processes currently employed in radiation oncology, many aspects of BEV imaging can be automated. Automation can serve an important function in safety improvement as seen in the *hierarchy of effectiveness* where automation and forcing functions are listed as the most effective means of intervention (Grout 2006). This conceptual model is part of a larger trend in health care that recognizes design and human factors engineering as a key component of safety (Gurses et al. 2009) and owes much to the early design work of Don Norman (Norman 1988) and subsequent research on human–computer interface. In the situations where BEV imaging can be implemented as a semiautomatic system (c.f. Chapter 11), it may be more reliable than other quality control measures that rely on human inspection. Automated safety nets are being explored in other applications in radiation oncology as well such as software-assisted plan verification (Yang and Moore 2012; Li et al. 2014; Noel et al. 2014; Xia et al. 2014) and automatic algorithms that assess plan quality against predictions of what is achievable (Moore et al. 2011; Wu et al. 2011; Good et al. 2013).

Several BEV related reports have appeared that reflect the points aforementioned. A landmark study (Mans et al. 2010) reported the experience with electronic portal imaging devices (EPIDs)-based dosimetry in clinical operations and demonstrated the ability to detect safety-critical problems in the delivery of beams. More information is available in Chapter 11.

In summary, BEV imaging offers a semiautomatic means to identify errors and thereby improve the quality and safety of radiotherapy. Because of its position in the overall workflow (i.e., at the very end of the chain of preparation and treatment), it is potentially sensitive to errors that occur throughout the whole process of care preparation and delivery. This is particularly attractive since it is known that errors often originate in the process of treatment planning (Clark et al. 2013). The following sections dissect the ways in which BEV imaging can serve a safety-critical role and its future potential in this regard.

7.2 CASE STUDIES: BEAM'S EYE VIEW AS A SAFETY CHECK IN ACTION

The following three case scenarios illustrate the use of BEV imaging to detect errors in treatments. Each case illustrates a slightly different operation of BEV imaging. All of these case studies are drawn from actual events from institutional incident learning systems (e.g., Nyflot et al. 2015). The events described here are near-miss events, that is, errors which were identified before they reached the patient, and so did not result in any harm. In each of these case scenarios, BEV imaging would have served to identify the error if other quality control measures failed.

7.2.1 Case 1: Incorrect field shape

A patient treatment plan was performed for a linear accelerator (LINAC) with a *beam modulator* device, that is, a multileaf collimator (MLC) for which there are no jaws (Elekta Inc., Crawley UK). As is standard with such plans, the jaws were automatically set to the largest field size (16 × 21 cm). It was then decided that the patient would actually be treated on another LINAC, and so the designated machine was switched in the treatment planning system (Pinnacle v.9.0, Phillips, Inc., Madison, WI). On this other LINAC, standard jaws are present and should be moved to block up to the edge of the MLC-defined field. However, the option in the planning software to *set jaws to block* was not checked, and so the jaws remained at their default location (16 × 21 cm setting). The end result was a field with proper MLC blocking but with a gap between MLC leafs that was not covered by a jaw. This gap (0.5 cm wide) extended over a critical structure.

In this case, this issue was identified by a physicist on plan review and was corrected. However, BEV imaging could identify this problem and others like it. The identification could happen either prior to treatment (e.g., port films or pretreatment EPID-based QA as discussed in Chapter 6) or potentially during the first treatment fraction if used *in vivo* (e.g., Chapter 6).

7.2.2 Case 2: Incorrect isocenter

Numerous variants of this error scenario have been noted in clinical practice, but this particular situation is extremely difficult to detect. In this case, a plan was generated in the treatment planning system (Pinnacle v.9.0, Phillips, Inc., Madison, WI) with all beams assigned to a particular point labeled as *isocenter*. However, the digitally reconstructed radiographs (DRRs) for this treatment were generated to a different point, *the calculation point*. A causal factor for this error is a human factors design issue: the treatment planning software uses the last point that was selected by the user as the location for the DRR and in this case it happened to be the *calculation point* and not the intended isocenter.

In this case, the discrepancy on the DRR was noted by a physicist on plan review prior to treatment and the problem was corrected. However, if the issue were not identified then the wrong DRR might have been used for localization and alignment, and the patient treatment could have proceeded to the incorrect location. This would be challenging to identify posttreatment. This case study bears similarities to another presented in the literature from our group (Ford et al. 2012a).

BEV imaging may serve a role in such situations. It would clearly need to be performed, however, during the patient treatment. Tests performed prior to treatment (c.f. Chapter 6) would not be sensitive to changes in the isocenter location as discussed further below.

7.2.3 Case 3: Bolus for sarcoma treatment

A patient receiving postoperative radiation therapy for a leg sarcoma was intended to be treated with 5-mm thick bolus placed over the surgical site in order to boost the dose to the skin that was considered at risk. The attending physician communicated this through the oncology information system. However, because of human–computer interface design issues, the relevant note was not apparent to staff. The treatment planning was performed as intended (i.e., with bolus in place) but the information was not communicated to the treatment team. The specialized *bolus alert* (part of the treatment field) was also not filled out in the oncology information system, preventing the warning to the treatment staff. In this case, an alert dosimetrist noticed the absence of bolus when called to check the first fraction of treatment and the issue was resolved.

BEV imaging could play a role in detecting problems such as this. It would need to be used during patient treatment for this to be effective. Pretreatment, phantom-based, BEV measurements would be unlikely to identify this problem.

7.3 QUANTIFYING THE POTENTIAL IMPACT OF BEAM'S EYE VIEW IMAGING AS A SAFETY TOOL

The case studies above illustrate the potential of BEV imaging as a safety tool, but to fully assess its potential impact, a quantitative study is needed. One tool to facilitate this is *quality control quantification* (QCQ) which has been explored by our group (Ford et al. 2012b). This technique takes incidents from clinical practice as its input. For each incident, one evaluates whether a particular quality control measure is potentially able to identify that incident and thereby disrupt the safety problem before it reaches the patient. In this way, one can quantify the potential sensitivity of each quality control measure.

There are some inherent biases in this type of analysis. First and foremost is the fact that it is based on voluntary incident reporting at the clinical level, and so only captures the incidents that occur in that clinic and only those that are reported into the system. One way to minimize this bias is to gather many hundreds or even thousands of incidents. This makes the analysis less subject to reporting bias. Bias can also be minimized by using multiinstitutional data. This is the approach taken in our studies.

Our studies suggest that there is a wide variation in the potential effectiveness of various common quality control measures (Ford et al. 2012b). Interestingly, *in vivo* EPID-based dosimetry appeared as one of the top-ranking quality measures in terms of its potential ability to identify safety-critical errors. *In vivo* EPID-based dosimetry is one type of BEV imaging and is discussed more in Chapter 11. Although this initial study was promising, further validation was needed. Recently we repeated this analysis on a completely separate incident learning database from another clinic and showed much the same pattern (Bojechko et al. 2015).

The data from this study showed an interesting effect: the potential effectiveness of EPID-dosimetry depends very much on how it was employed. This can be seen in Figure 7.1. This plot

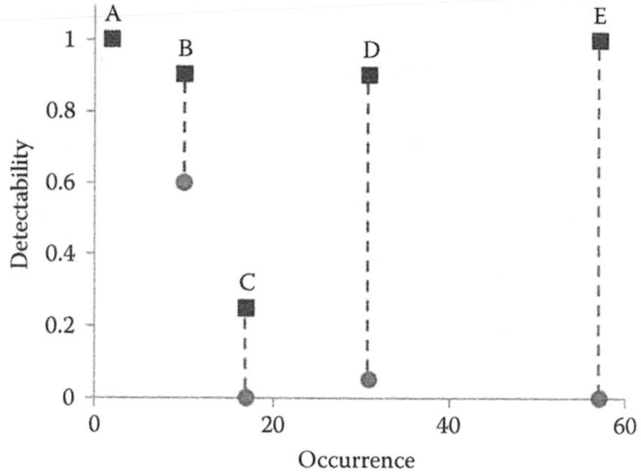

Figure 7.1 Detectability and occurrence for five different error scenarios drawn from incident reports. Circles are detectability based on EPID dosimetry performed prior to treatment, whereas squares are for EPID dosimetry performed *in vivo* at the first treatment fraction.

displays five different error scenarios according to the occurrence and detectability, a categorization scheme of risk assessment advocated by AAPM Task Group 100 (Huq et al. 2016). Here the relative occurrence score (arbitrary scale) is measured from the data in the incident learning database. The detectability of the events was evaluated under one of two scenarios: BEV-EPID dosimetry performed prior to treatment (circles) or EPID dosimetry performed during the first fraction (squares).

Five different error classes shown in Figure 7.1 are as follows: (A) error in physics dosimetry calculation, (B) corrupted treatment plan, (C) missing or incorrect documentation, (D) error in CT data, and (E) incorrect isocenter. Several patterns are immediately apparent. First, there are a number of error classes (e.g., D and E) that are not detectable at all with pretreatment measurement but which are potentially highly detectable (>90%) with *in vivo* EPID dosimetry. These error classes are relatively common in this clinical database. Second, there is one type of error class that has the same detectability regardless of which technique is used (e.g., A, both potentially 100%). It should be noted, however, that this error class occurs relatively more rarely. Finally, there is a class of errors which, although have higher detectability with *in vivo* dosimetry, have a detectability rate that still falls far short of 100% (e.g., C).

Overall, these data suggests that EPID dosimetry is an effective safety tool but only if employed *in vivo* with the patient present. There are thus technique-dependent trade-offs to how this system is utilized as summarized in Table 7.1. Table 7.1 also notes a potential disadvantage of *in vivo* EPID-based dosimetry, namely, the fact that the results are not available until just after the first treatment. For this reason a real-time solution may ultimately be required (Fuangrod et al. 2013; Monville et al. 2014) and this may be especially important for high-dose-per-fraction treatments.

In the aforementioned discussion the term *potential* appears frequently. In other words, the analyses considered an idealized situation whereby if a quality control check *could* identify an error it actually *does* identify the error. This is a significant assumption because it could alter the conclusions: a relatively sensitive test that does not perform well may actually be less effective than a less sensitive test that performs with high fidelity. EPID-based dosimetry is thought to be less prone to this problem because it can be implemented in a way that is semiautomatic and less reliant on

Table 7.1 Safety-critical aspects of beam's eye view imaging as employed under various scenarios

Technique	Advantages	Disadvantages
Imaging	• Provides localization information • Well established	• No dose measure • Field/beam information often not quantified
Pretreatment EPID dosimetry (Chapter 6)	• Provides dose measure • Provides check prior to first treatment • Well established	• Not sensitive to common errors
In vivo EPID dosimetry (Chapter 11)	• Provides dose measure • Sensitive to common errors	• No check prior to treatment • More false positives • In development

human performance (Olaciregui-Ruiz et al. 2013). However, it must be recognized that even EPID dosimetry will not be equally sensitive to all types of error scenarios that it may be applied to.

There is now emerging data that attempts to answer this question of actual reliability of a particular test. It relies on a technique in which deviations are artificially introduced into a plan. One then measures the ability of the quality assurance (QA) system to detect these deviations. This technique has been employed in the novel study by Carlone et al. (2013) where the authors introduced MLC calibration errors and then measured the ability of the IMRT QA to detect those errors. One can calculate both the sensitivity and specificity of the test and can create an ROC curve by varying the test parameters (in this case the cutoff value for the γ test).

This technique can be generalized to include many types of errors. Figure 7.2 shows an example as applied to *in vivo* EPID dosimetry, using the EPID system described in various reports (Mans et al. 2010; Olaciregui-Ruiz et al. 2013). The data in Figure 7.2 are drawn from six actual IMRT patient treatments. Errors were represented by introducing simulated deviations in the treatment plan. Figure 7.2 suggests that some deviations such as dose output changes are highly detectable (black line), whereas detection of other deviations is not better than random (gray line). The data shown in green represent a relatively large patient shift of 1 cm laterally. The conclusions are interesting, although more work is needed. Data like these begin to address the question of the sensitivity of this BEV imaging technique to various errors and the dependence on technique and measurement endpoint.

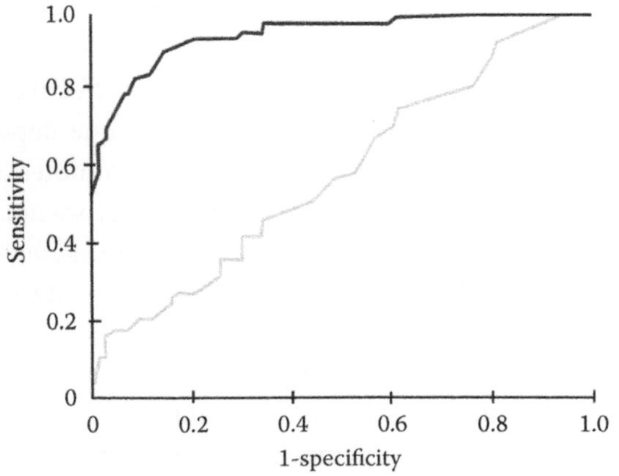

Figure 7.2 Receiver–operator curve (ROC) for *in vivo* EPID dosimetry measurements of two variations: a shift of the patient by 1 cm laterally (gray line) and an overall dose scaling by 6% (black line). AUC are 0.58 and 0.94, respectively. (Courtesy of Eric Ford, University of Washington.)

7.4 UNINTENDED CONSEQUENCES

In a complex environment like health care, error prevention efforts can have unintended consequences. A classic example of this is the experience with clinical provider order entry (CPOE) systems. One important motivation for implementing CPOE systems is to reduce error, that is, reduce adverse drug events related to medication error by eliminating transcription errors and other faults (Bates et al. 1999). However, such systems do not always have the intended effect. A landmark study from University of Pittsburgh Children's Hospital, Penn Ave, Pittsburgh, PA (Han et al. 2005) reported experience over at 18-month period in 2001–2003 with the implementation of a new CPOE system. This study actually showed an *increase* in the mortality of their hospitalized pediatric patients over this period from 2.8% to 6.6% (odds ratio 3.28, $p < 0.001$). This was attributed to human–computer interface design issues and other challenges during implementation. A survey by another group (Koppel et al. 2005) supports this, finding that one commonly used CPOE system facilitated error in 22 different types of error scenarios. Though such systems are intended to improve safety, it is important to be aware of possible unintended consequences.

One potential method for addressing this is to employ a formalized risk assessment to evaluate the new technology being considered. A recent study in this regard is directly relevant to BEV imaging (Sawant et al. 2010). The authors use the technique of failure mode and effects analysis to analyze the potential risk points in an MLC-based tumor-tracking system. They find, for example, that the system can fail by not asserting a beam-hold when the jaws are in an incorrect position. Through an examination of such failure modes, the authors were able to develop a quality management program to reduce the potential for error. Efforts like this can be applied more generally to the BEV technology and techniques in order to reduce the impact of unintended consequences.

7.5 CONCLUSIONS AND FUTURE DIRECTIONS

BEV imaging can serve as an important component of a patient safety and quality management program. This is demonstrated in a handful of studies both in operational practice and also from a more theoretical assessment of detectability. As technology and techniques continue to evolve, it will be important to consider and evaluate these effects. Several key questions may help guide the future work:

1. *What is the potential safety benefit of the BEV technology?*
   This question needs to be evaluated for each new technique, technology, or service in question relative to other existing techniques. Simply adding another layer of safety barrier may have little impact on the ultimate safety or quality of treatment. This is illustrated in Figure 7.3, which shows the impact of 15 separate quality control checks added in combination with one another. In these data there is little added value after the first five checks. This is because the previous checks have already been probed for the same types of errors. When considering whether to add additional layers of quality control, the new measures should either be sensitive to different error classes or should be more reliable in some way than the measures already in place.
2. *How can the BEV technology best be used to maximize its potential for safety improvement?*
   In the examples presented earlier it was suggested that the full potential for EPID-based dosimetry is only realized if it is used *in vivo* rather than prior to the treatment. This illustrates

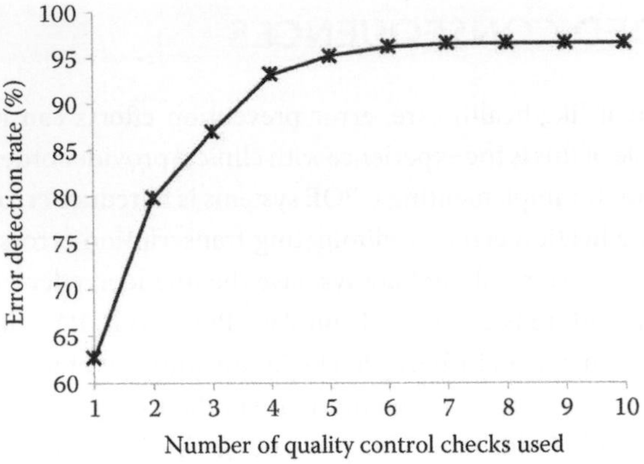

Figure 7.3 The sensitivity of multiple standard quality control checks when used in combination with one another. (Reprinted with permission from Ford, E. C., Terezakis, S. et al. *Int. J. Radiat. Oncol. Biol. Phys.*, 84, e263–269, 2012b.)

the fact that the potential of any given BEV technology may be very much related to how it is used. This must be carefully considered in evaluating these technologies and techniques.

It is likely that the maximum benefit will be derived from systems that are as automated as possible, that is, removing the reliance on human inspection. This need is understood and advocated by the experts in patient safety (Deming 1986; Grout 2006). It is being realized in some of the automated treatment plan quality check systems that are now emerging (Li et al. 2014; Noel et al. 2014; Xia et al. 2014; Dewhurst et al. 2015). In some applications, the BEV technology lends itself naturally to such automation.

3. *What are the risks in the technology or technique itself? What are the unintended consequences of introducing a given technology?*

   Any new technology or procedure change may have unintended consequences. These can be potentially very toxic or even lethal for the patient as illustrated by the examples aforementioned. It is therefore very important to address this issue in a systematic manner. At least one related study has appeared (Sawant et al. 2010), which does this prospectively using the failure mode and effects risk-estimate methodology. Future studies with such methods will be valuable. At the same time it must be realized that such risk-assessment methods are not able to capture all the error scenarios that one encounters in actual clinical practice as suggested by recent studies (Yang et al. 2015). It is important to gather information about error pathways from actual clinical operations. This should happen at the departmental level through a structured incident learning system, as specifically advocated by the American Society of Radiation Oncology (ASTRO) and other professional societies (Zeitman et al. 2012). At a more global scale, more information will hopefully be available through distributed incident learning systems such as radiation oncology incident learning system (RO-ILS) (TM) (Hoopes et al. 2015), SAFRON, ROSIS (Holmberg et al. 2010), or other systems. These systems may provide invaluable information about error pathways specific to BEV imaging, and this represents an important direction for shared learning.

Underpinning all of these questions and future progress is the need for quantitative measures of safety and quality. We have discussed a few methods for accomplishing this, but this itself is a work in development. In the future, more reliable and automatic means will be developed for quantifying the impact of these technologies with a meaningful endpoint. Ultimately this should be tied to clinical outcomes.

REFERENCES

Bates, D. W., J. M. Teich et al. (1999). The impact of computerized physician order entry on medication error prevention. *J Am Med Inform Assoc* **6**(4): 313–321.

Bogdanich, W. and K. Rebelo (2010). A pinpoint beam strays invisibly, harming instead of healing. *New York Times*, December 28, 2010.

Bogdanich, W. and R. R. Ruiz (2010). Radiation errors reported in Missouri. *New York Times*, February 24, 2010.

Bojechko, C., M. Phillps et al. (2015). A quantification of the effectiveness of EPID dosimetry and software-based plan verification systems in detecting incidents in radiotherapy. *Med Phys* **42**(9): 5363.

Carlone, M., C. Cruje et al. (2013). ROC analysis in patient specific quality assurance. *Med Phys* **40**(4): 042103.

Clark, B. G., R. J. Brown et al. (2013). Patient safety improvements in radiation treatment through 5 years of incident learning. *Pract Radiat Oncol* **3**(3): 157–163.

CoxHealth. (2010). CoxHealth announces some BrainLAB stereotactic radiation therapy patients received increased radiation dose. http://www.coxhealth.com/body.cfm?id=3701.

Deming, W. E. (1986). *Out of the Crisis: Quality, Productivity, and Competitive Position*. New York: Cambridge University Press.

Dewhurst, J. M., M. Lowe et al. (2015). AutoLock: A semiautomated system for radiotherapy treatment plan quality control. *J Appl Clin Med Phys* **16**(3): 5396.

Donaldson, L. (2008). *Radiotherapy Risk Profile: Technical Manual*. Geneva, Switzerland: World Health Organization.

Dunscombe, P., S. Evans et al. (2013). Introduction to quality. In *Quality and Safety in Radiotherapy*. Thomadsen, B., Dunscombe, P., Ford, E. et al. (Eds.). Madison, WI: Medical Physics Publishing. **36**: 1–30.

Executive, T. S. (2006). Report into unintended overexposure of Lisa Norris at Beatson, Glasgow. http://www.scotland.gov.uk/Publications/2006/10/27084909/0.

Fairchild, A., W. Straube et al. (2013). Does quality of radiation therapy predict outcomes of multicenter cooperative group trials? A literature review. *Int J Radiat Oncol Biol Phys* **87**(2): 246–260.

Ford, E. C., R. Gaudette et al. (2009). Evaluation of safety in a radiation oncology setting using failure mode and effects analysis. *Int J Radiat Oncol Biol Phys* **74**(3): 852–858.

Ford, E. C., K. Smith et al. (2012a). Prevention of a wrong-location misadministration through the use of an intradepartmental incident learning system. *Med Phys* **39**(11): 6968–6971.

Ford, E. C., S. Terezakis et al. (2012b). Quality control quantification (QCQ): A tool to measure the value of quality control checks in radiation oncology. *Int J Radiat Oncol Biol Phys* **84**(3): e263–269.

Fraass, B. A., L. B. Marks et al. (2011). Safety considerations in contemporary radiation oncology: Introduction to a series of ASTRO safety white papers. *Pract Radiat Onco* **1**(3): 188–189.

Fuangrod, T., H. C. Woodruff et al. (2013). A system for EPID-based real-time treatment delivery verification during dynamic IMRT treatment. *Med Phys* **40**(9): 091907.

Good, D., J. Lo et al. (2013). A knowledge-based approach to improving and homogenizing intensity modulated radiation therapy planning quality among treatment centers: An example application to prostate cancer planning. *Int J Radiat Oncol Biol Phys* **87**(1): 176–181.

Grout, J. R. (2006). Mistake proofing: Changing designs to reduce error. *Quality and Safety in Health Care* **15**(Suppl. 1): i44–i49.

Gurses, A. P., Y. Xiao et al. (2009). User-designed information tools to support communication and care coordination in a trauma hospital. *J Biomed Inform* **42**(4): 667–677.

Han, Y. Y., J. A. Carcillo et al. (2005). Unexpected increased mortality after implementation of a commercially sold computerized physician order entry system. *Pediatrics* **116**(6): 1506–1512.

Hendee, W. R. and M. G. Herman (2011). Improving patient safety in radiation oncology. *Med Phys* **38**(1): 78–82.

Holmberg, O., T. Knoos et al. (2010). *Radiation Oncology Safety Information System (ROSIS)*.

Hoopes, D. J., A. P. Dicker et al. (2015). RO-ILS: Radiation oncology incident learning system: A report from the first year of experience. *Pract Radiat Oncol* **5**(5): 312–318.

Huq, M. S., B. A. Fraass et al. (2008). A method for evaluating quality assurance needs in radiation therapy. *Int J Radiat Oncol Biol Phys* **71**(Suppl. 1): S170–S173.

Huq, M. S., B. Fraass et al. (2016). Application of risk analysis methods to radiation therapy quality management: Report of AAPM task group 100. *Med Phys* **43**(7): 4209–4262.

James, J. T. (2013). A new, evidence-based estimate of patient harms associated with hospital care. *J Patient Saf* **9**(3): 122–128.

Koppel, R., J. P. Metlay et al. (2005). Role of computerized physician order entry systems in facilitating medication errors. *J Am Med Inform Assoc* **293**(10): 1197–1203.

Li, H. H., Y. Wu et al. (2014). Software tool for physics chart checks. *Pract Radiat Oncol* **4**(6): e217–225.

Mans, A., M. Wendling et al. (2010). Catching errors with in vivo EPID dosimetry. *Med Phys* **37**(6): 2638–2644.

Marks, L. B., R. D. Adams et al. (2013). Enhancing the role of case-oriented peer review to improve quality and safety in radiation oncology: Executive summary. *Pract Radiat Oncol* **3**(3): 149–156.

Monville, M. E., Z. Kuncic et al. (2014). Simulation of real-time EPID images during IMRT using Monte-Carlo. *Phys Med* **30**(3): 326–330.

Moore, K., R. S. Brame et al. (2011). Quantitative metrics for assessing treatment plan quality. *Semin Radiat Oncol* **22**: 62–69.

Moran, J. M., M. Dempsey et al. (2011). Safety considerations for IMRT: Executive summary. *Pract Radiat Oncol* **1**(3): 190–195.

Noel, C. E., V. Gutti et al. (2014). Quality assurance with plan veto: Reincarnation of a record and verify system and its potential value. *Int J Radiat Oncol Biol Phys* **88**(5): 1161–1166.

Norman, D. A. (1988). *The Design of Everyday Things*. New York: Basic Books.

Nyflot, M. J., J. Zeng et al. (2015). Metrics of success: Measuring impact of a departmental near-miss incident learning system. *Pract Radiat Oncol* **5**(5): e409–416.

Ohri, N., X. L. Shen et al. (2013). Radiotherapy protocol deviations and clinical outcomes: A meta-analysis of cooperative group clinical trials. *J Natl Cancer Inst* **105**(6): 387–393.

Olaciregui-Ruiz, I., R. Rozendaal et al. (2013). Automatic in vivo portal dosimetry of all treatments. *Phys Med Biol* **58**(22): 8253–8264.

Peters, L. J., B. O'Sullivan et al. (2010). Critical impact of radiotherapy protocol compliance and quality in the treatment of advanced head and neck cancer: Results from TROG 02.02. *J Clin Oncol* **28**(18): 2996–3001.

Reasons, J. T. (1997). *Managing the Risks of Organizational Accidents*. Farnham, UK: Ashgate Publishing.

Sawant, A., S. Dieterich et al. (2010). Failure mode and effect analysis-based quality assurance for dynamic MLC tracking systems. *Med Phys* **37**(12): 6466–6479.

Thomadsen, B., P. Dunscombe et al., Eds. (2013). *Quality and Safety in Radiotherapy: Learning the New Approaches in Task Group 100 and Beyond*. Madison, WI: Medical Physics Publishing.

Wu, B. B., F. Ricchetti et al. (2011). Data-driven approach to generating achievable dose-volume histogram objectives in intensity-modulated radiotherapy planning. *Int J Radiat Oncol Biol Phys* **79**(4): 1241–1247.

Xia, J., C. Mart et al. (2014). A computer aided treatment event recognition system in radiation therapy. *Med Phys* **41**(1): 011713.

Yang, D. and K. L. Moore (2012). Automated radiotherapy treatment plan integrity verification. *Med Phys* **39**(3): 1542–1551.

Yang, F., N. Cao et al. (2015). Validating FMEA output against incident learning data: A study in stereotactic body radiation therapy. *Med Phys* **42**(6): 2777–2785.

Zeitman, A., J. Palta et al. (2012). *Safety is No Accident: A Framework for Quality Radiation Oncology and Care*. Fairfax, VA: ASTRO.

PART 3

INNOVATIONS

8

Beam's eye view imaging with low atomic number linear accelerator targets

JAMES L. ROBAR

8.1 INTRODUCTION: RATIONALE AND PHYSICAL MECHANISMS OF LOW-Z TARGET IMAGING

8.1.1 Rationale for low-Z target imaging

Beam's eye view (BEV) imaging has been applied to guidance of radiation therapy for decades (Rabinowitz et al. 1985; Rosenthal et al. 1992) and provides arguably the most useful view possible, that is, that of the tumor volume relative to the collimation of the treatment beam. However, since its inception, BEV imaging has been fundamentally limited by the beam's energy characteristics. Linear accelerator (LINAC) photon beams were designed specifically for therapeutic purposes (Podgorsak et al. 1975), providing appropriate depth-dose characteristics for typical patient geometries. Therapeutic energy spectra, for example, for a 6 MV beam, contain less than 0.5% of photons in the diagnostic energy range (Orton and Robar 2009). When applied to the task of radiographic imaging, this attribute presents two key disadvantages. First, the predominant interaction for the majority of primary photons in tissue is Compton scatter, the mass coefficient of which shows no dependence on the effective atomic number of the medium being imaged. A near absence of primary photons in the diagnostic energy range limits the proportion of photoelectric absorption occurring in the patient, and thus the strong dependence on atomic number (with a mass coefficient varying approximately with Z^3) of this interaction cannot be leveraged in producing differential attenuation between tissues, that is, subject contrast. The second limitation is the low efficiency of common detectors used for BEV imaging when used to detect therapeutic beams. For example, a typical Gd_2O_2S detector provides approximately 1% zero-frequency detective quantum efficiency (DQE) (Munro and Bouius 1998).

Accordingly, the motivation in the development and use of low atomic number (Z) target LINAC beams is to generate, within the usual beamline of a LINAC, a photon-energy spectrum that will be more useful for imaging. Specifically, the goal is to recover a significant proportion of photons in the diagnostic energy range, for example, in the range of 40%–50% (Orton and Robar 2009; Robar et al. 2009; Faddegon et al. 2010). This has the two-fold benefit of improving subject contrast by augmenting the proportion of photoelectric absorption in the patient, and increasing, for many flat panel, solid-state detectors, the efficiency of detection. Given that the low-Z target beams may be designed to replicate the geometry as a treatment beamline (e.g., with regard to source position and collimation), the approach may be combined with many of the other advantages and techniques of BEV imaging as described in Part 2 of this book.

8.1.2 Physical mechanisms of low-Z target beam generation

The concept of low-Z target beams for patient positioning is not new; in fact, the rationale for and experimental examples of *low-energy imaging* beams produced by thick, low-Z LINAC bremsstrahlung targets were described by Galbraith (1989). This work articulated the physics underlying the production of a photon beam containing an increased low-energy component, including the following points:

- Although diagnostic energy bremsstrahlung photons are created in high-Z and low-Z targets, the use of the latter reduces the absorption of low-energy photons, allowing a higher proportion of low-energy photons to escape the target. This is due to a reduced relative cross section for photoelectric absorption in the low-Z target.

- For any energy of electron beam incident on the target, according to thin target data (Motz 1955), the fractional yield of low-energy photons is highest for low-Z targets and the yield increases as the energy of the incident electron beam is decreased.
- Electron-electron bremsstrahlung is more significant in low-Z targets compared to high-Z targets. The spectrum produced has a lower peak energy than electron-nuclear bremsstrahlung (Motz 1955; Podgorsak et al. 1975).
- With regard to efficiency, while higher-Z targets give a greater yield of bremsstrahlung overall, over the forward 0°–15° angular range, that is, that subtended by a typical LINAC primary collimator in a LINAC, the yield is roughly independent of Z (Nordell and Brahme 1984).

This same early work described the experimental installation of beryllium and graphite targets within an AECL Therac-20 accelerator, demonstrating the production of a *soft* MV photon beam. A significant component of diagnostic energy photons was deduced by the rapid attenuation in water, and the fact that a more penetrating beam, that is, similar to the standard therapy beam, was recovered through the introduction of a sheet of lead that preferentially absorbed the low-energy (<150 keV) spectral component. Finally, Galbraith demonstrated that compared to imaging with a 6 MV therapy beam, low-Z targets offered the potential of substantially increasing image contrast, and that the relative advantage is reduced as a function of thickness of the patient (Figure 8.1).

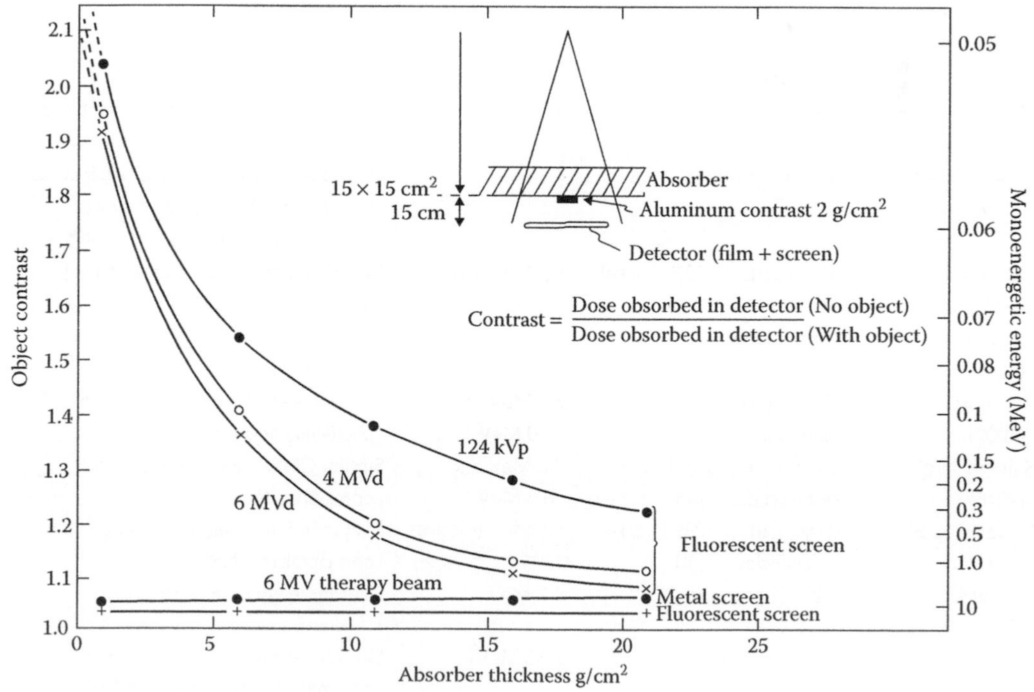

Figure 8.1 Measured contrast versus solid water absorber thickness for a 2 g/cm^2 disk in air. The 4 MVd and 6 MVd curves correspond to low-Z target beams, and are compared to contrast yielded by a 124 kVp and standard 6 MV therapy beam. (Reprinted from Galbraith, D. M., *Med. Phys.*, 16, 734–746, 1989. With permission.)

8.2 PARAMETERS IN DESIGN OF LOW-Z TARGET BEAMLINES

Compared to a standard therapeutic photon beam, three modifications are generally made to implement a low-Z imaging beamline. First, a low-Z target replaces the standard therapy target. The target typically has Z ≤ 13, which is low in comparison to conventional tungsten (Z = 74) or copper (Z = 29) therapy target materials. Second, the flattening filtration is removed to prevent selective absorption of low-energy photons downstream from the target. Third, as described in some (but not all) implementations, the mean energy of the incident electron beam is reduced compared to that typical of a therapeutic beam (Faddegon et al. 2008; Robar et al. 2009; Parsons and Robar 2012a) in order to increase the overall proportion of the low-energy photon population.

8.2.1 Choice of target material

Various low-Z target materials have been proposed, including beryllium, graphite, sintered diamond, and aluminum. Table 8.1 gives examples of materials and thicknesses used previously. It has been shown (Ostapiak et al. 1998; Tsechanski et al. 1998; Flampouri et al. 2005; Orton and

Table 8.1 Summary of low-Z target properties and reported effects on image quality

Author	Target thickness and material	Electron energy	Effect on imaging[a]
Galbraith (1989)	14.2 mm graphite	6 MeV	Contrast increase by up to a factor of 2
Tsechanski et al. (1998)	5 mm aluminum	4.0 MeV	Qualitative improvement of contrast in head-and-neck imaging
Ostapiak et al. (1998)	16.5 mm beryllium, 15.7 mm carbon	6.0 MeV	Contrast increase by up to a factor of 4
Flampouri et al. (2005)	6 mm aluminum	4 MeV	Contrast increase by up to a factor of 9.5 (thin objects)
Faddegon et al. (2008)	13.2 mm graphite	4.2 MeV	Contrast-to-Noise Ratio (CNR) increase by a factor of 3, improvement of spatial resolution by a factor of 2
Roberts et al. (2008)	20 mm carbon (28%), nickel exit window (71%)	5.6 MeV (mean)	Contrast increase by factor of 1.3 for thick objects and 4.6 for thin objects
Orton et al. (2009)	10 mm aluminum	6.0 MeV	Contrast increase by a factor between 1.6 and 2.8
Robar et al. (2009)	6.7 mm and 10.0 mm aluminum	3.5 MeV and 7.0 MeV	CNR increase by up to a factor of 2.4 (7.0 MeV) and 4.3 (3.5 MeV)
Sawkey et al. (2010)	13.2 mm graphite, 13.2 mm sintered diamond	4.6 MeV and 6.4 MeV	Similar CNR/dose with diamond compared to graphite
Roberts et al. (2011)	20 mm carbon (28%), nickel exit window (71%)	5.6 MeV (mean)	Factor of 5 to 7 less dose required for comparable CNR
Fast et al. (2012)	13.2 mm graphite	4.2 MeV	CNR increase by a factor of 2.6
Robar et al. (2012)	7.6 mm graphite	2.35 MeV	Greater reduction of dose with field collimation compared to 6 MV
Parsons and Robar (2012a)	7.6 mm carbon, 6.7 mm aluminum	1.85 to 2.35 MeV	CNR increase by factor ranging from 2.2 to 9.7

[a] All studies compared to 6 MV, with the exception of Tsechanski et al. who compared to 10 MV.

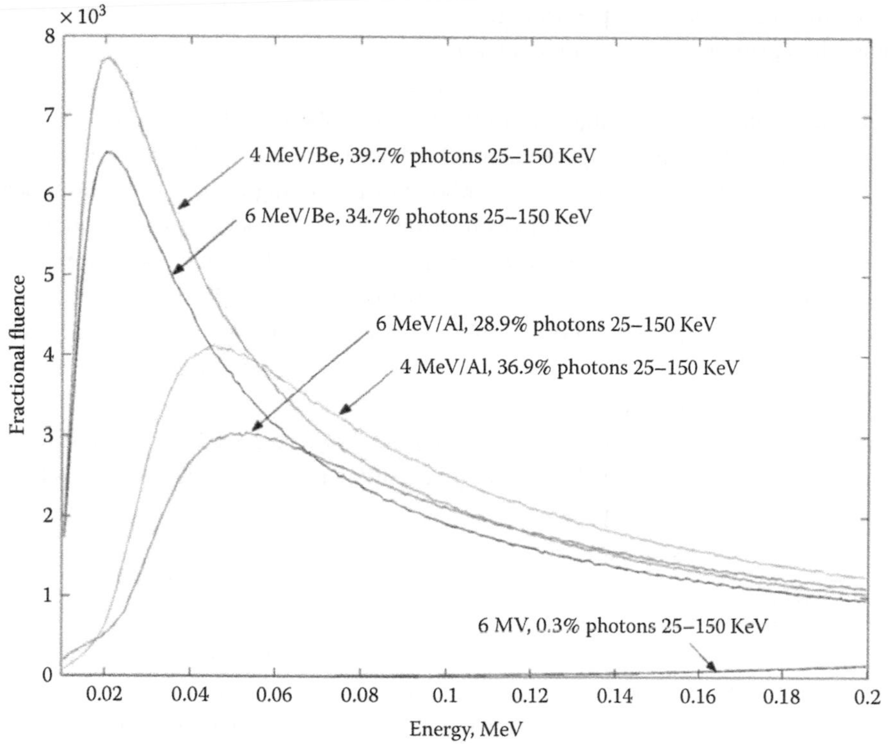

Figure 8.2 Monte Carlo calculated photon spectra for a standard 6 MV therapy beam, compared to beryllium and aluminum targets for electron energies of 4 MeV and 6 MeV. In this work, almost 40% of photons generated have energies between 25 and 150 keV, compared to 0.3% for the 6 MV beam. (Reprinted from Orton, E. J. and Robar, J. L., *Phys. Med. Bio.*, 54, 1275–1289, 2009. With permission.)

Robar 2009; Parsons and Robar 2012a; Parsons et al. 2014) that use of lower-Z target materials will increase the total proportion of low-energy photons in the spectrum produced. For example, Figure 8.2 compares energy spectra from aluminum (Z = 13) and beryllium (Z = 4) targets over the energy range below 200 keV. As shown, the low-energy photon population between 25 keV and 150 keV varies from approximately 29% to 40%, in comparison to just 0.3% for a clinical 6 MV beam (Orton and Robar 2009). The proportion in this useful energy range increases modestly, from 36.9% to 39.7% when aluminum is replaced by beryllium, for the 4 MeV beam. An important observation is that while a lower target atomic number increases fractional fluence, much of this gain occurs at very low energies, for example, 40 keV. Although this may be advantageous in imaging at low separation (e.g., breast or extremities), for more common patient geometries (e.g., head-and-neck, lung, and pelvis) this component will increase patient dose without contributing to image formation.

If the energy-spectral characteristics may be controlled by varying the target Z, one would expect that image quality will be affected. In a Monte Carlo (MC) modeling study, Flampouri et al. (Flampouri et al. 2005) demonstrated that when imaging thin subjects, contrast depends significantly on the atomic number of the target used. For example, contrast is increased by approximately 30% when Z is lowered from 20 to 4, for objects within a 10 cm thick water phantom. However, when the subject thickness increases to 20 cm, over the same range of target materials contrast varies by less than 10%. This finding is consistent with the spectral variation shown in Figure 8.2; for larger separation geometries, the additional very low-energy fractional fluence has a less pronounced effect on contrast.

Aside from atomic number, various other physical properties are important in selecting an appropriate target material. Although aluminum has been shown to have advantages with regard to contrast-to-noise ratio (CNR) improvement (Orton and Robar 2009; Robar et al. 2009), it has a comparatively low melting point (660.3°C). Although the melting point of beryllium is approximately twice as high, it is associated with low neutron activation energy. In addition, the machining of beryllium is inconvenient due to its toxicity if inhaled in particulate form. This leaves carbon as a viable candidate. Carbon offers the advantage of no melting point and a very high sublimation temperature of 5530°C. Both graphite and diamond have been used as LINAC targets, with the latter offering the advantages of (i) a higher thermal conductivity by a factor of at least seven, allowing dissipation of heat during beam production, and (ii) a higher density, facilitating a thinner target that may yield better spatial resolution characteristics (Connell and Robar 2010). Sawkey has shown that graphite and diamond give similar results with regard to CNR as a function of imaging dose (Sawkey et al. 2010).

8.2.2 Target thickness

Tsechanksi (1998) suggested that it may be preferable to use a *thin* target, that is, with a thickness less than the continual slowing down approximation (CSDA) range of the electron in the target medium. The reasoning behind this suggestion was that the majority of bremsstrahlung events occur in the superficial target layers, and thus, thicker targets would simply harden the beam. This contradicted the earlier suggestion by Galbraith that thick targets are appropriate, given the contribution of straggling electrons to low-energy bremsstrahlung output (Galbraith 1989). In Tsechanski's work, it was demonstrated that the integrated photon fluence below 150 keV was indeed higher, for example, for a 1.5 mm compared to a 4 mm target; however, this was demonstrated for an intermediate Z (copper), not for a lower atomic number material. The thin-target approach suffers from the drawback of requiring a filter, for example, a plastic sheet in the accessory tray of the LINAC to prevent excessive superficial dose to the patient by transmitted electrons (Tsechanski et al. 1998; Robar 2006; Orton and Robar 2009). Using a similar thin-target approach, Ostapiak et al. (1998) measured an increase of dose in the build-up region by a factor of 3 if the filter were omitted, for a beam generated by 6 MeV electrons on a 1.7 cm carbon target. Robar (2006) showed that removal of a polystyrene filter in an aluminum or beryllium target beamline results in a very high superficial dose over the first several millimeters, caused by transmitted electrons, followed by a more deeply penetrating dose due to the bremsstrahlung photon component. Other than the inconvenience of requiring a removable accessory for imaging, this approach is also limited in that the tray attenuator will itself act as a source of scatter close to the patient and detector, potentially compromising image quality. Following the observation by Flampouri et al. (2005) that contrast actually varies minimally as a function of low-Z target thickness, most subsequent implementations employed full-thickness targets, that is, greater than the CSDA range of incident electrons.

8.2.3 Target location

Conceptually, the low-Z target can be placed in a similar location as the therapy target, for example, within the usual target arm of the LINAC. However, for practical reasons, most experimental investigations have placed the target in air, downstream from the primary collimation, for example, in place of the scattering foil or flattening filter (FF). This offers a convenient experimental setup, since one or more targets can be installed, for example, without removing vacuum from the target assembly and primary collimator. Installed low-Z targets are simply moved into position through manual collimator rotation (Tsechanski et al. 1998; Robar 2006; Robar et al. 2009). In this arrangement, the LINAC may be operated in electron mode in order to retract the usual therapy target. Although electron modes typically involve beam currents that are two to three orders of magnitude lower than photon modes, this generally does not pose a problem for imaging because high currents are not required. Finally, if heat dissipation is of concern, unlike conventional LINAC targets in vacuum, air cooling is an option.

The placement of a low-Z target in air, downstream from the vacuum assembly has been used by Faddegon in a Siemens LINAC (Faddegon et al. 2008). Roberts in an Elekta unit (Roberts et al. 2008) and Robar in Varian platforms (Robar et al. 2009). In the Siemens Primus unit, the target was located in place of the scattering foil normally used to scatter 18 to 21 MeV beams. In the Elekta Primus unit, the target was located at the bottom of the primary collimator, supported by an aluminum cone that was attached to the high-energy difference filter mountings. In the Varian high-energy clinical linear accelerator (Clinac) unit, low-Z targets have been placed in vacant ports of the carousel, that is, replacing the FF.

Common to all of these arrangements is the potential for broadening of the electron beam through scatter in air after emerging from the vacuum assembly, upstream of the low-Z target. The concern here is compromise of spatial resolution resulting from enlargement of the focal spot. Orton et al. (2009) experimented with various target positions by systematically raising the beam-side surface of the target above the carousel in which it was installed. On a high-energy Varian LINAC, this group was able to locate the target within 9 mm of the beryllium exit window of the primary collimator assembly without obstructing rotation of the carousel. When the target was located at the level of the carousel, the full width half maximum (FWHM) of the incident electron beam was shown to be 2.4 mm; moving the upper surface to within 9 mm of the beryllium exit window of the primary vacuum reduced this width to 1.9 mm. In a Varian high-energy Clinac unit, Connell and Robar (2010) placed radiochromic film at the location of the upper surface of the low-Z target, which allowed direct measurement of the electron beam profile as it impinged on the target surface, showing FWHM values of 2.3 mm and 2.7 mm for 7.0 MeV and 4.5 MeV beams, respectively. For these experimental arrangements with low-Z targets in air, focal spots thus were broadened slightly, for example, by approximately 1 mm FWHM (Keall et al. 2003) over those for therapy targets in vacuum. This has been found to not compromise with spatial resolution significantly; in fact, as described in Section 8.4.2, the spatial resolution can be slightly improved over that for a standard therapy beam (Connell and Robar 2010). In contrast, for an Elekta unit, Roberts determined that the electron fluence distribution was 8 cm wide at the level of the carbon target (Roberts et al. 2008). The reason that imaging was still possible without severe degradation of spatial resolution was due to the fact that more than 70% of photons were actually produced in the nickel exit window of the vacuum structure, not the low-Z target.

8.2.4 Removal of flattening filtration

There are few examples of low-Z target beamlines where flattening filtration was left in place. This is not surprising, given that filtration is commonly composed of stainless steel, copper, or lead, which would simply remove the useful, low-energy component of the beam. It is clear, however, that if the FF is removed while the LINAC target is unmodified, marked improvements of image quality are realized. For example, Christensen et al. (2013) demonstrated that the increased fluence rate, the softer energy spectrum, and the removal of scatter from the flattening filtration contribute to improved image quality in fluoroscopic imaging. In this work, the relative modulation transfer function (RMTF) was shown to increase by 40% at a spatial frequency of 0.75 line pairs/mm and CNR was increased over that yielded by flattened beams by up to a factor of 4. Some of this improvement was simply due to the increased fluence rate of the flattening filter free (FFF) beam with a fixed acquisition frame rate. Faddegon demonstrated that the major source of extrafocal scattered fluence is the FF, comprising 2% of treatment beamline fluence. These photons have lower energies than the primary beam and are detected at higher efficiency, and thus removal of the FF serves to improve spatial resolution (Faddegon et al. 2008; Faddegon et al. 2010).

8.2.5 Incident electron energy

Although a low-Z target beamline vastly increases the proportion of low-energy photons produced, typical energy spectra will contain <50% of photons within the diagnostic energy range, and a high energy *tail* persists to the maximum energy of electrons incident on the target (Figure 8.2). These photons are undesirable for several reasons. First, they will interact through Compton scatter and thus will contribute little to subject contrast. Second, for higher energy photons a larger proportion of the photon energy is transferred to the recoil electron, resulting in a higher patient dose per interaction. Third, this population will not be efficiently absorbed by most detectors. An approach to proportionally lower this high-energy population is to reduce the mean energy of the incident electron beam.

The effects of lowering the electron beam energy on image quality have been described in several studies. For sintered diamond targets, Sawkey performed low-Z target cone-beam computed tomography (CBCT) imaging using both 6 MV and 4 MV beams (Sawkey et al. 2010). When imaging at lower doses (<10 cGy), the lower energy yielded improved CNR by 25%. In order to further reduce the overall proportion of photons in the high-energy tail, it is advantageous to lower the electron beam energy below that used typically in therapy. This approach has been used, albeit with a high-Z target, in the tomotherapy unit (Accuray, Incorporated), where the energy was lowered to 3.5 MeV (Yartsev et al. 2007). To some extent, lowering the energy below that typically used for therapy is also possible on other LINAC platforms. In aluminum-target CBCT imaging, for example, the electron energy was reduced from 7.0 to 3.5 MeV, which produced an improvement in CNR by a factor of approximately 1.8 (Robar et al. 2009). Parsons lowered the energy further on a Varian high-energy Clinac unit, performing aluminum and carbon target imaging over the energy range from 1.85 MeV to 2.35 MeV (Parsons and Robar 2012a). Beam production remained stable over this range; however, output from the waveguide/bending magnet dropped by approximately 40% as energy was lowered from 2.2 to 1.9 MeV. The output declined precipitously when energy was lowered below this limit. Adjusting energy over this range yielded an increase of the relative spectral population in the diagnostic energy range, from 48.5% to 54% as energy was reduced from 2.35 to 1.9 MeV. However, the spectral change resulted in only modest improvements in CNR as a function of dose for realistic subject geometries. The authors, therefore, concluded that low-Z imaging at 2.35 MeV was preferred given the higher dose rate achievable (Parsons and Robar 2012a).

8.3 ENERGY AND DOSE DISTRIBUTIONS OF LOW-Z TARGET BEAMS

8.3.1 FEATURES OF LOW-Z TARGET ENERGY SPECTRA

Typical clinical therapy beams contain almost no photons in the diagnostic energy range. For example, Figure 8.2 shows almost no photons below 100 keV. Further examples of low-Z target energy spectra are given in Figures 8.3, 8.5, and 8.6, for Varian, Siemens, and Elekta treatment units, respectively. Figure 8.3 demonstrates that, at least within a narrow range of electron energies, the mode of the spectrum is set predominantly by the target material, for example, with carbon giving a peak near 20 keV and aluminum at approximately 45 keV (Parsons and Robar 2012a). In this example, the fraction of diagnostic photons ranges from 46% to 54%, with higher proportions yielded by lower electron energy beams and lower Z targets. Noting that very low-energy photons will only increase patient dose without ameliorating image quality, this work also demonstrated that a carbon target can be laminated with a very thin copper layer to selectively filter photons below 25 keV (Figure 8.4). Figure 8.5 compares energy spectra between a carbon target imaging beamline (IBL) and treatment beamline (TBL) in a Siemens unit (Faddegon et al. 2008). The example also demonstrates that although the TBL spectrum coincides only marginally with the detector response, the IBL is comparatively well matched. This feature introduces the benefit of improving the CNR without increasing patient dose. A similar set of spectra is shown in Figure 8.6 for an Elekta unit with XVI and iView/GT detectors, for a mean energy of 5.6 MeV and a 20 mm carbon target (Roberts et al. 2008). This beamline differed from other examples in that more than 70% of photons arose from bremsstrahlung in the moderate-Z (nickel) exit window, not from the low-Z target below.

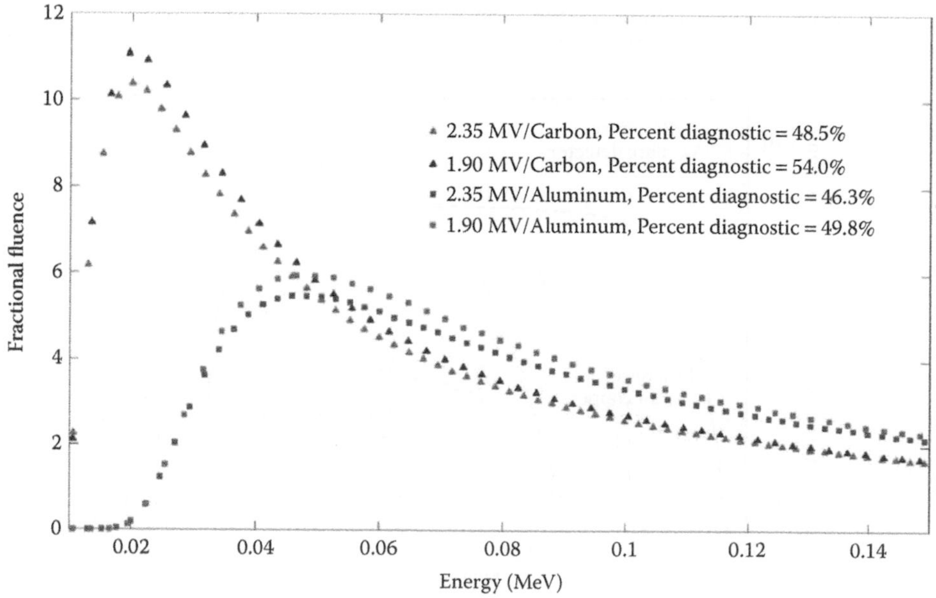

Figure 8.3 Energy spectra for carbon and aluminum targets below 150 keV, for incident electron beam energies of 1.90 and 2.35 MeV. (Reprinted from Parsons, D. and Robar, J. L., *Med. Phys.*, 39, 4568–4611, 2012. With permission.)

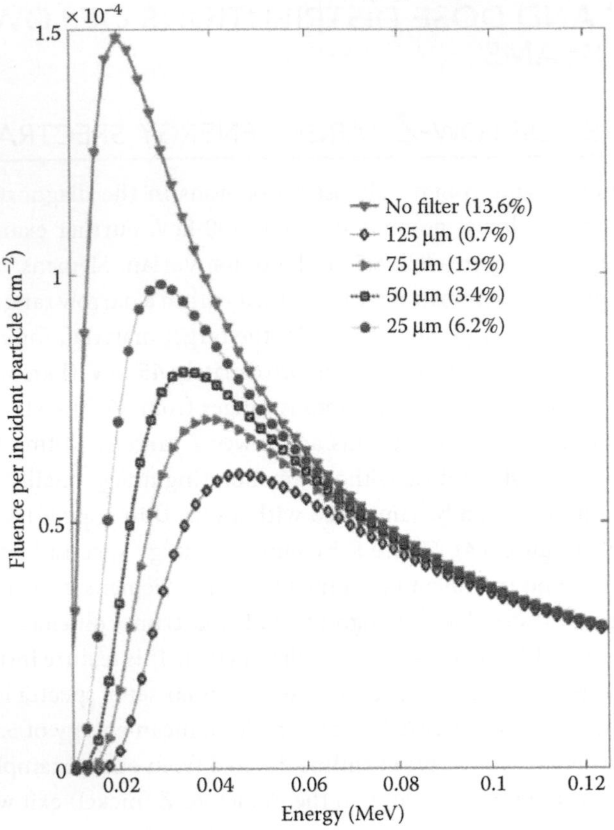

Figure 8.4 Energy spectra for 2.35 MeV electrons incident on a carbon target backed with various thin layers of copper for filtration of the very low-energy component. (Reprinted from Parsons, D. and Robar, J. L., *Med. Phys.*, 39, 4568–4611, 2012. With permission.)

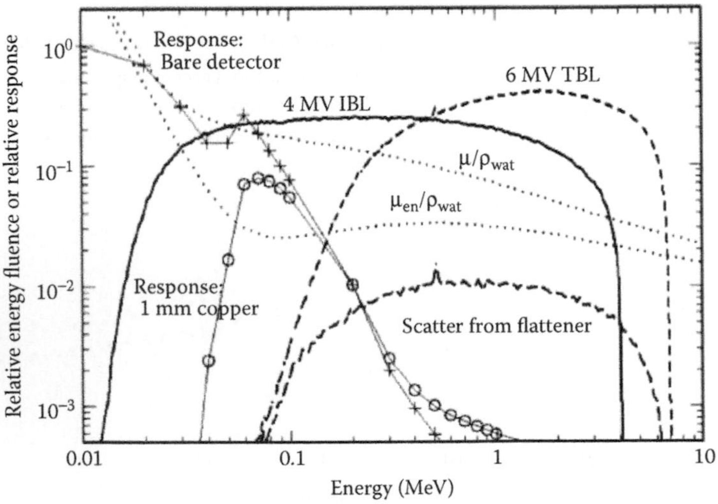

Figure 8.5 Calculated energy spectra for the treatment beamline (TBL) and low-Z target imaging beamline (IBL) for a Siemens treatment unit. Also shown are the energy response of the detector with and without copper layer and variation of mass attenuation and mass energy attenuation coefficients for water. (Reprinted from Faddegon, B. A., Wu, V., Pouliot, J., Gangadharan, B., and Bani-Hashemi, A., *Med. Phys.*, 35, 5777–5810, 2008. With permission.)

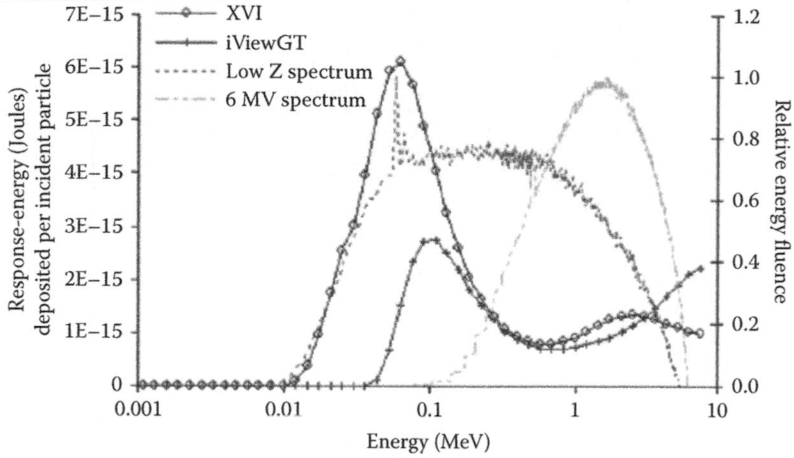

Figure 8.6 Comparison of calculated low-Z target and 6 MV therapeutic spectra for an Elekta treatment unit, compared to the energy response of XVI and iViewGT detectors. (Reprinted from Roberts, D A, Hansen, V. N., Niven, A. C., Thompson, M. G., Seco, J., and Evans, P. M., *Phys. Med. Bio.*, 53, 6305–6319, 2008. With permission.)

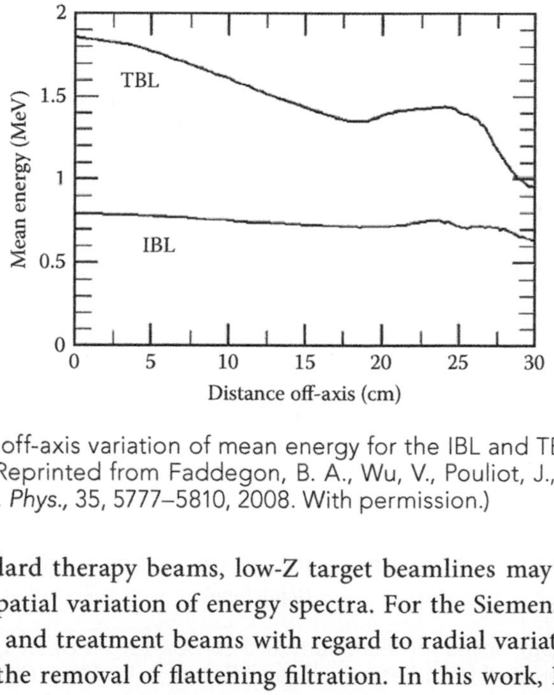

Figure 8.7 Calculated off-axis variation of mean energy for the IBL and TBL for the Siemens carbon target beamline. (Reprinted from Faddegon, B. A., Wu, V., Pouliot, J., Gangadharan, B., and Bani-Hashemi, A., *Med. Phys.*, 35, 5777–5810, 2008. With permission.)

Compared to standard therapy beams, low-Z target beamlines may also introduce differences with regard to spatial variation of energy spectra. For the Siemens beamline, Figure 8.7 compares the imaging and treatment beams with regard to radial variation of mean energy, a change largely due to the removal of flattening filtration. In this work, Faddegon et al. (2008) noted that reducing the off-axis variation of energy spectrum serves to reduce cupping artifacts in raw CBCT images.

8.3.2 Dose distributions from low-Z target beams

As described earlier, low-Z targets produce softer energy spectra and therefore depth-dose curves will exhibit less penetration in water compared to high-Z target beams with corresponding electron energies. Figure 8.8 shows, for example, that for a carbon target beamline, the dose at 10 cm is approximately 10% less than the treatment beamline (Faddegon et al. 2008).

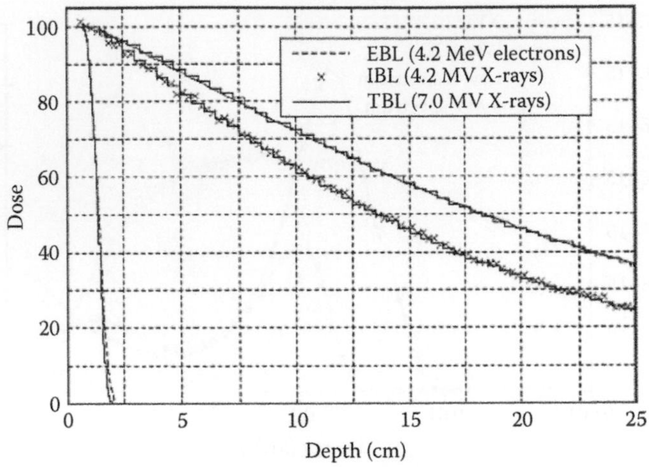

Figure 8.8 Depth ionization curves for electron beamline (EBL), low-Z target imaging beamline (IBL), and treatment beamline (TBL) for the Siemens unit. (Reprinted with permission from Faddegon, B. A., Wu, V., Pouliot, J., Gangadharan, B., and Bani-Hashemi, A., *Med. Phys.*, 35, 5777–5810, 2008.)

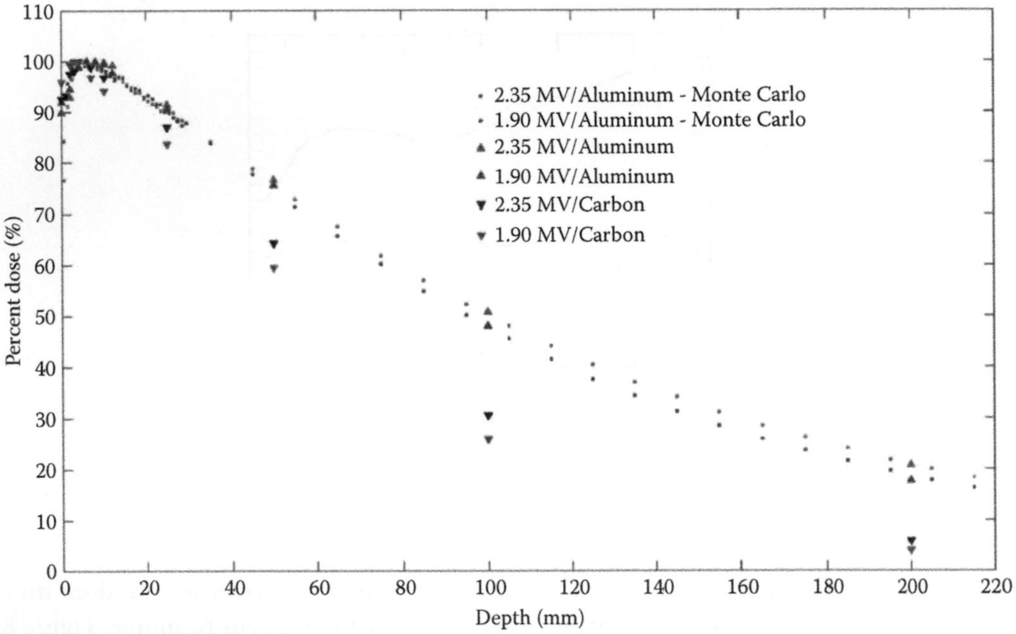

Figure 8.9 Percent depth doses for aluminum and carbon target beams, for 1.90 and 2.35 MeV electron energies. (Reprinted from Parsons, D. and Robar, J. L., *Med. Phys.*, 39, 4568–4611, 2012. With permission.)

Figure 8.9 shows a more pronounced example, where 1.9 MV and 2.35 MV carbon and aluminum target beams show PDD_{10cm} values of 30% and 50%, that is, 37% and 16% lower than for a typical 6 MV therapeutic beam (Parsons and Robar 2012a).

Similar to FFF therapy beams (Titt et al. 2006), low-Z target beamlines are forward peaked along the radial dimension, rather than being flat or exhibiting beam horns. Figure 8.10 compares

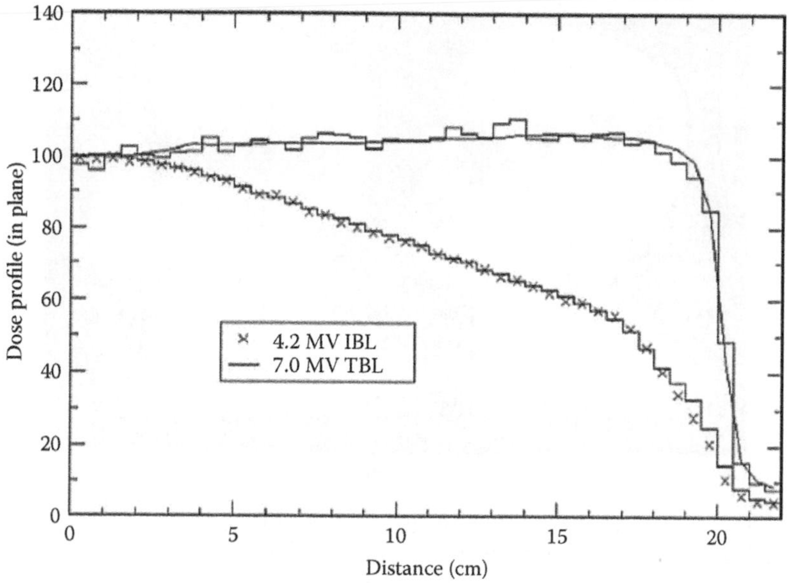

Figure 8.10 Off-axis dose profiles for a 4.2 MV carbon target imaging beamline (IBL) compared to that for a 7.0 MV treatment beamline (TBL). Solid lines and crosses show measured data, compared with Monte Carlo simulated results (steps). (Reprinted with permission from Faddegon, B. A., Wu, V., Pouliot, J., Gangadharan, B., and Bani-Hashemi, A., *Med. Phys.*, 35, 5777–5810, 2008.)

profiles at shallow depths in water for the Siemens carbon target beamline (Faddegon et al. 2008). Profiles will be more forward peaked at higher electron energies, according to the angular distribution of the Larmor relationship describing bremsstrahlung intensity. Profiles will also be more forward peaked for lower atomic number targets due to the variation of mass angular scattering power, which is roughly proportional to Z. Forward-peaked fluence profiles do not present a problem for imaging, as flood-field corrections will account for the nonuniformity of beam fluence. In addition, the fluence distribution is quite compatible with CBCT imaging for typical subjects, with higher fluence near the central axis compared to the periphery, that is, similar to that produced by a bow-tie filter for a full-fan acquisition.

For CBCT acquired with a 6 MV therapeutic beam, Gayou et al. measured dose distributions and modeled the image acquisition using a commercial treatment planning system (Gayou et al. 2007). Given the changes in characteristics of dose profiles shown earlier, one can expect differences in imaging dose distributions for patients when low-Z target beams are employed. In order to visualize and account for imaging dose distributions, Robar et al. (2014) commissioned the Eclipse anisotropic analytical algorithm (AAA) (Varian Medical Systems, Incorporated) to model photon distributions with an accuracy of better than 5%, despite the fact that the algorithm is usually configured for higher quality therapeutic beams. This allowed generation of three-dimensional dose distributions for low-Z target CBCT protocols, comparison of volume-of-interest acquisition with full-field imaging, and almost complete compensation for the presence of the imaging dose distribution through optimization of volumetric modulated arc therapy (VMAT) plans (Figure 8.11).

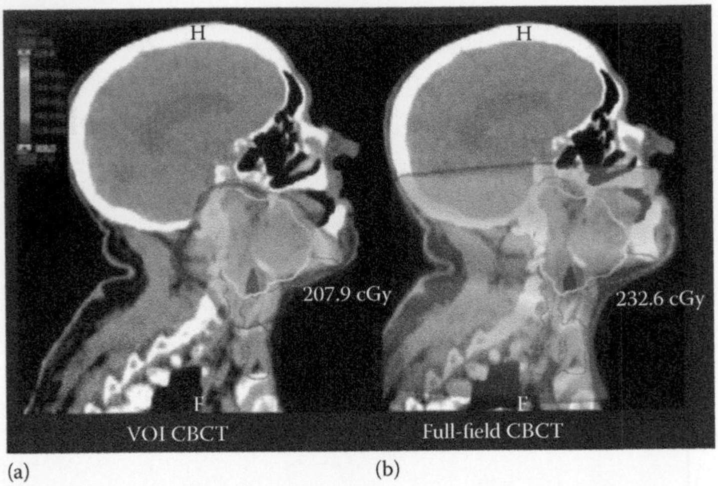

Figure 8.11 CBCT imaging dose distributions calculated within a commercial treatment planning system for a 2.35 MV/carbon target beam, where the MLC has been used to conform to the target volume during image acquisition (a) VOI CBCT and (b) for full-field imaging. (From Robar, J. L., Leary, D., and Anderson, C., *Radia. Oncol. Bio.*, 90, S884, 2014.)

8.4 IMAGE QUALITY PRODUCED BY LOW-Z TARGET BEAMS

8.4.1 Contrast, contrast-to-noise ratio, and imaging dose

As described earlier, recovery of diagnostic energy photons in the energy spectrum benefits both subject contrast and noise characteristics. A review of the literature shows a wide range of reported values, given the variability of target designs, beam energies used, phantom geometry, and imaging geometry. Table 8.1 summarizes the findings: for example, previous studies indicate that contrast improves by a factor ranging from 1.3 to 4.6, compared to a 6 MV therapeutic beam. CNR will increase by a factor ranging from 2.4 to 4.3 for beamlines in which the electron energy is comparable to that used for a typical therapeutic beam. However, if the energy is lowered, for example, below 2.4 MeV, substantial further improvements in CNR are seen. For example, as shown in Figure 8.12 for BEV planar imaging, Parsons and Robar (2012a) demonstrated a factor increase of CNR ranging from 3.7 to 4.3, from 5.0 to 6.0, and from 7.2 to 10.0 for cortical bone, inner bone, and brain, respectively, where objects were placed within a 15 cm thick solid water phantom. For the low-contrast breast object within this phantom, no advantage was seen below approximately 0.1 cGy imaging dose; however, for a very thin (3 cm) phantom, a factor up to 2.7 improvement in CNR was observed.

Figure 8.12 illustrates the increase of CNR with dose due to the reduction of noise, which should vary roughly as $1/\text{dose}^2$. Thus one can also express the benefit of low-Z target beams in terms of dose reduction, for equivalent CNR, compared to a standard therapy beam. For example, for the Siemens IBL, the same CNR was obtained with approximately half of the dose for muscle and liver objects (Faddegon et al. 2008). For adipose, the required imaging dose was reduced by approximately a factor of 5. For the same beamline, Fast et al. (2012) also showed a dose reduction

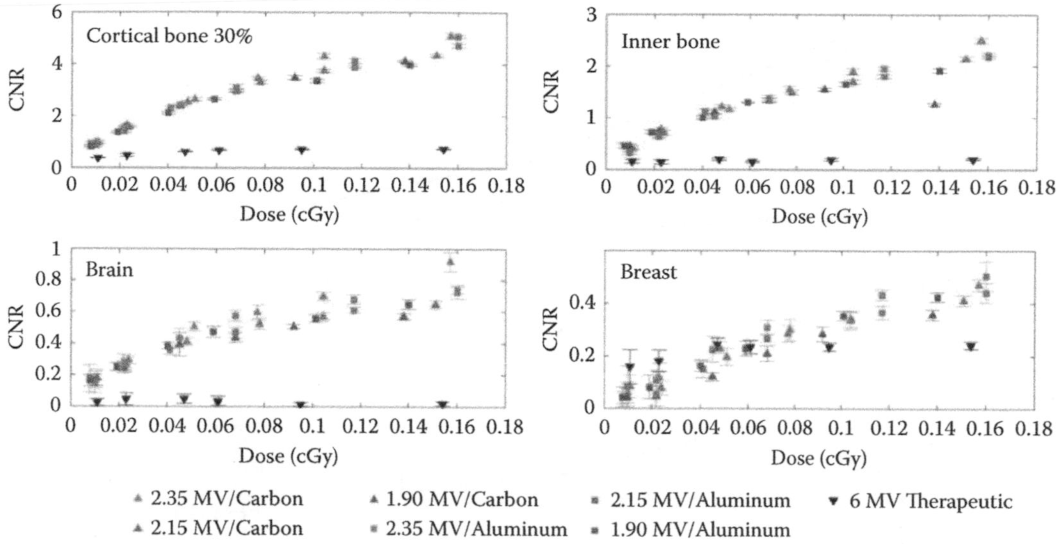

Figure 8.12 CNR versus dose curves for planar imaging using carbon and aluminum target beams with electron beam energies of 1.90, 2.15, or 2.35 MeV, compared to 6 MV therapy beam. (Reprinted from Parsons, D. and Robar, J. L., *Med. Phys.*, 39, 4568–4611, 2012. With permission.)

by a factor of 5. For the 3.5 MV/aluminum beamline described by Robar, for bone and lung objects, the dose is reduced by a factor of approximately 8, for the same CNR (Robar 2006; Robar et al. 2009).

8.4.2 Spatial resolution

Although the main rationale for low-Z targets beamlines is improved CNR, several studies of spatial resolution have been conducted. Changes in resolution relative to therapy beams may be expected because low-Z target beamline designs (i) may cause some broadening of the electron beam upstream of the low-Z target in air, for example, compared to targets in vacuum, which would degrade resolution, and (ii) usually involve removal of flattening filtration, which would serve to improve resolution by eliminating the primary source of extrafocal scatter. In CBCT carbon target imaging, Faddegon et al. (2008) reported a two-fold improvement in spatial resolution, with resolvable spatial frequency improving from 0.2 line pairs/mm for the therapy beam to 0.4 line pairs/mm for the carbon target beamline. Improved resolution in patient imaging was also evident (Faddegon et al. 2010). Connell and Robar (2010) performed a detailed study of spatial resolution as a function of target material (beryllium, aluminum, and tungsten) and thickness (20%, 60%, or 100% CSDA range). Figure 8.13 shows measured f50 values, that is, the spatial frequency at which the MTF drops to 0.5. For the experimental targets, higher energies produced slightly better spatial resolution; however, the variation is small and therefore this factor would not outweigh the CNR gains realized by lowering energy. Beryllium yielded slightly better resolution than aluminum. This study also showed that for a *full-thickness* target, the spatial resolution may be either slightly improved or degraded compared to that for a 6 MV therapy beam, depending on the low-Z target beamline parameters chosen.

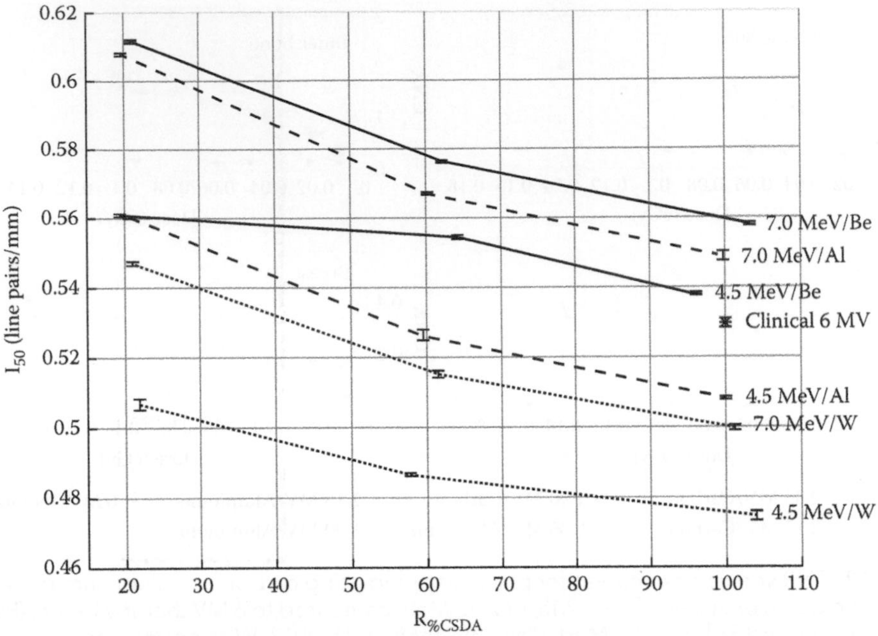

Figure 8.13 Variation of spatial resolution, quantified by F50, that is, the spatial frequency where MTF is reduced to 0.5, as a function of target thickness, expressed as percent CSDA range of the electron in the target medium. (Reprinted from Connell, T. and Robar, J. L., *Med. Phys.*, 37, 124–128, 2010. With permission.)

8.5 CONSIDERATIONS FOR DETECTOR DESIGN

8.5.1 Modifying common detectors for low-Z target beams

Typical electronic portal imaging device (EPID) detectors include a metal conversion plate, composed of copper or steel, a scintillating phosphor, and a photodiode array switched by an array of thin-film transistors. In the aS500 and aS1000 detectors (Varian Medical, Incorporated), for example, the conversion plate is a 1 mm copper layer located on the beam side of the phosphor layer. The purpose of the conversion plate is to increase the efficiency of absorption of MeV photons in standard therapy beams. Electrons set into motion, that is, Compton recoil electrons, then deposit energy in the adjacent phosphor scintillator. In detection of a low-Z target beam, however, it is evident that the presence of a conversion plate may be suboptimal because the range of electrons set into motion by 25–150 keV photons will be insufficient to allow transport to the phosphor layer. In this case, the diagnostic energy photons would be simply absorbed and eliminated from the detection process. For 2.35 MV/carbon and 7.0 MV/carbon beams, Parsons studied the effect of copper layer thickness on both CNR and spatial resolution over the range between 0 and 1.6 mm, in 0.2 mm thickness increments (Parsons and Robar 2012b). Compared to the standard 1.0 mm thickness, removal of copper produced a CNR improvement by factors of 1.4 to 4 during planar imaging, depending on the material imaged (Figure 8.14). In low-Z target CBCT imaging, CNR improved by factors ranging from 1.3 to 2.1 with the conversion plate removed.

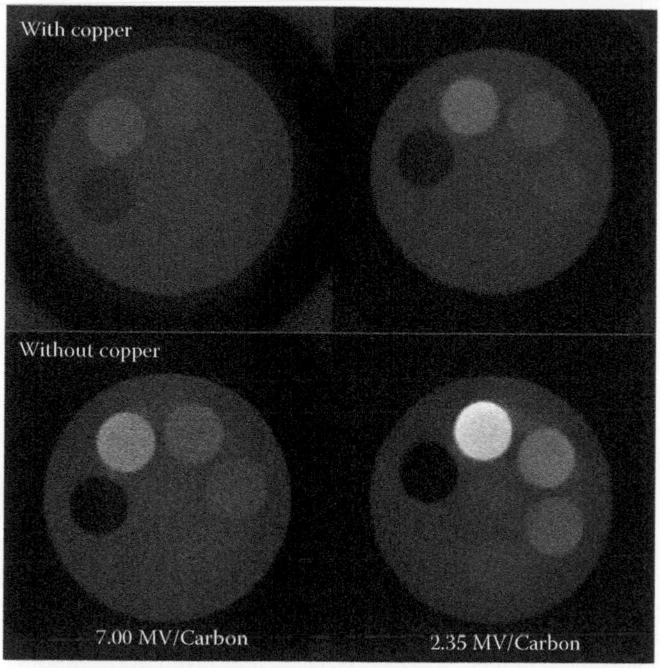

Figure 8.14 Images of a CNR phantom with and without the copper layer on the detector, for 7.0 MV/carbon target and 2.35 MV/carbon target beams. (Reprinted from Parsons, D. and Robar, J. L., *Med. Phys.*, 39, 5362–5410, 2012b. With permission.)

8.5.2 Combining low-Z target beams with high efficiency detectors

Figures 8.5 and 8.6 demonstrate that typical low-Z target energy spectra are well matched to relative detector response compared to standard therapy beams. A further approach to improving detection efficiency is by modifying the detector element in addition to the spectrum, for example, by replacing the standard Gd_2O_2S EPID scintillator by a thicker, segmented scintillator, for example, composed of CsI or BGO (Seppi et al. 2003; Sawant et al. 2005; Wang et al. 2009) or a sintered pixelated array (SPA) composed of Gd_2O_2S ceramic. In the work by Breitbach et al. (2011), a standard 1 mm Cu layer/0.26 mm Kodak Lanex Fast B scintillator was replaced by 2.46 mm Al sheet/1.8 mm thick SPA. Assuming full absorption of energy of all keV photons in the spectrum and complete detection of scintillation light, upper bounds of detective quantum efficiency (DQE_{UB}) were calculated. When combined with a 4.2 MV/carbon target beam, the high-efficiency detector increased DQE_{UB} by factors ranging from 3.6 to 4, with slightly greater improvements realized when imaging larger water phantoms. Experimentally, use of the high-DQE detector improved CNR by a factor of 1.6 on an average, when compared to that for the standard EPID. Combining the low-Z target beam with the SPA detector showed substantial improvement in patient CBCT image quality (Figure 8.15). The authors indicated that the coupling of the SPA array to the photodiode array was not optimized and thus improved performance may be realized in future versions.

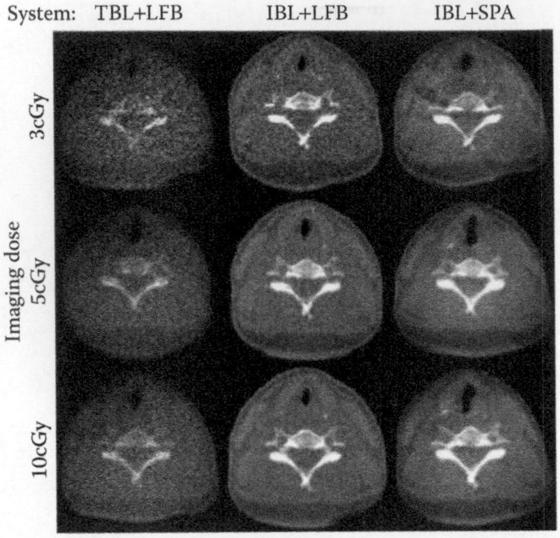

Figure 8.15 Example head-and-neck CBCT data for the treatment beamline (TBL) with Lanex Fast B (LFB) detector, the imaging beamline (IBL) with LFB and the IBL with the sintered pixelated array detector (SPA). (Reprinted from Breitbach, E. K., Maltz, J. S., Gangadharan, B., Bani-Hashemi, A., Anderson, C. M., Bhatia, S. K., Stiles, J., Edwards, D. S., and Flynn, R. T., *Med. Phys.*, 38, 5969–5979, 2011. With permission.)

8.6 FUTURE DIRECTIONS

8.6.1 Commercial implementations of low-Z targets

Although Table 8.1 demonstrates a depth of experience with low-Z target beams in research settings, similar beamlines are now available on commercial Clinac. The TrueBeam platform (Varian Medical Systems, Inc.) offers a *low-x imaging 2.5 MV* option that allows improved CNR characteristics compared to a standard therapy beam. Akin to previous experimental approaches, the imaging target is located in air. An MC study using the VirtuaLinac computational resource demonstrated that this beam contains approximately 22% of photons between 25 keV and 150 keV, compared to 40% to 50% using a 2.35 MV/carbon target beam on a TrueBeam or Clinac platform (Parsons et al. 2014). Therefore, although image quality improvement is not as dramatic as for previous experimental approaches, contrast almost doubles compared to that produced by the FFF 6 MV beam on the TrueBeam, for a 20 cm thick phantom (Figure 8.16).

8.6.2 Continuously variable waveguides

As discussed earlier, previous investigations have demonstrated the value of lowering the incident electron beam energy in increasing CNR per unit dose. Parsons et al. demonstrated this below 2.4 MeV using a standard high-energy waveguide; however the same study also reported that it was challenging to produce beam current below 1.85 MeV (Parsons and Robar 2012a). Roberts et al. have described an experimental, continuously variable waveguide, in which a rotovane is used (Roberts et al. 2012). The rotovane consists of an off-axis cell within the waveguide containing a rotatable vane. When adjusted, the vane modifies RF coupling between adjacent on-axis cells and thus both amplitude and polarity of electric field. This allows electrons in the waveguide to be decelerated

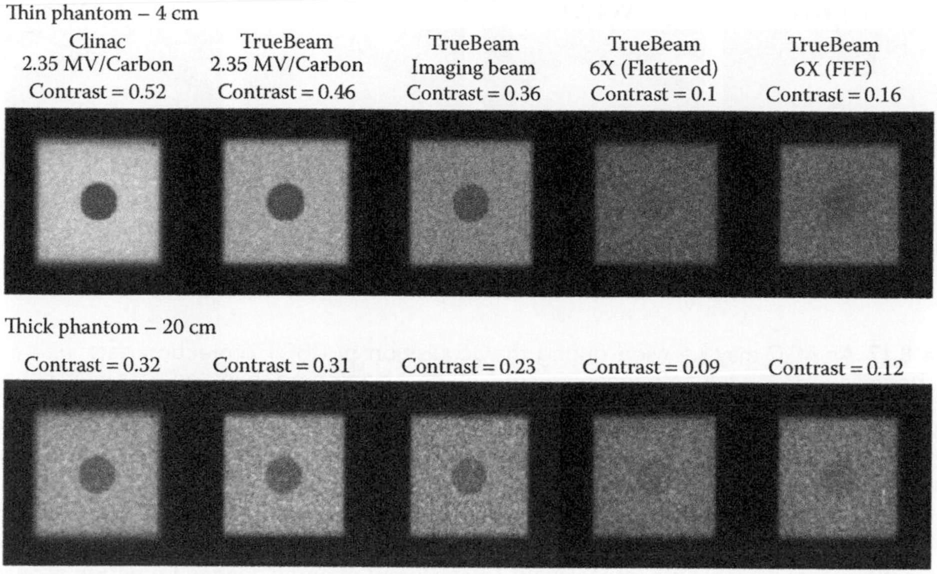

Figure 8.16 Monte Carlo calculated images generated by scoring dose deposited in the phosphor layer of an IDU20 detector for the TrueBeam imaging beam, compared to a standard 6 MV beam, a 6 MV flattening filter-free beam, and previously-investigated carbon target beamline. Although the imaging mode does not produce the contrast of the low-energy carbon target beams, it more than doubles the contrast produced by either therapy beam. (Reprinted from Parsons, D., Robar, J. L., and Sawkey, D., *Med. Phys.*, 41, 021719–021727, 2014. With permission.)

yielding a significant current of electrons with energies lower than those usually produced by waveguides in Clinacs. The investigation showed fine control of energy over the range of 1.4 to almost 7 MeV. This waveguide thus represents a promising new technology in the context of low-Z target imaging: by imaging at 1.4 MeV with a carbon target, the authors demonstrated that CBCT image data were acquired at doses less than 2 cGy, with the same CNR as gantry-mounted kV on-board imaging system. These findings, therefore, point to the possibility of inclusion of waveguides offering imaging modes that rival kV imaging systems while constraining imaging doses within acceptable limits in the context of radiotherapy guidance. Moreover, unlike gantry mounted kV on-board imaging systems, the approach would allow either CBCT or BEV imaging using a single beamline, while streamlining the overall design of the treatment delivery/image guidance platform.

8.6.3 Dynamic MLC sequences and low-Z targets

In contrast to a kV on-board imaging system, the low-Z target beamline includes a device for detailed shaping of imaging beams, that is, the multileaf collimator (MLC). The presence of this device is somewhat fortuitous but introduces interesting options for image acquisition. For example, during CBCT imaging, the MLC may be adjusted continuously as a function of gantry angle to collimate to a chosen volume of interest (VOI), for example, based on the target volume to be aligned for therapy. This has the effect of reducing dose within the VOI due to reduction of scatter, and outside of the VOI, due to collimation. Robar et al. (2012) demonstrated dose reductions by 15% and 75% within, and outside of an imaged VOI, respectively. Leary and Robar (2014) extended this concept with a 2.35 MV/carbon target beam, showing that a dynamic MLC sequence may be used during continuous gantry rotation, allowing a central VOI to be imaged at higher CNR compared to a surrounding anatomical region imaged at lower CNR (and dose). This suggests the

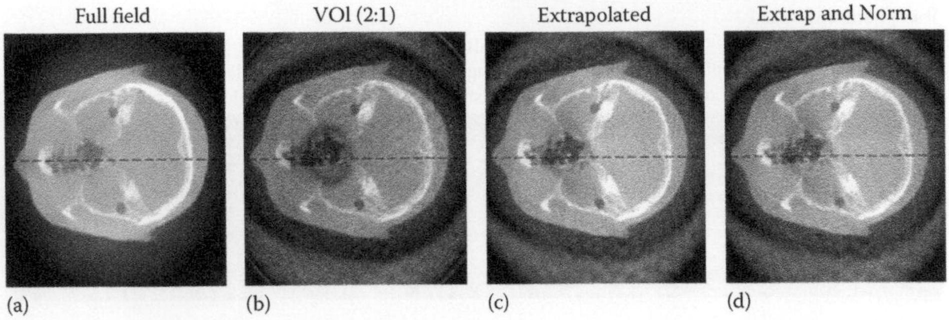

Figure 8.17 An MLC may be used during the acquisition of CBCT projection data with a low-Z target beamline. Here a full-field reconstructed image is compared to image data where the dose is roughly doubled between a volume of interest (VOI), visible on the sinuses compared to the peripheral tissues. Preprocessing projection data allows minimization of artifacts between the central VOI and the surrounding regions, and normalization of intensities. (Reprinted with permission from Leary, D. and Robar, J. L., *Med. Phys.*, 41, 011909–011918, 2014.)

potential for a more generalized *fluence-field optimization* (Bartolac et al. 2011), for example, allowing a practitioner to control both image quality metrics and imaging dose as a function of anatomic location within the patient (Figure 8.17).

8.6.4 High-Z/low-Z switching targets

One of the strengths of BEV imaging is the capacity to perform imaging during treatment and to visualize the tumor (or surrogates thereof) relative to the collimation of the treatment beam. Although innovative marker-less lung-tracking methods have been developed, for example, the robustness of these methods is compromised by low-CNR characteristics of image data acquired (Bryant et al. 2014). Assuming the low-Z target beam will not be used for treatment, a mechanism for rapidly switching between a high-Z therapy target and low-Z imaging target would be useful for near real-time BEV imaging. Figure 8.18 shows a prototype developed recently by the authors at

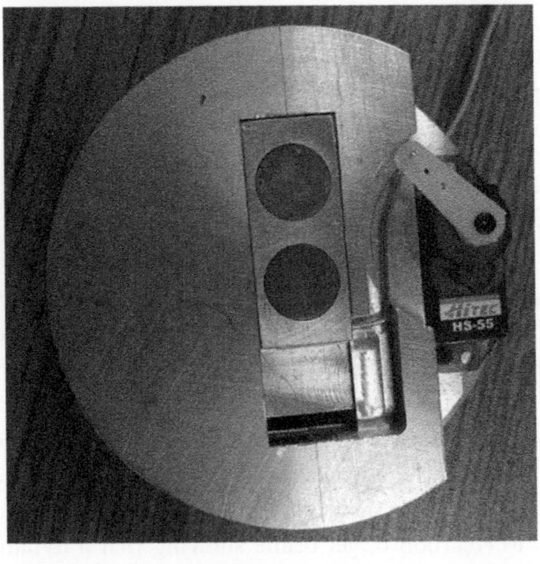

Figure 8.18 Prototype of a switching carbon/tungsten target installed experimentally in a port of a High-energy Clinac carousel. (Courtesy of Varian Medical, Inc.)

Dalhousie University, Halifax, Canada consisting of adjacent tungsten/copper and carbon targets. The entire assembly can be installed within the usual port of the carousel on a Varian high-energy accelerator. This technology could allow intrafractional low-Z target imaging, that is, the tungsten target therapy beam would be switched momentarily and at user-defined moments to acquire a high-quality, low-Z target image. The approach could be used either for live monitoring or, for example, to update models describing periodic target volume motion.

REFERENCES

Bartolac, S., S. Graham, J. Siewerdsen, and D. Jaffray. 2011. Fluence field optimization for noise and dose objectives in CT. *Medical Physics* 38 (S1): S2–S17. doi:10.1118/1.3574885.

Breitbach, E. K., J. S. Maltz, B. Gangadharan, A. Bani-Hashemi, C. M. Anderson, S. K. Bhatia, J. Stiles, D. S. Edwards, and R. T. Flynn. 2011. Image quality improvement in megavoltage cone beam CT using an imaging beam line and a sintered pixelated array system. *Medical Physics* 38 (11): 5969–5979. doi:10.1118/1.3651470.

Bryant, J. H., J. Rottmann, J. H. Lewis, P. Mishra, P. J. Keall, and R. I. Berbeco. 2014. Registration of clinical volumes to beams-eye-view images for real-time tracking. *Medical Physics* 41 (12): 121703. doi:10.1118/1.4900603.

Christensen, J. D., A. Kirichenko, and O. Gayou. 2013. Flattening filter removal for improved image quality of megavoltage fluoroscopy. *Medical Physics* 40 (8): 081713–081717. doi:10.1118/1.4812678.

Connell, T. and J. L. Robar. 2010. Low-Z target optimization for spatial resolution improvement in megavoltage imaging. *Medical Physics* 37 (1): 124–128. doi:10.1118/1.3267040.

Faddegon, B. A., M. Aubin, A. Bani-Hashemi, B. Gangadharan, A. R. Gottschalk, O. Morin, D. Sawkey, V. Wu, and S. S. Yom. 2010. Comparison of patient megavoltage cone beam CT images acquired with an unflattened beam from a carbon target and a flattened treatment beam. *Medical Physics* 37 (4): 1737–1745. doi:10.1118/1.3359822.

Faddegon, B. A., V. Wu, J. Pouliot, B. Gangadharan, and A. Bani-Hashemi. 2008. Low dose megavoltage cone beam computed tomography with an unflattened 4 MV beam from a carbon target. *Medical Physics* 35 (12): 5777–5810. doi:10.1118/1.3013571.

Fast, M. F., T. Koenig, U. Oelfke, and S. Nill. 2012. Performance characteristics of a novel megavoltage cone-beam-computed tomography device. *Physics in Medicine and Biology* 57 (3): N15–N24. doi:10.1088/0031-9155/57/3/N15.

Flampouri, S., H. A. McNair, E. M. Donovan, P. M. Evans, M. Partridge, F. Verhaegen, and C. M. Nutting. 2005. Initial patient imaging with an optimised radiotherapy beam for portal imaging. *Radiotherapy and Oncology* 76 (1): 63–71. doi:10.1016/j.radonc.2005.04.006.

Galbraith, D. M. 1989. Low-energy imaging with high-energy bremsstrahlung beams. *Medical Physics* 16 (5): 734–746.

Gayou, O., D. S. Parda, M. Johnson, and M. Miften. 2007. Patient dose and image quality from mega-voltage cone beam computed tomography imaging. *Medical Physics* 34 (2): 499–508. doi:10.1118/1.2428407.

Keall, P. J., J. V. Siebers, B. Libby, and R. Mohan. 2003. Determining the incident electron fluence for Monte Carlo-based photon treatment planning using a standard measured data set. *Medical Physics* 30 (4): 574–582.

Leary, D. and J. L. Robar. 2014. CBCT with specification of imaging dose and CNR by anatomical volume of interest. *Medical Physics* 41 (1): 011909–011918. doi:10.1118/1.4855835.

Motz, J. W. 1955. Bremsstrahlung differential cross-section measurements for 0.5- and 1.0-Mev electrons. *Physical Review* 100 (6): 1560–1571.

Munro, P. and D. C. Bouius. 1998. X-ray quantum limited portal imaging using amorphous silicon flat-panel arrays. *Medical Physics* 25 (5): 689–702.

Nordell, B. and A. Brahme. 1984. Angular distribution and yield from bremsstrahlung targets (for radiation therapy). *Physics in Medicine and Biology* 29 (7): 797–810. doi:10.1088/0031-9155/29/7/004.

Orton, E. J. and J. L. Robar. 2009. Megavoltage image contrast with low-atomic number target materials and amorphous silicon electronic portal imagers. *Physics in Medicine and Biology* 54 (5): 1275–1289. doi:10.1088/0031-9155/54/5/012.

Ostapiak, O. Z., P. F. O'Brien, and B. A. Faddegon. 1998. Megavoltage imaging with low z targets: Implementation and characterization of an investigational system. *Medical Physics* 25 (10): 1910–1918. doi:10.1118/1.598380.

Parsons, D. and J. L. Robar. 2012a. Beam generation and planar imaging at energies below 2.40 MeV with carbon and aluminum linear accelerator targets. *Medical Physics* 39 (7): 4568–4611. doi:10.1118/1.4730503.

Parsons, D. and J. L. Robar. 2012b. The effect of copper conversion plates on low-Z target image quality. *Medical Physics* 39 (9): 5362–5410. doi:10.1118/1.4742052.

Parsons, D., J. L. Robar, and D. Sawkey. 2014. A Monte Carlo investigation of low-Z target image quality generated in a linear accelerator using varian's virtualinaca. *Medical Physics* 41 (2): 021719–021727. doi:10.1118/1.4861818.

Podgorsak, E. B., J. A. Rawlinson, and H. E. Johns. 1975. X-ray depth doses from linear accelerators in the energy range from 10 to 32 Mev. *The American Journal of Roentgenology, Radium Therapy, and Nuclear Medicine* 123 (1): 182–191.

Rabinowitz, I., J. Broomberg, M. Goitein, K. McCarthy, and J. Leong. 1985. Accuracy of radiation field alignment in clinical practice. *Radiation Oncology Biology* 11 (10): 1857–1867.

Robar, J. L. 2006. Generation and modelling of megavoltage photon beams for contrast-enhanced radiation therapy. *Physics in Medicine and Biology* 51 (21): 5487–5504. doi:10.1088/0031-9155/51/21/007.

Robar, J. L., T. Connell, W. Huang, and R. G. Kelly. 2009. Megavoltage planar and cone-beam imaging with low-Z targets: Dependence of image quality improvement on beam energy and patient separation. *Medical Physics* 36 (9): 3955–3959. doi:10.1118/1.3183499.

Robar, J. L., D. Leary, and C. Anderson. 2014. Compensation for imaging dose in low-Z target volume-of-interest CBCT. *Radiation Oncology Biology* 90 (S): S884. doi:10.1016/j.ijrobp.2014.05.2521.

Robar, J. L., D. Parsons, A. Berman, and A. MacDonald. 2012. Volume-of-interest cone-beam CT using a 2.35 MV beam generated with a carbon target. *Medical Physics* 39 (7): 4209–4210. doi:10.1118/1.4728977.

Roberts, D. A., V. N. Hansen, A. C. Niven, M. G. Thompson, J. Seco, and P. M. Evans. 2008. A low zlinac and flat panel imager: Comparison with the conventional imaging approach. *Physics in Medicine and Biology* 53 (22): 6305–6319. doi:10.1088/0031-9155/53/22/003.

Roberts, D. A., V. N. Hansen, M. G. Thompson, G. Poludniowski, A. Niven, J. Seco, and P. M. Evans. 2011. Comparative study of a low- zcone-beam computed tomography system. *Physics in Medicine and Biology* 56 (14): 4453–4464. doi:10.1088/0031-9155/56/14/014.

Roberts, D. A., V. N. Hansen, M. G. Thompson, G. Poludniowski, A. Niven, J. Seco, and P. M. Evans. 2012. Kilovoltage energy imaging with a radiotherapy linac with a continuously variable energy range. *Medical Physics* 39 (3): 1218–1310. doi:10.1118/1.3681011.

Rosenthal, S. A., J. M. Galvin, J. W. Goldwein, A. R. Smith, and P. H. Blitzer. 1992. Improved methods for determination of variability in patient positioning for radiation therapy using simulation and serial portal film measurements. *Radiation Oncology Biology* 23 (3): 621–625.

Sawant, A., L. E. Antonuk, Y. El-Mohri, Q. Zhao, Y. Li, Z. Su, Y. Wang et al. 2005. Segmented crystalline scintillators: An initial investigation of high quantum efficiency detectors for megavoltage x-ray imaging. *Medical Physics* 32 (10): 3067–3083.

Sawkey, D., M. Lu, O. Morin, M. Aubin, S. S. Yom, A. R. Gottschalk, A. Bani-Hashemi, and B. A. Faddegon. 2010. A diamond target for megavoltage cone-beam CT. *Medical Physics* 37 (3): 1246–1248. doi:10.1118/1.3302831.

Seppi, E. J., P. Munro, S. W. Johnsen, E. G. Shapiro, C. Tognina, D. Jones, J. M. Pavkovich et al. 2003. Megavoltage cone-beam computed tomography using a high-efficiency image receptor. *Radiation Oncology Biology* 55 (3): 793–803.

Titt, U., O. N. Vassiliev, F. Pönisch, L. Dong, H. Liu, and R. Mohan. 2006. A flattening filter free photon treatment concept evaluation with Monte Carlo. *Medical Physics* 33 (6): 1595–1598. doi:10.1118/1.2198327.

Tsechanski, A., A. F. Bielajew, S. Faermann, and Y. Krutman. 1998. A thin target approach for portal imaging in medical accelerators. *Physics in Medicine and Biology* 43 (8): 2221–2236. doi:10.1088/0031-9155/43/8/016.

Wang, Y., L. E. Antonuk, Q. Zhao, Y. El-Mohri, and L. Perna. 2009. High-DQE EPIDs based on thick, segmented BGO and CsI:Tl scintillators: Performance evaluation at extremely low dose. *Medical Physics* 36 (12): 5707–5718. doi:10.1118/1.3259721.

Yartsev, S., T. Kron, and J. V. Dyk. 2007. Tomotherapy as a tool in image-guided radiation therapy (IGRT): Theoretical and technological aspects. *Biomedical Imaging and Intervention Journal* 3 (1): e16. doi:10.2349/biij.3.1.e16.

9

Real-time tumor tracking

JOERG ROTTMANN

9.1 INTRODUCTION AND CLINICAL MOTIVATION: TUMOR MOTION, TREATMENT MARGINS, AND TRACKING

9.1.1 TUMOR MOTION

Tumor motion is typically classified as one of two types: (i) *interfractional motion* refers to differences between planned and actual tumor position observed during patient setup for daily treatment delivery and (ii) *intrafractional motion* refers to differences between daily setup position and actual tumor position during therapy delivery (Keall et al. 2006). As interfractional motion can be corrected through daily setup imaging, the main concern for tumor tracking is intrafractional tumor position uncertainty. Sources of intrafractional tumor motion include respiration, heartbeat, gas passing through the bowel, and voluntary motion related to uncomfortable setup positions. Large respiration-induced motion is often observed for lesions located in the thoracic and upper abdominal areas including lung (Ekberg et al. 1998; Seppenwoolde et al. 2002; Bissonnette et al. 2009), pancreas (Whitfield et al. 2012), and liver (Case et al. 2010). Tumors attached or close to the diaphragm often show the largest motion amplitudes. Figure 9.1 illustrates this behavior for lung tumors.

9.1.2 TREATMENT MARGINS

Intrafractional motion generally causes a blurring of the target dose distribution. In the current clinical standard of care intrafractional motion is, therefore, accounted for by irradiating an enlarged tissue volume encompassing all tumor positions expected to occur during treatment delivery. For treatment sites with respiratory motion present, an estimate for the expected tumor motion range

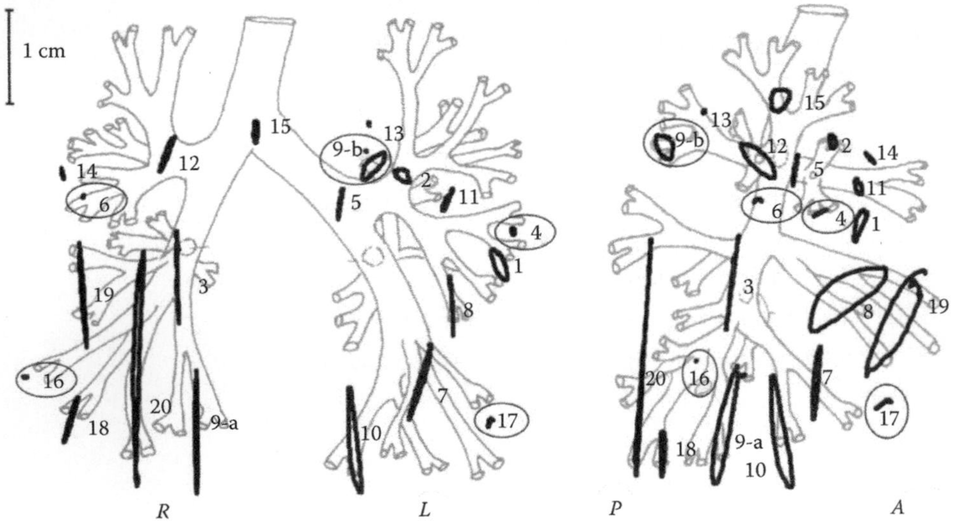

Figure 9.1 Lung tumor mobility as one of the most prominent examples for respiration-induced intrafractional tumor motion. Lesions attached to bony anatomy are circled. It can be seen that tumors closer to the diaphragm exhibit larger mobility on average. (From Seppenwoolde, Y. et al., *Inter. J. Radiat. Oncol. Biol. Phys.*, 53, 822–834, 2002.)

is typically derived from a 4D computed tomography (4DCT), that is, a temporal sequence of CT image volumes in which each CT image corresponds to a different phase in the patient's breathing cycle (Vedam et al. 2003). The 4DCT is used to define appropriately extended treatment margins, namely, the internal target volume (ITV), which is the union of all observed tumor positions in the 4DCT sequence (ICRU 2010 and van Herk 2004). However, it has been shown that 4DCT-derived motion ranges may not accurately predict tumor motion observed during treatment delivery due to the short observation length utilized for 4DCT generation and elapsed time between acquisition and treatment start of typically several days (St. James et al. 2012). Furthermore, the ITV concept necessitates the irradiation of additional healthy tissues (surrounding the tumor) to prescription dose, a potentially limiting factor particularly for hypofractionated treatments such as stereotactic body radiation therapy (SBRT) that deliver large doses in few (≤ 5) fractions. This especially applies to large target volumes and targets adjacent to or surrounded by organs-at-risk (OAR), for example, lung, pancreas, and liver.

Intrafractional motion in radiotherapy of the left breast can increase the heart dose, which may lead to coronary artery disease. For deep inspiration breath hold (DIBH) techniques the patient is asked to inhale and then hold the breath while the radiation is delivered. DIBH can maximize the distance between the treatment target and the heart and therefore minimize the heart dose (Yeung et al. 2015). Tracking the chest wall during treatment delivery has been shown to be a useful tool to ensure that the planned motion tolerances are met during therapy delivery (Jensen et al. 2014).

Although intrafractional motion is most pronounced in targets moving with the patient's respiration, other treatment sites may exhibit baseline drifts. Prostate motion during radiation delivery can lead to 3D mean displacements (per fraction) of up to 9 mm were observed in supine setup (Kupelian et al. 2007; Colvill et al. 2015). These unpredictable drifts mostly occur due to gas passing through the patient's bowel and are more pronounced in prone setup (Kitamura et al. 2002).

9.1.3 Tracking

Tumor tracking can be used to verify and mitigate the negative effects of tumor motion on the delivered dose distribution while improving the necessary treatment margins. The essential ingredients for real-time tumor tracking during radiotherapy delivery are

1. Robust real-time localization of the target during radiotherapy delivery.
2. Adaptation of the delivery geometry (radiation beam or patient support) to compensate for the observed motion.
3. Motion prediction algorithm to overcome the system latency, that is, the delay between tumor location observation and hardware adjustment to compensate for tumor motion.

In the context of this chapter the focus will be on the first point, that is, real-time *target localization* techniques. Localization can either be accomplished in 2D, that is, in the plane of view or in 3D via 3D point reconstructions (e.g., of fiducial markers) or 3D volume reconstruction. The term *real time* will be used in the context of the application, that is, tumor motion. Respiratory motion typically has a cycle time of ≈2.1–5s (Rodríguez-Molinero et al. 2013) and latencies smaller than about 0.5s have been considered manageable for prediction (Rottmann and Berbeco 2014; Krauss et al. 2011).

9.1.4 The beam's eye view perspective

The beam's eye view (BEV) perspective offered by the electronic portal imaging devices (EPIDs) is particularly well suited for assessment and compensation of intrafractional motion as it captures the dosimetrically most relevant directions of motion, that is, both directions of the steep dose falloff. In comparison: the kV on-board imager (OBI) is typically mounted with an imaging axis perpendicular to the treatment beam and can therefore only capture one of the two directions of steep dose gradient (Figure 9.2). For tracking purposes the EPID is operated in cine mode acquiring a continuous stream of images. Typically achievable frame rates are on the order of 10 frames per second (fps) with currently deployed clinical hardware (cf. also Chapter 2).

9.1.5 General hardware considerations

A general physical limit to real-time motion estimation from portal imagery is imposed by the information content encoded in the image. Compared to kV X-ray imaging, MV imaging offers substantially less image contrast due to the energy dependence of the photon interactions via photoelectric and Compton effect. Photons in the kV energy range interact predominantly via the photoelectric effect with contrast proportional to the effective atomic number ratio of the imaged materials ($C \propto Z_{\mathrm{eff}}^3$), whereas contrast in portal images is mainly formed through Compton interactions of the incident photons with the electrons of the imaging object, that is, proportional to the electron density ratio ($C \propto \rho_{e^-}$) (Motz and Danos 1978; Herman et al. 2001). The EPID technology currently in clinical use furthermore suffers from a low detective quantum efficiency (DQE) of 1%–2% at zero frequency (Antonuk et al. 1990), which limits the achievable noise characteristics. As a figure of comparison kV on-board imaging systems typically achieve a zero frequency DQE on the order of ≈70%. However, it is expected that future technology advancements will improve the performance of MV imaging and with it the tracking capabilities (cf. Chapter 12) (Rottmann et al. 2016). Another important consideration is the EPID imaging frame rate which for tracking applications must be kept above 4.5 fps to enable acceptable residual localization uncertainties (Yip et al. 2014).

As the EPID is a gantry-mounted system, mechanical precision with gantry rotation (sagging) has to be taken into account through careful assessment and correction with a calibration curve, as it has a direct impact on tracking accuracy (Rowshanfarzad et al. 2012). Furthermore, collisions

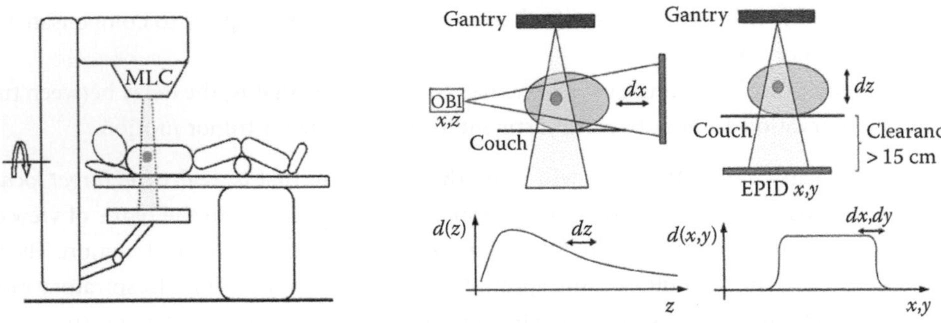

Figure 9.2 Illustration of the beam's eye view (BEV) image acquisition with the EPID (a) and comparison of its acquisition geometry (b) relative to the standard kV on-board imaging (OBI) device. It is apparent that the beam's eye view perspective captures the dosimetrically more relevant information.

of the EPID cassette with the patient or the treatment couch are a concern, particularly for noncoplanar beam setups.

Another consideration is the acquisition frame rate that has been in the range of ≈1 fps until recently. There is a lower threshold for successful BEV tracking which for markerless applications has been found to be about 4 fps (Yip et al. 2014). The latest generation of BEV imaging devices support > 10 fps, which is considered sufficient for both markerless tracking and motion mitigation (cf. also Section 9.3.1).

The delivery of variable dose rates, often used in intensity modulated treatment deliveries, can affect the image quality introducing stripe artifacts during beam-pulse synchronized cine readout (cf. Chapter 2).

The source to imager position (SID) of the EPID cassette needs to be chosen carefully for BEV tracking applications. In general, for distances (<15 cm) between EPID surface and patient surface, backscatter radiation generated in the EPID is not negligible and may lead to erythema for the patient when exposed throughout the entire treatment course. However, there are also other reasons to maximize the SID including minimization of scattered radiation from the patient and treatment support (couch) reaching the detector and the improved magnification—even at SID = 180 cm a treatment aperture of 3 cm diameter would only measure ≈150 pixels on the detector (assuming a pixel pitch ≈ 0.33 mm).

9.2 TARGET LOCALIZATION WITH BEAM'S EYE VIEW IMAGES

A multitude of tumor-localization methods utilizing the EPID have been developed over the past two decades. All methods face the aforementioned challenges of a low-image contrast and a restricted field of view (FOV) imposed by the treatment beam aperture. The FOV limitations are particularly pronounced in tumor sites exhibiting large motion ranges, such as tumors located in the thorax or upper abdomen, which can lead to periodic temporary occlusions of the tracking target. For intensity-modulated radiation therapy (IMRT) and volumetric modulated arc therapy (VMAT), the multileaf collimator (MLC) leafs moving through the treatment field cause further temporary FOV occlusions. In addition, the dose rate is often varied for intensity-modulated delivery types. This typically leads to stripe artifacts on the portal images (cf. Chapter 2). Although the boundary between cancerous and normal tissues may provide sufficient contrast in some situations such as early stage lung tumors, many other treatment sites do not provide sufficient soft tissue contrast for target localization. In these situations alternative surrogate structures may be tracked, for example, the diaphragm for tumors in proximity. When no natural surrogates are available, implanting radiopaque fiducial markers (typically made from gold) inside or adjacent to the tumor volume may offer an alternative (cf. Figure 9.3). However, the percutaneous implantation procedure may cause clinical concern over side effects tied to the invasiveness of the procedure (e.g., spread of microscopic disease along the insertion path, infection, and pneumothorax). For lung tumors, bronchoscopic implantation may also be an option—however, not all locations in the lung are reachable through this technique.

In this Section, a brief overview of available techniques for target location tracking shall be given, distinguishing between approaches to derive 2D localization and 3D localization of the treatment target (tumor). Most algorithms only provide an estimate for the center of mass motion of the target over time and disregard target deformation or surrounding anatomy and OAR. Some algorithms designed for kV imaging can be applied on MV images if the image contrast is sufficient.

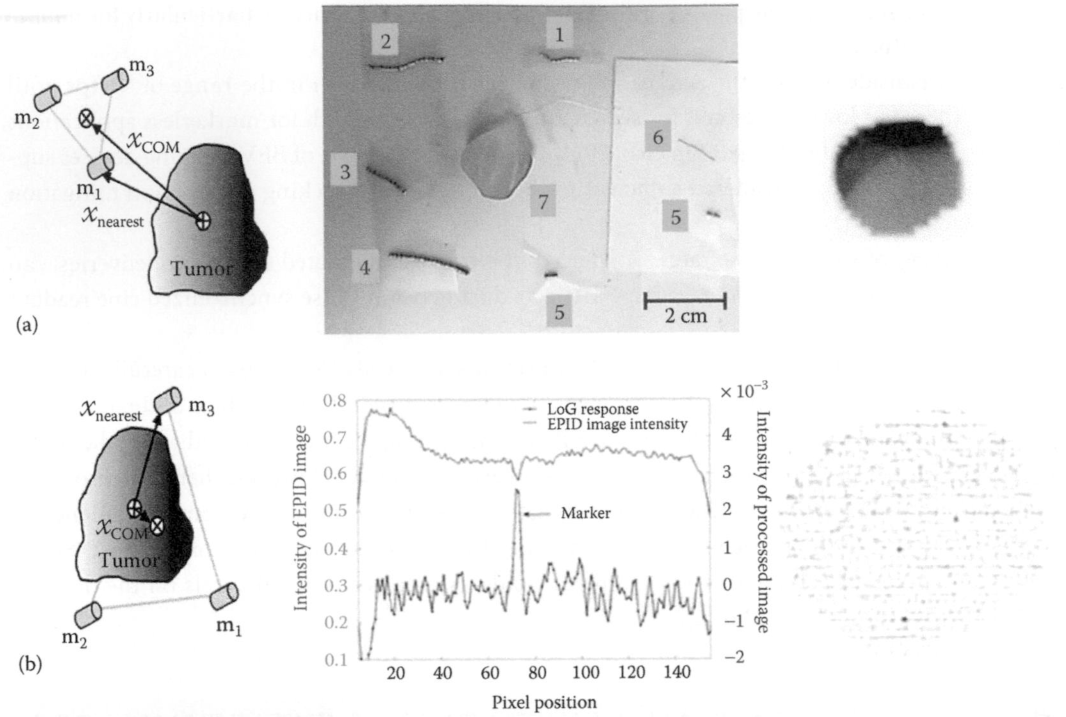

Figure 9.3 Fiducial implantation and its limitations. (From Seppenwoolde, Y., Wunderink, W., Veen, S. R. W.-V., Storchi, P., Romero, A. M., and Heijmen, B. J. M., *Phys. Med. Biol.*, 56, 5445–5468, 2011.) Samples of clinically used fiducial markers. A marker similar to type #5 was used for the case of liver SBRT shown on the right (raw EPID also displaying the diaphragm and cropped/enhanced image showing the location of the three implanted fiducial markers). (From Park, S.-J., Ionascu, D., Hacker, F., Mamon, H., and Berbeco, R., *Med. Phys.*, 36, 4536–4546, 2009.)

9.2.1 Tracking methods based on template matching

The most intuitive method for motion estimation in a sequence of images (typically 2D projections) is to choose one or several easily identifiable landmarks and locate them on subsequent images in the sequence; this strategy is commonly referred to as template matching. Since the technique works for both anatomical and fiduciary landmarks and is used by many algorithms as one of the center pieces, it will be discussed here as an introduction.

Finding suitable trackable landmarks for template matching can be accomplished manually or in an automated fashion. A small image area T_i surrounding each landmark on a reference image R is cropped and used as template. To find the location of these templates T_i on a subsequent image each possible position of the template on the subsequent image is evaluated in terms of a similarity measure. Depending on the application, similarity measures can include normalized cross correlations (NCC), sum of squared difference (SSD), or normalized mutual information (NMI). The *best matching* position is typically found by using the global maximum of similarity for all possible positions of the template on the matching image. To limit computational cost for the similarity evaluation and the chance for ambiguity, a search region is defined on the matching image, which is much smaller than the original image dimension. For EPID–EPID similarity, that is, both template and matching image are EPID acquisitions, NCC has some pronounced advantages over SSD: (1) it can be very efficiently implemented as a convolution operation (Lewis 1995), (2) potentially even using graphics processing unit (GPU) acceleration, and (3) unlike SSD it is not sensitive to global changes in illumination between template and matching image. Illumination changes over time are commonly observed with respiratory-induced motion, particularly in the lung due to compression/expansion of tissues changing the radiological path length for the traversing X-ray photons.

Template matching results in positions $p_k(t_i)$ of maximum similarity for each template T_i on each image $I(t_i)$. The use of a global maximum for matching can be problematic if several local maxima exist or if the maxima are not well defined (think of a maximum ridge—a straight line of maxima). Prior knowledge can be utilized to decrease mismatches. This includes optimal selection of search region and the utilization of the landmark positions relative to each.

Most tracking algorithms, including template matching, work in three steps: (1) image preprocessing to enhance visibility of relevant landmarks, (2) cropping to an adequate region of interest (ROI), and (3) identification of tracking landmarks on each image in the sequence and postprocessing including the calculation of the target position from the landmarks. In the preprocessing step image filters such as median, Gaussian, Laplacian of Gaussian (LoG), or histogram equalization are used to reduce noise levels and enhance visibility.

9.2.2 Tracking with fiducials

After the introduction of the first *cine*-capable electronic portal imaging systems it was considered to use implanted fiducial markers to overcome limitations in target localization due to the inherently low EPID image contrast. Several early studies showed that the (automatic) localization of fiducial markers placed adjacent to or within the prostate gland is feasible and may be used for setup and intrafractional motion assessment (Balter et al. 1994; Vigneault et al. 1997; Nederveen et al. 2001; Buck et al. 2003). Expanding on this idea it was proposed to track fiducial markers in real time for treatment adaption with the therapy beam (Keall et al. 2004). Despite the advantages of MV-EPID imaging listed in Sections 9.1.4 and 9.1.5, it is currently not used routinely in the clinic for automated target tracking or treatment adaptation. This is partially due to the increase in IMRT and VMAT but also due to the much better image quality achievable with readily available kV on-board imaging technology.

In general, at least three implanted fiducial markers are needed for reliable tumor position estimation. The precision of the estimate depends on the placement of the markers relative to the point of interest. Ideally the markers are placed close to the tumor on either side spanning an equilateral triangle. However, this is typically not achievable due to the access route via percutaneous needle or bronchoscope and often limited visualization capacity during the implantation procedure. Markers can migrate over time or even be lost (e.g., in the lung due to cough) necessitating careful monitoring of their interfractional location stability. The target registration error (TRE), that is, the difference between estimated and real target location, is usually not easily estimated, particularly if the actual target is not visible in the EPID images due to limited contrast. The fiducial registration error (FRE) can be calculated in retrospect and is often used as a predictor for the TRE. However, this may not always be appropriate as TRE and FRE are poorly correlated (Datteri and Dawant 2012). Figure 9.3 illustrates this and depicts various fiducial marker models along with a typical liver stereotactic body radiation therapy (SBRT) demonstrating a best-case scenario for visibility. It has been reported that a diameter of at least 0.75 mm is required for sufficient visibility (Chan et al. 2015).

Robust fiducial marker localization in clinical scenarios relies on reliable marker segmentation and occlusion identification (for instance, by bony anatomy or the field edge). In order to improve marker visibility and reduce bony anatomy occlusions, detector-based spectral imaging with the BEV imager has been recently suggested. This will be illustrated in the following example of liver SBRT with a technique complementary to template matching. In-treatment cine-EPID images for liver SBRT typically exhibit a homogenous background allowing exploitation of gradient techniques to enhance the contrast between fiducial marker and background. Using a LoG filter with subsequent thresholding of the standard deviation can be used as a blob detector identifying potential marker positions (cf. Figure 9.3). It can be implemented very efficiently by utilizing kernel techniques enabling computation in the Fourier domain. To remove false positives from the group of candidate positions and identify occluded markers, prior knowledge about the spatial relationship between the markers (identified on pretreatment imaging) can be used. Submillimeter fiducial localization accuracy in the BEV plane can be achieved with this technique on patient data (Park et al. 2009). 3D localization can be achieved by either utilizing a second imaging panel (e.g., OBI) or using information from the previously irradiated treatment field. However, the latter technique is not immediately suitable for recovering intrafractional motion. Employing additional prior knowledge from pretreatment imaging such as cone-beam computed tomography (CBCT) can alleviate this issue.

In the case that fiducial marker implantation is a viable option, visibility can be enhanced by adjusting the MLC sequence for maximum fiducial visibility. It has been shown that plans generated in this fashion do not necessarily suffer degraded dose distributions (Ma et al. 2009; Zhao et al. 2009).

9.2.3 Markerless tracking

The basic idea in markerless BEV tracking is to utilize the soft tissue contrast found at the border between tumor and surrounding healthy tissue and/or possible differences in texture between cancerous and healthy tissues to estimate the tumor location. Due to the dismal soft tissue contrast in liver, pancreatic, and prostate tumors, research on markerless BEV tracking has focused on lung tumors. The main benefits of markerless tracking over using implanted fiducials are the direct target observation (i.e., any uncertainties associated with the correlation of the fiducial positions and the actual tumor are eliminated) and avoiding the risks associated with the fiducial implantation procedure. Especially the latter point has drawn much interest to the further development of markerless techniques of tumor tracking for both BEV and kV imaging.

It is important to note that the noncoplanar beam arrangements typically seen in SBRT treatments do not affect the efficacy of BEV markerless tumor tracking. This is in contrast to kV images, which typically suffer from the increased radiation path length (Yip et al. 2014).

Various algorithms have been proposed for markerless BEV lung tumor tracking including template matching (Richter et al. 2010; Rottmann et al. 2010a, 2013a), level sets (Schildkraut et al. 2010; Zhang et al. 2015), and scale invariant feature transforms (SIFT) (Xie et al. 2013). We will discuss some of these algorithms briefly in Sections 9.2.3.1 and 9.2.3.2.

9.2.3.1 STiL ALGORITHM

The soft tissue localization (STiL) algorithm is based on multitemplate matching and is the only markerless algorithm, so far for which real-time beam tracking with the MLC for tumor motion mitigation has been experimentally demonstrated (Rottmann et al. 2013b). The STiL algorithm identifies suitable landmarks on the reference image (e.g., first of the sequence) automatically by texture analysis with a variance filter. Local maxima indicate strong texture. Landmark bunching can be avoided by demanding a minimal distance between landmarks. The uniqueness of each landmark within its search region is maximized by calculating an NCC-based autosimilarity map S_{auto} for each landmark and the Gaussian curvature at the origin—if S_{auto} falls off omnidirectionally from the origin, this indicates uniqueness. A scoring function is used to weigh texture against uniqueness and the best scoring landmarks are selected for tracking. Tracking of all landmarks is performed using NCC as a similarity measure (as described earlier). Temporary occlusions, attachment to static structures (e.g., bony anatomy), or poor landmark identification due to deformations or image quality problems are addressed by regularizing the set of landmarks on each image. The relative position of all landmarks to each other on the reference image is used as a signature for this purpose and deviation from this signature on subsequent images is penalized by expunging landmarks contributing the most to the deviation until an acceptable level is reached or not enough landmarks remain. The output of the algorithm for each image $I(t_i)$ acquired at time t_i is a 2D offset vector $p(t_i)$ of the average template offset from the reference image. This is illustrated in Figure 9.4.

9.2.3.2 TRACKING WITH LEVEL-SET METHODS

Level-set methods (LSM) were originally developed to track wave fronts in oceanography and burning flames (Osher 1988), but gained popularity for medical image segmentation due to their robust performance on images featuring high levels of noise and/or gray value fluctuations. A level-set is defined for a real-valued function $\phi:\mathbb{R}^n \longmapsto \mathbb{R}$ as the set of all points that map to a constant level $c \in \mathbb{R}$. Consider, for example, a 2D plane described by the function $\phi:(x, y) \longmapsto z$ and its zero level set $C = [(x, y)|\phi(x, y) = 0]$. Then C is the set of zero-level points and it describes a contour line as depicted in Figure 9.5c. The idea is now to demand for this contour line to inscribe the outline of the tumor. Since the contour line is defined by the surface function ϕ, the task of tumor tracking on a sequence of fluoroscopic images can be reformulated to finding a time-dependent surface function $\phi(\boldsymbol{x}, t)$ (with $\boldsymbol{x} = [x(t), y(t)]^T$) whose zero level set at time t yields the tumor contour at that same time. The main benefit of this somewhat unintuitive approach is that there are no assumptions made for the shape of the tumor contour itself, that is, over time, even splitting, or merging of two separate lesions would not pose a problem, which would otherwise be difficult to implement if tracking the tumor outline itself.

Since ϕ has to be zero for all times on the contour line, by definition its time derivative has to vanish there as well. So, ϕ may be calculated by solving the partial differential equation (PDE) obtained from writing out the time derivative and choosing an initial value $\phi(\boldsymbol{x}, t = 0)$:

$$\frac{\partial\phi(\boldsymbol{x},t)}{\partial t} + \boldsymbol{v}(\boldsymbol{x},t)\left|\nabla\phi(\boldsymbol{x},t)\right| = 0 \tag{9.1}$$

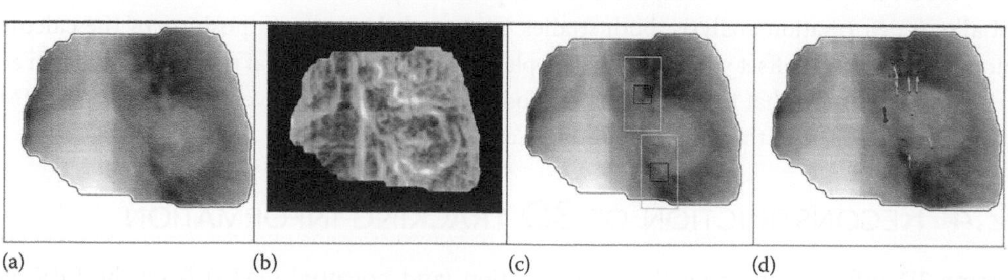

Figure 9.4 Template identification and tracking with the STiL algorithm. From left to right: (a) reference EPID image; (b) variance filtered reference image; (c) landmark candidates identified (templates marked in red, search region marked in green)—only two templates are shown for clarity; (d) tracking—each green arrow indicates a template offset, the red arrow at the center indicates the average offset in the BEV.

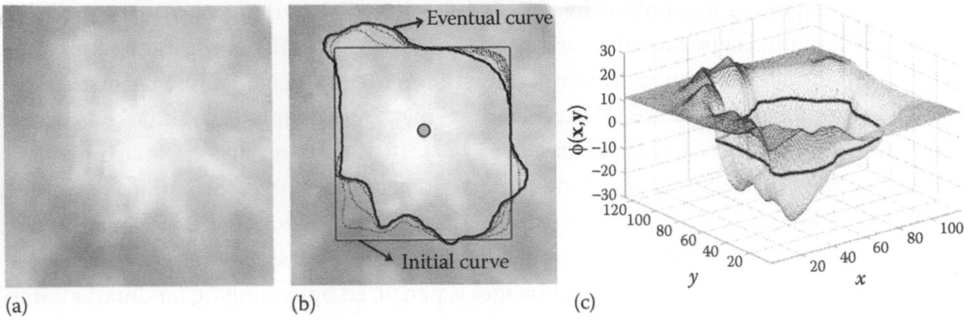

Figure 9.5 Illustration of the deformable contour definition: (a) shows the original EPID image captured during a lung tumor treatment, (b) shows the initial contour (blue) and the final contour (black) and several iterations in between (red)—the centroid used for racking is marked in green, and (c) shows a surface plot of the surface function $\phi(\mathbf{x},\mathbf{y})$ and the final contour. (From Zhang, X., Homma, N., Ichiji, K., Takai, Y., and Yoshizawa, M., *Med. Phys.*, 42, 2510–2523, 2015.)

This equation is often referred to as the *level-set equation*. Here $\boldsymbol{v}=(d/dt)\boldsymbol{x}$ denotes the speed function—also often called force function as it drives the motion of the contour. Solving the above PDE numerically is not trivial as the typically used finite difference approaches tend to fail quickly. There are many other numerical methods available for solving the above level-set equation—however, they need to be chosen carefully and are typically problem specific. Finding the tumor contour depends therefore on finding an appropriate speed function v that drives the zero level set to conform to the tumor boundary. LSMs can be classified by their choice of speed function into edge-based models and region-based models. Although edge-based models utilize boundary metrics including image intensity derivatives to define the velocity function, region-based models utilize metrics such as texture, motion, or image intensity. The latter may have advantages due to the noise present in EPID images that may affect the estimation of accurate image intensity gradients. Implementations of the LSM are usually iterative, that is, the initial value $\phi(\boldsymbol{x},\boldsymbol{y}=0)$ can be chosen arbitrarily given that solutions converge to the actual tumor contour. This is illustrated in Figure 9.5b, where the initial contour is chosen as a (blue) rectangle converging via subsequent intermediate contour iterations marked in red to the final contour marked in black.

In the context of real-time tumor tracking in the BEV, two independent studies reported the implementation of an LSM for lung tumor tracking (Schildkraut et al. 2010; Zhang et al. 2015) each using a different speed function for the driving of the level set. Both reported similar execution times (≈500 ms/f) and errors comparable to the previously described algorithms. Also, while the LSM allows deformation analysis, both studies only used the centroid position of the calculated contours, that is, a 2D offset vector in the BEV plane relative to the initial tumor position. An additional consideration is that in the context of modulated treatment delivery techniques (i.e., IMRT and VMAT) the LSM method may not be applicable.

9.2.4 Reconstruction of 3D tracking information

Inferring 3D information about the target location (and potential OARs) from the BEV alone requires either utilization of several portal images from different gantry angles and/or the use of prior knowledge in conjunction with these images. The first approach is limited by the amount of target motion and the angular separation between the portal images (Park et al. 2009). The performance improvements with respect to accuracy that are attainable by incorporating additional prior knowledge strongly depend on the validity and extent of the used prior, for example, population based data versus patient specific data (Yue et al. 2011). Alternatively, a mix of kV-OBI imaging and MV-EPID imaging may be used for instant triangulation that eliminates the error sensitivity to tumor motion (Wiersma et al. 2009). However, the target area (or fiducials) may not be visible from all gantry angles on kV projection images due to increased X-ray attenuation in bony anatomy and the associated additional imaging dose may be a potential concern.

Since 3D information is generally reconstructed from 2D information (plus a potential prior), the main challenges are similar to 2D localization, that is, the restricted FOV and limited image contrast. Intensity-modulated treatment deliveries pose a particularly challenging problem in this context.

In Sections 9.2.4.1 and 9.2.4.2, we will discuss two methods more in detail that illustrate the aspects of 3D point reconstructions and full volumetric reconstructions from one or several BEV images with or without prior knowledge.

9.2.4.1 3D POINT RECONSTRUCTION

For 3D point reconstructions 2D information from at least two directions is required, ideally separated by an observation angle of 90° to minimize triangulation errors. A simple geometric reconstruction can yield the 3D position $\boldsymbol{x}=(x,y,z)$ by back projecting both 2D localization $\boldsymbol{u}_1=(u_1,v_1,\theta_1)$ and $\boldsymbol{u}_2=(u_2,v_2,\theta_2)$ and evaluating the intersection point (cf. Figure 9.6) (Park et al. 2009; Yue et al. 2011). However, if the target is not static during the acquisition of

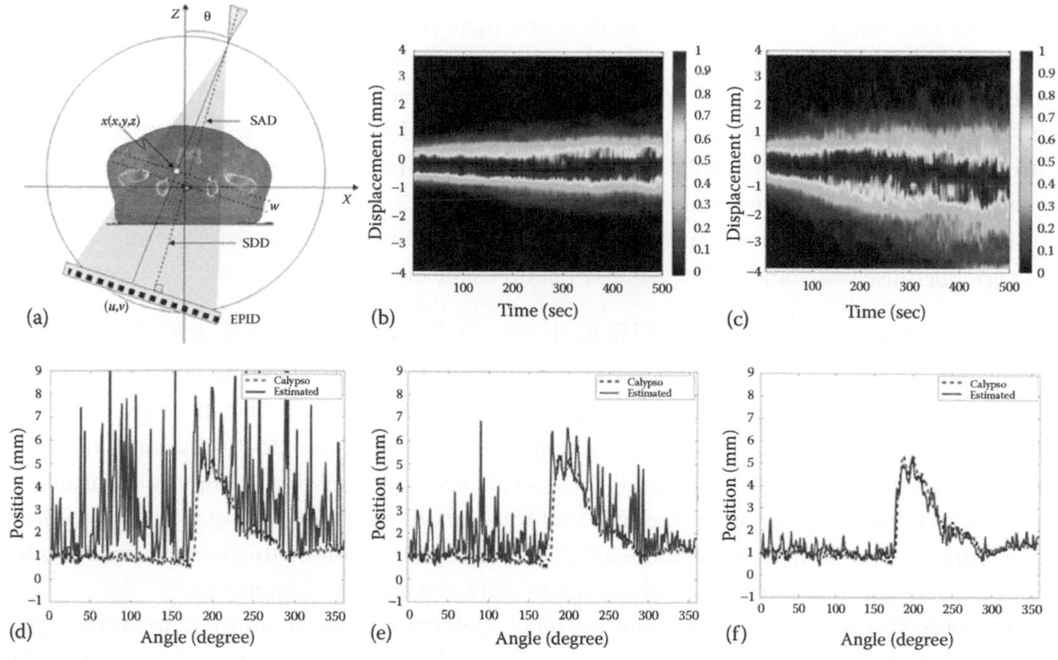

Figure 9.6 Incorporating prior knowledge to estimate 3D fiducial locations during volumetric modulated arc treatment (VMAT) deliveries in real time: (a) the setup geometry is shown noting the fiducial location **x**, its coordinates in the imager plane $\boldsymbol{u}=(\mathbf{u},\mathbf{v})$, gantry angle θ, and isocenter offset *w*. The accumulated histograms for prostate motion probability as a function of time are shown for lateral (b) and longitudinal (c). On the bottom row the 3D displacement magnitude for prostate motion estimated from seven gantry angles is shown for the purely geometric method (d), maximum likelihood (e), and maximum a posteriori (f). Note: Dotted line, Calypso; red line, estimated. (From Yue, Y., Aristophanous, M., Rottmann, J., and Berbeco, R. I., *Med. Phys.* 38, 3222–3231, 2011.)

positions (u_1, v_1, θ_1) and (u_1, v_1, θ_1), errors in the 3D reconstruction are introduced by the target displacement between the two image acquisitions (Mao et al. 2009; Park et al. 2009).

The easiest strategy to reduce this error is to utilize BEV images from more than two gantry angles, for example, BEV images from the gantry angles around the actual point of interest $\{\theta_{-n},\ldots,\theta_0,\ldots,\theta_n\}$. The fiducial location during acquisition at gantry angle θ_0 can then be estimated as the median of the reconstructed positions from all permutations (θ_i,θ_j) (Yue et al. 2011).

A more reliable 3D position reconstruction may be achieved by employing statistical methods. The displacement $\boldsymbol{d} = \boldsymbol{x}_2 - \boldsymbol{x}_1$ between the landmark positions observed on BEV image acquisitions at gantry angle θ_1 and θ_2 may be modeled in the framework of maximum a posteriori probability (MAP) as

$$\hat{\boldsymbol{d}} = arg\max_{\boldsymbol{d}}\left[\log p(f_1 \mid f_2, \boldsymbol{d}) + \log p(\boldsymbol{d})\right] \quad (9.2)$$

Here $f_i : \boldsymbol{x}_i \longmapsto \boldsymbol{u}_i$ denotes the function projecting the landmark (or fiducial) position from patient coordinates to EPID coordinates and $\hat{\boldsymbol{d}}$ denotes the estimated fiducial displacement. The first term in Equation 9.2 is maximum likelihood estimation for $\hat{\boldsymbol{d}}$ and the second term a prior (which may be omitted if not available). The maximum likelihood estimation can either be driven by incorporating several landmarks (or fiducials) per projection or by utilizing a number of projections. The prior can be estimated by evaluating representative motion in a patient population or from previous treatment days. In Figure 9.6a–c this is illustrated for prostate motion, the time axis refers to the time difference between the BEV image acquisitions. As expected it can be observed that the displacement probability increases with elapsed time. Using Equation 9.2 is a particular advantage when images from only very few gantry angles are available and there is motion present—it has been found to consistently outperform the purely geometric reconstruction, independent of the number of utilized gantry angles. Although the MAP method performs best even when no prior motion is available and one needs to resort to maximum likelihood estimation, the statistical methods outperform the purely geometric approach. On a dataset of prostate motion from 17 patients, submillimeter maximum error in all directions was only achievable with the MAP method (Yue et al. 2011).

This method of 3D point reconstruction from 2D BEV images is particularly useful for VMAT deliveries because landmarks (fiducials) may not be visible on each image.

9.2.4.2 3D VOLUMETRIC RECONSTRUCTION—THE MOTION-MODELING APPROACH

A recently developed method allows the reconstruction of real-time 3D volumetric image data from a single portal image in conjunction with prior knowledge in form of a patient-specific respiratory motion model (Soehn et al. 2005; Zhang et al. 2007; Li et al. 2010; Mishra et al. 2014). The technique provides 3D target and OAR localization information including information on deformations, a clear advantage over 2D information, or 3D point reconstructions described (cf. Figure 9.3 and Sections 9.2.2 and 9.2.4.1). A natural extension of this method from tumor motion tracking applications is the calculation of delivered dose (cf. Chapter 10).

The basic idea is to optimize the deformation of a reference image volume CT_0 to produce projection images with maximum similarity to an actually observed portal image $I(t)$ at time t. This is done via a motion model that may be built as follows: each phase of a pretreatment 4DCT is registered to a reference phase (e.g., end-of-exhale) via deformable image registration (DIR). The resulting deformation vector fields $DVF(t_i)$ describe the voxel motion and deformation of the reference CT_0 to yield the $CT(t_i)$ volumes at times t_i. The DVFs can be parameterized using principal component analysis (PCA). By using the PCA eigenvectors to span an orthogonal basis for the DVFs, the DVF may be approximated for any time point, that is, not only for the times t_i observed during the 4DCT acquisition:

$$\mathrm{DVF}(t) \approx \mathrm{DVF} + \sum_k w_k(t)\boldsymbol{u}_k \tag{9.3}$$

Here DVF is the mean DVF over all 4DCT phases, $\boldsymbol{u}_k$ represent an orthogonal basis of principal components to the DVFs, and $w_k(t)$ are time-dependent scalar weighting factors (i.e., coordinates in the DVF basis). It has been shown that using only 3 base vectors $\boldsymbol{u}_k$ is sufficient to describe the patient motion during respiration. Therefore, $\mathrm{CT}(t)$ may be described with only 3 time-dependent scalar parameters $w_k(t)$; $k=1,2,3$. To find these parameters in real time an iterative optimization approach is employed that seeks to minimize a cost function describing the difference between the observed EPID image and an MV digitally reconstructed radiograph (DRR) projected with $\mathrm{CT}(t)$:

$$\min_{w} \mathrm{J}(\boldsymbol{w},\lambda) = \left\| \mathrm{P} \cdot \mathrm{CT}\left(\mathrm{CT}_0, \langle \mathrm{DVF} \rangle, \boldsymbol{u}_k, w_k(t)\right) - \lambda \cdot I(t) \right\|_2^2 \tag{9.4}$$

where:

J is the cost function

P is a projector creating an MV-DRR in EPID geometry from a CT volume

λ is a scalar parameter accounting for differences in the illumination between the observed EPID image $I(t)$ and the MV-DRR

$\boldsymbol{w}$ is a vector of the PCA coordinates $w_k(t)$

$\|*\|$ is the Euclidian vector norm

The algorithm can be implemented efficiently using the compute unified device architecture (CUDA) on Nvidia Inc. (NVIDIA) graphics cards utilizing parallel computing with GPUs (Figure 9.7).

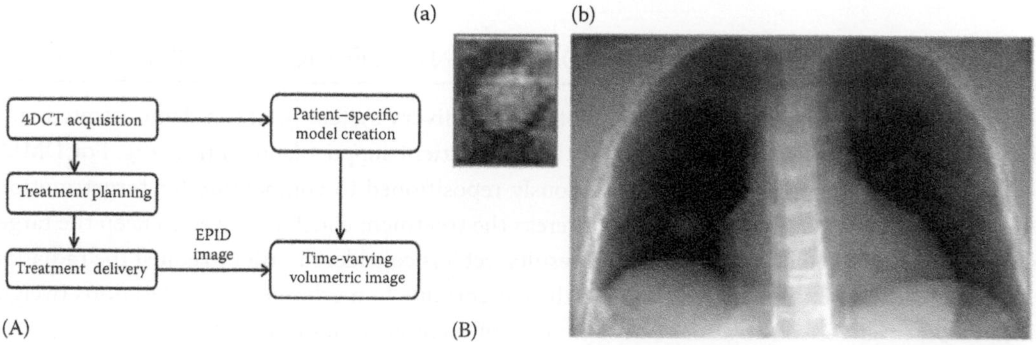

Figure 9.7 (A) Flowchart of the 3D fluoroscopy generation procedure and (B) an EPID (a) in comparison with an MV-DRR (b). Note the small FOV of the EPID images. (From Mishra, P. et al., *Med. Phys.* 41, 081713, 2014.)

9.3 OPTIONS FOR THE ADAPTION OF DELIVERY GEOMETRY AND RESPIRATORY MOTION PREDICTION

In Section 9.2, the extraction of 2D or 3D information from BEV images to estimate tumor displacement was discussed. In this section a brief overview of available techniques for real-time treatment adaptation to mitigate the dosimetric impact of these dispositions will be given. Both system latency and hardware adjustment will be discussed.

9.3.1 TIMING AND MOTION PREDICTION

To leverage the information derived from BEV imaging for real-time treatment adaptation, computational efficiency is important. However, there is always a time delay between the observed tumor position and the hardware adjustment to account for it. Major contributions to this system latency are the image acquisition, tumor localization, calculation of hardware adjustment coordinates, and the actual hardware motion (Poulsen et al. 2010). The latency is mainly a concern for periodic motion induced by respiration. The problem can be formulated as finding the tumor position at a future time from the N_p previous positions:

$$f_p : \left[x(t_{i-N_p}, \ldots, x(t_i)\right]^T \longmapsto \hat{x}(t_i + \tau_p) \tag{9.5}$$

Here f_p is the prediction function, τ_p is the system latency, and x and $\hat{x}$ are the observed and predicted tumor positions, respectively. Many techniques have been proposed and implemented, including linear predictors (LP), support vector machines (SVN), and artificial neural networks (ANN). All methods require some training data and most of them also require some history (noted above as the N_p previous observations). However, this can be a challenge in the case of BEV imaging as one cannot acquire MV images for training as, for example, in kV imaging. Utilizing other data sources representing the patient's organ motion can be a solution for this problem, for example, chest motion can be recorded with a camera or pressure-sensitive belt worn by the patient. The typically observed unstable phase shift between tumor motion and surrogate motion does not have an impact in this context (Rottmann und Berbeco 2014).

Another potential problem is the necessity of smooth, well-sampled input data. This translates into the requirement of an adequate frame rate, for example, 10 Hz, and possibly some smoothing of the real-time position input data.

9.3.2 REAL-TIME HARDWARE ADJUSTMENT FOR MOTION MITIGATION

Real-time hardware adjustment of the radiotherapy delivery system has been demonstrated with dynamic multileaf collimator (DMLC) tracking and patient support (couch) tracking. For DMLC tracking, the treatment aperture is continuously repositioned to compensate for target motion. For couch tracking the aperture is static, whereas the treatment couch is moved to keep the target at iscoenter. Both methods yield equivalent results, yet it is considered easier to adjust the radiation beam aperture rather than the couch (with the patient) due to mechanical considerations (weight lift), the potential for couch/gantry collisions and concerns about motion sickness.

DMLC tracking has been experimentally validated with BEV imaging (Rottmann et al. 2013b) and clinically demonstrated in patient treatments with other motion inputs (Calypso) (Keall et al. 2016) both using the Varian hardware platform. The general technical capacities for DMLC tracking are available for all commercial platforms from Varian, Elekta, and Siemens (Sawant et al. 2008; Tacke et al. 2010; Davies et al. 2013). The major benefits for using the BEV to drive motion

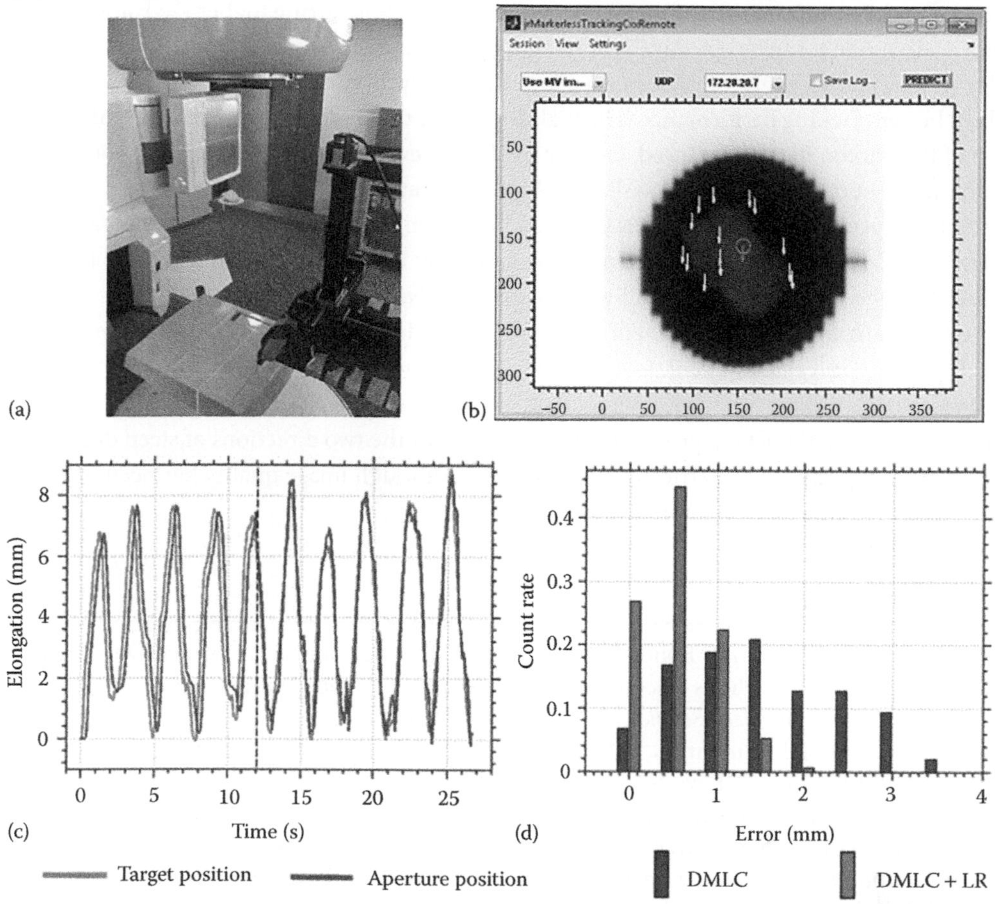

Figure 9.8 DMLC tracking with real-time markerless BEV localization: (a) and (b) Experimental setup on a clinical LINAC with a 3D printed tumor model on a slab of solid water mounted on the Washington University 4D Phantom and a screenshot of the graphical user interface, the user sees in the control room. (c) and (d) motion of a target and aperture demonstrating the effect of system latency and linear regression (LR) prediction. The dashed line shows the point when $\boldsymbol{N_p}$ has been acquired (cf. Section 9.3.1).

compensation are (1) the availability on most clinical LINACs (Clinac), (2) the passive nature (no additional dose), and (3) the ability to capture the most relevant dosimetric information even in 2D (cf. Section 9.1.4). However, there are also some limitations with BEV imaging in combination with DMLC tracking. If the imaging is interrupted by a beam-off event (e.g., due to a machine interlock) the history information needed for latency prediction is lost (cf. Figure 9.8c).

9.4 BEV TRACKING—CLINICAL IMPLEMENTATION AND OUTLOOK

In current clinical practice in-treatment cine-BEV imaging can be used for real-time visualization during treatment delivery (Rottmann et al. 2010b). These data may also be used retrospectively to assess potential issues such as tumor migration or to estimate delivered dose (Aristophanous et al. 2011). There are additional applications under development that are expected to find their way into the clinic in the near future. The development of better EPIDs featuring higher DQE and a modular (layered) design that can be optimized for tracking performance (Rottmann et al. 2016) is currently under way. A higher DQE translates into a lower noise level at the same exposure per image that leads to better tracking performance (Hu et al. 2016). The use of detector-based spectral decomposition of the photon fluence captured with the EPID may even enable the removal of bony anatomy or enhancement of fiducial marker visibility (Myronakis et al. 2016).

Another future application that has already been demonstrated in an experimental setting is the real-time adaptation of treatment margins driven by DMLC tracking in the BEV (Rottmann et al. 2014). The idea behind this technique is to adaptively grow or shrink margins depending on tumor localization uncertainty. For this application-delivered dose reconstruction is particularly important and can be facilitated with the BEV images as they record both tumor and MLC positions.

In summary, BEV tracking provides 2D information of the two directions of steep dose gradient making the perspective dosimetrically most valuable. Although image quality has been a challenge, work research is underway to improve the situation.

BIBLIOGRAPHY

Antonuk, L. E., J. Yorkston, J. Boudry, M. J. Longo, J. Jimenez, and R. A. Street. Development of hydrogenated amorphous silicon sensors for high energy photon radiotherapy imaging. *IEEE Transactions on Nuclear Science* 37(2): 165–170, (1990).

Aristophanous, M., J. Rottmann, L. E. Court, and R. I. Berbeco. EPID-guided 3D dose verification of lung SBRT. *Medical Physics* 38: 495–503, (2011).

Balter, J., K. Lam, H. F. Sandler, F. Littles, and R. T. Haken. Automatic localization of the prostate at the start of the treatment using implanted radio-opaque markers. *Proceedings of the Third International Workshop on Portal Imaging.* San Francisco, CA: Manitoba Cancer Treatment and Research Foundation. (1994).

Bissonnette, J.-P. et al. Quantifying interfraction and intrafraction tumor motion in lung stereotactic body radiotherapy using respiration-correlated cone beam computed tomography. *International Journal of Radiation Oncology, Biology, Physics* 75(3): 688–695, (2009).

Buck, D., M. Alber, and F. Nüsslin. Potential and limitations of the automatic detection of fiducial markers using an amorphous silicon flat-panel imager. *Physics in Medicine and Biology* 48(6): 763–774, (2003).

Case, R. B. et al. Interfraction and intrafraction changes in amplitude of breathing motion in stereotactic liver radiotherapy. *International Journal of Radiation Oncology, Biology, Physics* 77(3): 918–925, (2010).

Chan, M. F., G. N. Cohen, and J. O. Deasy. Qualitative evaluation of fiducial markers for radiotherapy imaging. *Technology in Cancer Research & Treatment* 14: 298–304, (2015).

Colvill, E. et al. Multileaf collimator tracking improves dose delivery for prostate cancer radiation therapy: Results of the first clinical trial. *International Journal of Radiation Oncology, Biology, Physics* 92(5): 1141–1147, (2015).

Cui, Y., J. G. Dy, G. C. Sharp, B. Alexander, and S. B. Jiang. Multiple template-based fluoroscopic tracking of lung tumor mass without implanted fiducial markers. *Physics in Medicine and Biology* 52(20): 6229–6242, (2007).

Datteri, R. D. and B. M. Dawant. Estimation and reduction of target registration error. *International Conference on Medical Image Computing and Computer-Assisted Intervention* 15(3): 139–146, (2012).

Davies, G. A., P. Clowes, J. L. Bedford, P. M. Evans, S. Webb, and G. Poludniowski. An experimental evaluation of the agility MLC for motion-compensated VMAT delivery. *Physics in Medicine and Biology* 58: 4643, (2013).

Ekberg, L., O. Holmberg, L. Wittgren, G. Bjelkengren, and T. Landberg. What margins should be added to the clinical target volume in radiotherapy treatment planning for lung cancer? *Radiotherapy and Oncology* 48(1): 71–77, (1998).

Herman, M. G. et al. Clinical use of electronic portal imaging: Report of AAPM radiation therapy committee task group 58. *Medical Physics* 28: 712–737, (2001).

Hu, Y. H., J. Rottmann, M. Myronakis, and R. I. Berbeco. Using fractal dimension analysis to distinguish lung tumors in MV imaging. *International Journal of Radiation Oncology, Biology, Physics* 96: E644–E645, (2016).

ICRU. 4. Definition of volumes. *Journal of the ICRU* 10(1): 41–53, (2010).

Jensen, C. et al. Cine EPID evaluation of two non-commercial techniques for DIBH. *Medical Physics* 41(2): 021730, (2014).

Keall, P. J. et al. On the use of EPID-based implanted marker tracking for 4D radiotherapy. *Medical Physics* 31(12): 3492–3499, (2004).

Keall, P. J. et al. The management of respiratory motion in radiation oncology report of AAPM task group 76. *Medical Physics* 33(10): 3874–3900, (2006).

Keall, P. J. et al. Real-time 3D image guidance using a standard LINAC: Measured motion, accuracy, and precision of the first prospective clinical trial of kilovoltage intrafraction monitoring-guided gating for prostate cancer radiation therapy. *International Journal of Radiation Oncology, Biology, Physics* 94: 1015–1021, (2016).

Kitamura, K. et al. Three-dimensional intrafractional movement of prostate measured during real-time tumor-tracking radiotherapy in supine and prone treatment positions. *International Journal of Radiation Oncology, Biology, Physics* 53(5): 1117–1123, (2002).

Krauss, A., S. Nill, and U. Oelfke. The comparative performance of four respiratory motion predictors for real-time tumour tracking. *Physics in Medicine and Biology* 56: 5303–5317, (2011).

Kupelian, P. et al. Multi-institutional clinical experience with the calypso system in localization and continuous, real-time monitoring of the prostate gland during external radiotherapy. *International Journal of Radiation Oncology, Biology, Physics* 67: 1088–1098, (2007).

Langen, K. M. et al. Observations on real-time prostate gland motion using electromagnetic tracking. *International Journal of Radiation Oncology, Biology, Physics* 71: 1084–1090, (2008).

Lewis, J. P. Fast template matching. *Proceedings of Vision Interface 95*. Quebec City, Canada: Canadian Image Processing and Pattern Recognition Society. 120–123, (1995).

Li, R. et al. Real-time volumetric image reconstruction and 3D tumor localization based on a single x-ray projection image for lung cancer radiotherapy. *Medical Physics* 37(6): 2822–2826, (2010).

Liu, W., R. D. Wiersma, W. Mao, G. Luxton, and L. Xing. Real-time 3D internal marker tracking during arc radiotherapy by the use of combined MV-kV imaging. *Physics in Medicine and Biology* 53(24): 7197–7213, (2008).

Ma, Y., L. Lee, O. Keshet, P. Keall, and L. Xing. Four-dimensional inverse treatment planning with inclusion of implanted fiducials in IMRT segmented fields. *Medical Physics* 36: 2215–2221, (2009).

Mao, W. et al. Image-guided radiotherapy in near real time with intensity-modulated radiotherapy megavoltage treatment beam imaging. *International Journal of Radiation Oncology, Biology, Physics* 75: 603–610, (2009).

Meyer, J., A. Richter, K. Baier, J. Wilbert, M. Guckenberger, and M. Flentje. Tracking moving objects with megavoltage portal imaging: A feasibility study. *Medical Physics* 33(5): 1275–1280, (2006).

Mishra, P. et al. An initial study on the estimation of time-varying volumetric treatment images and 3D tumor localization from single MV cine EPID images. *Medical Physics* 41(8): 081713, (2014).

Motz, J. W. and M. Danos. Image information content and patient exposure. *Medical Physics* 5: 8–22, (1978).

Myronakis, M. et al. WE-DE-BRA-07: Megavoltage spectral imaging with a layered detector. *The International Journal of Medical Physics Research and Practice* 43(6): 3813, (2016).

Nederveen, A. J., J. J. Lagendijk, and P. Hofman. Feasibility of automatic marker detection with an a-Si flat-panel imager. *Physics in Medicine and Biology* 46(4): 1219–1230, (2001).

Osher, S. and J. A. Sethian. Fronts propagating with curvature dependent speed: Algorithms based on Hamilton–Jacobi formulations. *Journal of Computational Physics* 79: 12–49, (1988).

Park, S.-J., D. Ionascu, F. Hacker, H. Mamon, and R. Berbeco. Automatic marker detection and 3D position reconstruction using cine EPID images for SBRT verification. *Medical Physics* 36(10): 4536–4546, (2009).

Poulsen, P. R., B. Cho, A. Sawant, D. Ruan, and P. J. Keall. Detailed analysis of latencies in image-based dynamic MLC tracking. *Medical Physics* 37(9): 4998–5005, (2010).

Richter, A., J. Wilbert, K. Baier, M. Flentje, and M. Guckenberger. Feasibility study for markerless tracking of lung tumors in stereotactic body radiotherapy. *International Journal of Radiation Oncology, Biology, Physics* 78: 618–627, (2010).

Rodríguez-Molinero, A., L. Narvaiza, J. Ruiz, and C. Gálvez-Barrón. Normal respiratory rate and peripheral blood oxygen saturation in the elderly population. *Journal of the American Geriatrics Society* 61(12): 2238–2240, (2013).

Rottmann, J., M. Aristophanous, A. Chen, L. Court, and R. Berbeco. A multi-region algorithm for markerless beam's-eye view lung tumor tracking. *Physics in Medicine and Biology* 55(18): 5585–5598, (2010a).

Rottmann, J. and R. Berbeco. Using an external surrogate for predictor model training in real-time motion management of lung tumors. *Medical Physics* 41(12): 121706, (2014).

Rottmann, J., P. Keall, and R. Berbeco. Markerless EPID image guided dynamic multi-leaf collimator tracking for lung tumors. *Physics in Medicine and Biology* 58(12): 4195–4204, (2013a).

Rottmann, J., P. Keall, and R. Berbeco. Real-time soft tissue motion estimation for lung tumors during radiotherapy delivery. *Medical Physics* (AAPM) 40: 091713, (2013b).

Rottmann, J., P. Keall, and R. Berbeco. TU-F-17A-07: Real-time personalized margins. *Medical Physics* 41(6): 474, (2014).

Rottmann, J., P. Keall, A. Chen, D. Sher, Y. Yue, and R. Berbeco. Dynamic treatment margin reduction for lung SBRT. *International Journal of Radiation Oncology, Biology, Physics* 81(2): S769–S770, (2010b).

Rottmann, J., D. Morf, R. Fueglistaller, G. Zentai, J. Star-Lack, and R. Berbeco. A novel EPID design for enhanced contrast and detective quantum efficiency. *Physics in Medicine and Biology* 61(17): 6297–6306, (2016).

Rowshanfarzad, P., M. Sabet, D. J. O'Connor, P. M. McCowan, B. M. C. McCurdy, and P. B. Greer. Detection and correction for EPID and gantry sag during arc delivery using cine EPID imaging. *Medical Physics* 39(2): 623–635, (2012).

Sawant, A. et al. Management of three-dimensional intrafraction motion through real-time DMLC tracking. *Medical Physics* 35: 2050–2061, (2008).

Schildkraut, J. S. et al. Level-set segmentation of pulmonary nodules in megavolt electronic portal images using a CT prior. *Medical Physics* 37(11): 5703–5710, (2010).

Seppenwoolde, Y. et al. Precise and real-time measurement of 3D tumor motion in lung due to breathing and heartbeat, measured during radiotherapy. *International Journal of Radiation Oncology, Biology, Physics* 53(4): 822–834, (2002).

Seppenwoolde, Y., W. Wunderink, S. R. W.-V. Veen, P. Storchi, A. M. Romero, and B. J. M. Heijmen. Treatment precision of image-guided liver SBRT using implanted fiducial markers depends on marker-tumour distance. *Physics in Medicine and Biology* 56(17): 5445–5468, (2011).

Soehn, M., M. Birkner, D. Yan, and M. Alber. Modelling individual geometric variation based on dominant eigenmodes of organ deformation: Implementation and evaluation. *Physics in Medicine and Biology* 50(24): 5893–5908, (2005).

St James, S., P. Mishra, F. Hacker, R. I. Berbeco, and J. H. Lewis. Quantifying ITV instabilities arising from 4DCT: A simulation study using patient data. *Physics in Medicine and Biology* 57(5): L1–L7, (2012).

Tacke, M. B., S. Nill, A. Krauss, and U. Oelfke. Real-time tumor tracking: Automatic compensation of target motion using the Siemens 160 MLC. *Medical Physics* 37: 753–761, (2010).

van Herk, M. Errors and margins in radiotherapy. *Seminars in Radiation Oncology* 14: 52–64, (2004).

Vedam, S. S., P. J. Keall, V. R. Kini, H. Mostafavi, H. P. Shukla, and R. Mohan. Acquiring a four-dimensional computed tomography dataset using an external respiratory signal. *Physics in Medicine and Biology* 48(1): 45–62, (2003).

Vigneault, E., J. Pouliot, J. Laverdière, J. Roy, and M. Dorion. Electronic portal imaging device detection of radio opaque markers for the evaluation of prostate position during megavoltage irradiation: A clinical study. *International Journal of Radiation Oncology, Biology, Physics* 37: 205–212, (1997).

Whitfield, G. et al. Quantifying motion for pancreatic radiotherapy margin calculation. *Radiotherapy and Oncology: Journal of the European Society for Therapeutic Radiology and Oncology* 103(3): 360–366, (2012).

Wiersma, R. D., N. Riaz, S. Dieterich, Y. Suh, and L. Xing. Use of MV and kV imager correlation for maintaining continuous real-time 3D internal marker tracking during beam interruptions. *Physics in Medicine and Biology* 54(1): 89–103, (2009).

Xie, Y., L. Xing, J. Gu, and W. Liu. Tissue feature-based intra-fractional motion tracking for stereoscopic x-ray image guided radiotherapy. *Physics in Medicine and Biology* 58(11): 3615–3630, (2013).

Yeung, R. et al. Cardiac dose reduction with deep inspiration breath hold for left-sided breast cancer radiotherapy patients with and without regional nodal irradiation. *Radiation Oncology* 10: 200, (2015).

Yip, S., J. Rottmann, and R. Berbeco. The impact of cine EPID image acquisition frame rate on markerless soft-tissue tracking. *Medical Physics* 41(2): 021730, (2014).

Yip, S. S. F., J. Rottmann, and R. I. Berbeco. Beam's-eye-view imaging during non-coplanar lung SBRT. *Medical Physics* 42: 6776–6783, (2015).

Yue, Y., M. Aristophanous, J. Rottmann, and R. I. Berbeco. 3-D fiducial motion tracking using limited MV projections in arc therapy. *Medical Physics* 38(6): 3222–3231, (2011).

Zhang, Q. et al. A patient-specific respiratory model of anatomical motion for radiation treatment planning. *Medical Physics* 34(12): 4772–4781, (2007).

Zhang, X., N. Homma, K. Ichiji, Y. Takai, and M. Yoshizawa. Tracking tumor boundary in MV-EPID images without implanted markers: A feasibility study. *Medical Physics* 42(5): 2510–2523, (2015).

Zhao, B., J. Dai, and C. C. Ling. Considering marker visibility during leaf sequencing for segmental intensity-modulated radiation therapy. *Medical Physics* 36: 3906–3916, (2009).

10

Beam's eye view imaging for in-treatment delivered dose estimation in photon radiotherapy

JOHN H. LEWIS

10.1 INTRODUCTION AND MOTIVATION

In this chapter, the use of electronic portal imaging devices (EPIDs) for delivered dose verification will be discussed. The most common clinical use of EPID is for patient imaging prior to treatment delivery, with the imager retracted during beam delivery. In-treatment beam's eye view (BEV) imaging is accomplished by leaving the EPID extended during treatment delivery in order to capture the exit radiation from the treatment beam. Images are captured in *cine* mode, giving a sequence of images resembling fluoroscopy, though generally with lower image quality and frame rate. The images can be used to verify patient positioning during treatment, and as described here, to calculate the actual delivered dose distribution. Delivered dose is estimated from BEV images by using the EPID to measure changes in anatomy during treatment delivery (intrafraction motion), and incorporating the real-time updates in anatomical configurations to modify or recalculate the delivered dose at each time point.

The motivation for in-treatment delivered dose estimation is to detect discrepancies between the dose distribution that is prescribed or planned, and the dose distribution that is actually delivered in the presence of intrafraction deformations. Accurate determination of delivered dose distributions can improve the ability of clinical studies to correlate patient outcomes with radiation doses. Currently, clinical studies on dose-response are generally based on comparisons of planned dose distributions to patient outcomes. As stated earlier, there are often differences between the planned and actually delivered doses. It is reasonable to assume that patient outcomes are more directly related to the delivered dose, and accurate methods of determining the delivered dose could improve the ability of clinical studies to derive dose-response relationships. Determination of delivered dose distributions could also improve individual patient care. Underdosing of the target or overdosing of critical structures can degrade the quality of a patient's treatment. Calculating the dose as treatment is delivered allows treatment plans to be modified to correct dose errors, either during treatment delivery or before the next treatment fraction is delivered. This type of system is called convergent radiotherapy (CRT), in that corrections will allow the delivered dose to converge on the prescribed dose (Berbeco et al. 2008).

Intrafraction motion arises from a variety of sources, including the respiratory, cardiac, gastrointestinal, and skeletal muscular systems. Each type of motion can change the delivered dose distribution to different extents. A great deal of literature has been devoted to assessing the magnitude of this motion for various treatment sites (Keall et al. 2006; Marchant et al. 2008; Minn et al. 2009; Kron et al. 2010; Ramakrishna et al. 2010; Quon et al. 2012; Floriano et al. 2013; Glide-Hurst et al. 2015; Hamamoto et al. 2015; Yorke et al. 2015; Han et al. 2016), and the corresponding dosimetric consequences (George et al. 2003; Naqvi and D'Souza 2005; Seco et al. 2008; Adamson et al. 2011; Waghorn et al. 2011; Olsen et al. 2012; Belec and Clark 2013; Yin et al. 2013; Lovelock et al. 2015; Zhuang 2015; Rico et al. 2016). The magnitude of dosimetric changes is dependent on several factors, including the magnitude and type of motion, the treatment modality, and the clinical equipment used. For example, a treatment delivered in three fractions will be more sensitive to intrafraction motion variations than a treatment delivered in 30, where dosimetric discrepancies can average out over time. St. James et al. (2013) reported errors in the volume of lung tumor receiving prescription dose of up to 20.9% in a single fraction when the actual motion during treatment was considered. Cai et al. (2015, 2016) reported changes in lung tumor D95 (the minimum dose received by 95% of the volume) of up to 25.5% in a single fraction.

EPID-BEV imaging has several characteristics that make it well suited for delivered dose verification in photon radiotherapy. The images are acquired using the treatment beam as the X-ray source, providing essentially *free* information, without increasing treatment time or imaging dose (the backscatter from the imager during treatment has been shown to be negligible [Kilby and Savage 2003]). Most modern clinical linear accelerators (Clinac) come equipped with an EPID, and no additional hardware is required to capture *cine* images during treatment. In addition to these benefits, BEV imaging provides spatial information in the most important plane for target tracking in photon radiotherapy. This point is illustrated in Figure 10.1. BEV imaging provides information in the plane with the sharpest dose falloff (Rottmann et al. 2010). The dose falloff is much shallower along the central axis of the treatment beam, because of the shape of photon percent depth dose (PDD) curves. Thus, a positional error of a given magnitude could have a much larger dosimetric effect if it is perpendicular to the beam's central axis, than if it is along the beam's central axis.

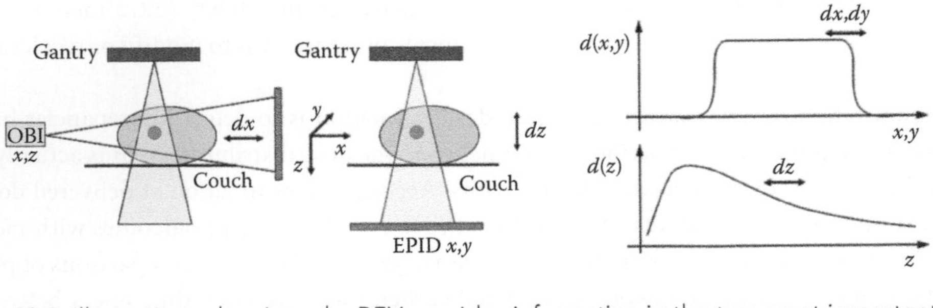

Figure 10.1 Illustrations showing why BEV provides information in the two most important directions for delivered dose verification in photon radiotherapy. The on-board imaging (OBI) system represents a typical kV X-ray imaging system mounted perpendicular to the treatment beam on a LINAC gantry. BEV imaging captures the directions with the larger potential dose errors for a given positional shift. (From Rottmann, J., Aristophanous, M., Chen, A., Court, L., and Berbeco, R., *Phys. Med. Biol.*, 55, 5585–5598, 2010.)

10.2 COMPONENTS OF A DELIVERED DOSE VERIFICATION FRAMEWORK

Various methods of delivered dose verification using the EPID have been developed, but each shares two key required components. These components are (1) a method of estimating target and/or tissue motion based on EPID images and (2) a method of incorporating the estimated motion into calculations of the delivered dose.

Estimation of target and normal tissue motion can be achieved either based directly on tissue visible in EPID images, or with the assistance of implanted fiducial markers. Providing real-time target localization for tracking or gating-based treatment has motivated the most previous works in this area, and an in-depth description of motion tracking in EPID images is provided in Chapter 9. For tracking treatments, knowledge of the target's location in the BEV provides enough information to follow the target with the treatment field. In the case of delivered dose verification, it becomes more important to estimate the motion of both the tumor and other tissues, so that accurate estimates of the 3D delivered dose distribution can be achieved.

The second component of delivered dose verification is a method of incorporating EPID-measured motion information into the calculation of delivered dose. In general, this requires synchronization of information about the treatment machine (gantry angle, field shape, dose rate, etc.) with the estimated motion at each time point. The most direct method is to determine the instantaneous fluence being delivered by the machine, and use the patient anatomy estimated at the same time point to perform a completely new dose calculation. The final delivered dose distribution is achieved by accumulating the doses delivered at all time points during treatment onto a representative reference image. This brute-force method requires a large number of dose calculations and registration tasks, and is a challenging task for routine patient care. Researchers have attempted to address this issue through a combination of automated scripts and simplifying approximations. One method of simplifying the process is to incorporate motion through convolution of either the planned 3D dose distribution or (for better accuracy) the 2D photon fluence patterns with a probability density function (PDF) derived from the target position (Lujan et al. 1999; Chetty et al. 2003; Craig et al. 2003; Li et al. 2008; Waghorn et al. 2010; Adamson et al. 2011; Aristophanous et al. 2011; Bharat et al. 2012). Convolution with a motion PDF allows for faster results, but it cannot account for relative motion between the normal tissue and target. Convolution methods also do not account for interplay effects between the target and MLC motions, which can cause deviations in the delivered dose. A review of published work on both brute-force and convolution-based methods is provided in Section 10.3.

10.3 METHODS OF USING MEASURED MOTION TO CALCULATE DELIVERED DOSE

Several methods for calculating delivered dose distributions based on motion tracked in EPID images have been presented in the literature. The utility of each method varies based on the treatment modality and anatomical site. In this section, some of the key works published on this topic are reviewed.

In 2008, Berbeco et al. published a method of calculating delivered dose distributions based on intrafraction motion derived from *cine*-EPID images. The study focused on liver metastasis targets with implanted fiducial markers, treated with 3D conformal radiotherapy. This approach used EPID images acquired at 0.7 Hz during treatment delivery. The position of the implanted fiducial markers were manually defined in each EPID image, and compared to digitally reconstructed radiographs (DRRs) generated from the planning CT to determine the intrafraction target shifts. Dose was calculated by dividing each treatment beam into a number of subbeams corresponding to the number of EPID images acquired during that beam's delivery. For each subbeam, the isocenter was shifted in the BEV by an amount equal to the fiducial marker motion in the corresponding EPID image, and dose was recalculated using the originally planned fluence. The final delivered dose was determined by adding together the doses from each subbeam. This method was applied to four patients, one of which showed a substantial underdosing during one fraction of a three-fraction treatment (Figure 10.2).

The method published in Berbeco et al. (2008) was one of the first publications in this area, but requires a substantial amount of manual effort, both in tracking the fiducial markers and in creating and calculating the dose for each subbeam. Later improvements on this work have focused on developing methods to automate and hasten the process (Aristophanous et al. 2011). Automated tracking of fiducial markers or tumor motion have been discussed in Chapter 9. The method also accounts for rigid shifts of the target in the BEV. Motion along the beam axis is not accounted for, though as described previously this is the least important direction for photon treatment verification. Deformation of the tumor and surrounding tissue is not accounted for, and the method will not capture the changes in delivered dose to normal tissues that move relative to the tumor. This has motivated other methods of delivered dose verification, as will be discussed later in this section. Due to the lack of a ground truth delivered dose distributions in this study, it is not possible to assess the accuracy of the method. So long as the tumor motion measured from the fiducial markers is accurate, it seems reasonable to expect that the dose calculated with this method is a better estimate of the delivered dose than the originally planned dose distribution is.

In 2012, Poulsen et al. published a method similar to the shift-and-add approach employed by Berbeco et al. (2008), but incorporating an automated script to eliminate some of the manual labor. Treatment plans were initially developed in the clinical treatment planning system (TPS), and then were exported for modification by an in-house MATLAB® (MathWorks, Natick, MA) script. The

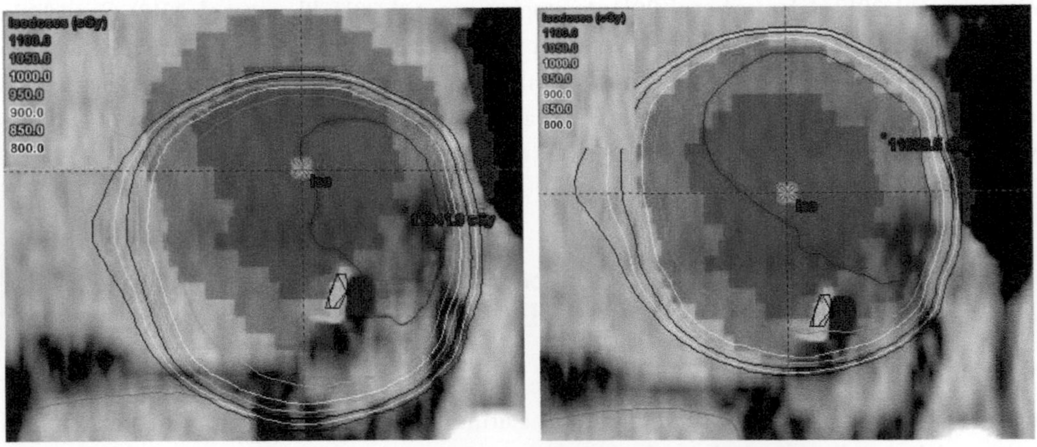

(a) Day 1 (delivered) (b) Day 1 (planned)

Figure 10.2 An example of planned (b) and delivered (a) dose distributions for a single fraction of liver SBRT treatment, as reported in Berbeco et al. (2008). A shift in patient anatomy caused substantial underdosing of the tumor that was detected with EPID. (From Berbeco, R. I., Hacker, F., Zatwarnicki, C., Park, S. J., Ionascu, D., O'Farrell, D., and Mamon, H. J., *Med. Phys.*, 35, 3225–3231, 2008.)

script divided target positions measured during treatment into 1 mm bins, and then created subbeams derived from the initial treatment fields that corresponded to the part of the treatment that was delivered while the target was within each bin. The positional shifts of the target were modeled by a shift in the isocenter for each subbeams. All subbeams were concatenated into a DICOM treatment plan, which was reimported to the clinical treatment planning system for dose calculation. This process is shown in Figure 10.3. Poulsen et al.'s method was developed as a general solution appropriate for any motion input, but was tested using EPID images of a physical respiratory phantom.

Aristophanous et al. published an updated version of the work of Berbeco et al. (2008), where the manual shift-and-add technique was replaced by convolution with a PDF of the target's position (Berbeco et al. 2008; Aristophanous et al. 2011). The planned fluence of each beam was convolved to

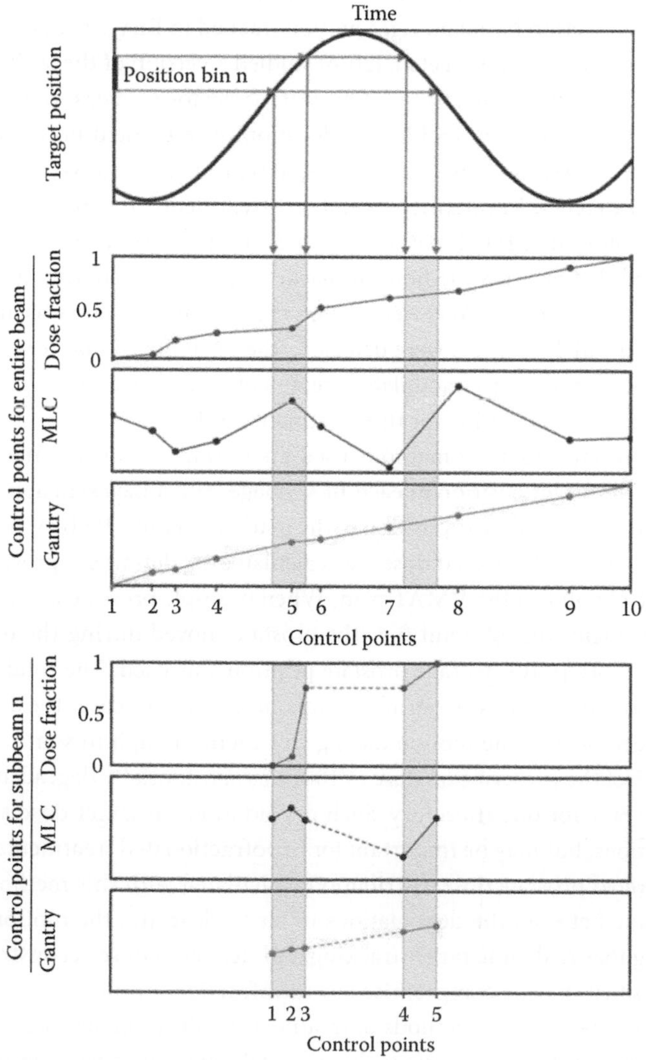

Figure 10.3 Graphs explaining the process of subbeam construction. The gray regions show pieces of a delivered treatment that were grouped into a subbeam based on the target position at the time at which those pieces of the initial plan were delivered. Delivered dose is calculated by accumulating the contributions from each control point. (From Poulsen, P. R., Schmidt, M. L., Keall, P., Worm, E. S., Fledelius, W., and Hoffmann, L., *Med. Phys.*, 39, 6237–6246, 2012.)

a PDF derived from the tumor motion tracks measured with EPID during the delivery of that beam, yielding a *super fluence*. Tumor motion tracking was performed using an automated markerless tracking algorithm (Rottmann et al. 2010). The method was applied to data from a physical phantom driven by motion trajectories from lung cancer radiotherapy patients. Excellent agreement was found between the calculated and measured *tumor* dose in the physical phantom. The use of motion convolution instead of manual isocenter shifting, and an automated tracking method make this delivered dose verification method more suitable for clinical implementation. Convolution of the motion track with the fluence is similar to manual isocenter shifting in that it accounts only for 2D shifting of the target in the BEV. Similar to the previous work from Berbeco et al. (2008), the method may not be appropriate for calculating dose delivered to normal tissues that move relative to the target. Deformations of the patient surface or tissue heterogeneities may also have an effect on the delivered dose distribution, and are not accounted for with this method.

Delivered dose verification based on rigid motion tracked in BEV images has also been studied for prostate cancer treatment. Azcona et al. (2014) studied a version of the shift-and-add technique in 2013. This study was performed using 32 prostate trajectories measured from eight patients. Prostate motion was measured by tracking the location of implanted fiducial markers with *cine* EPID. Unlike the previously described studies, these treatments were volumetric modulated arc therapy (VMAT) treatments. In intensity-modulated treatments, the anatomical motion is more difficult to track because the MLC leaves can obstruct large portions of the BEV image. Azcona et al. (2014) reported that at least one fiducial marker was visible, on average, 56.2% of the time (range 10.9%–90.2%). When no markers were visible, the prostate position was interpolated between the nearest images in which markers were detected. The visibility of markers or anatomical landmarks in BEV images for intensity-modulated treatments remains one of the main challenges in BEV-based delivered dose calculation for this treatment modality.

Azcona et al.'s prostate tracking methods used an automatic fiducial marker detection algorithm to determine the target position in each BEV image, and a Bayesian approach to computing the 3D prostate location based on the position of the markers in each 2D BEV image (Li et al. 2011a; Azcona et al. 2013a, 2013b). Delivered dose was calculated by defining a prostate position associated with each control point in the VMAT plan. When multiple prostate positions were measured corresponding to a single control point (i.e., the prostate moved during the delivery of a control point's monitor unit [MU]), the average prostate position was used. The total delivered dose was estimated by accumulating the dose from all control points. In most cases the dose to the target was only slightly degraded by prostate motion during treatment, though in some cases the target volume receiving 100% of the prescription dose (V100) was substantially degraded. In the worst case, the V100 fell below 60% for one trajectory. Such degradations in target dose are likely to average out over many fractions, but may be important for hypofractionated treatments. Figure 10.4 shows an example of delivered prostate dose distributions calculated with this method. Interestingly, no correlation was found between the degradations in target dose and the motion magnitude of the prostate, suggesting that real-time target tracking and delivered dose reconstruction is necessary to detect treatments that may be compromised by intrafraction motion.

Recently, motion-model based methods of treatment verification have been developed that are capable of accounting for 3D deformations during treatment. The basic approach is to develop a respiratory motion model from images acquired prior to treatment delivery that capture respiratory motion, and to combine the prior information provided by these models with *cine*-EPID images measured during treatment delivery to estimate 3D images for each delivery time point. That is, for each single frame of a *cine*-EPID sequence, a corresponding 3D image is generated. This

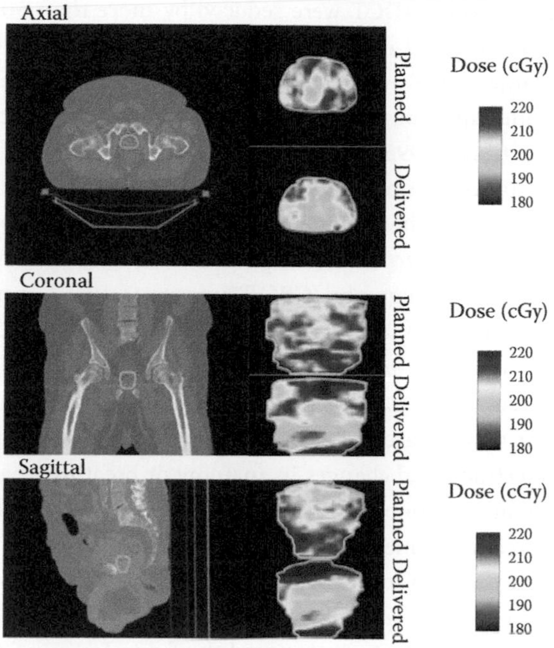

Figure 10.4 Images showing how the delivered prostate dose was altered by motion during treatment delivery for a single fraction of conventionally fractionated treatment. The prescribed dose was 200 cGy. (From Azcona, J. D., Li, R., Mok, E., Hancock, S., and Xing, L., *Int. J. Radiat. Oncol. Biol. Phys.*, 86, 762–768, 2014.)

method of generating time-varying volumetric treatment imaging is referred to as *3D fluoroscopy*. For each time point, the beam geometry and output is used to calculate dose using the external contour and electron densities derived from the corresponding 3D image, and the final delivered dose is determined by accumulating the dose from all time points. The final step of dose accumulation onto a reference image is accomplished by inverting the displacement vectors used to generate each corresponding 3D fluoroscopic image. Since this method uses updated 3D images for each dose calculation, it has the ability to account for deformations in the target and other anatomical structures, and has the potential to provide both tumor and normal tissue delivered dose estimations. A potential drawback is that the estimation of 3D images at each time point is more complex than simply measured rigid shifts in the target, and relies on motion models derived from prior information. If the anatomical shapes and deformations measured in the prior imaging set do not represent the anatomy at the time of treatment delivery, the delivered dose estimation accuracy will be unreliable.

In 2015, Cai et al. (2015) published a treatment verification method based on 3D fluoroscopic images generation from motion models derived from four-dimensional computed tomography (4DCT). Motion models were derived by registering each phase of a 4DCT dataset to the end-of-exhale phase from that set. Principal component analysis was performed on the set of displacement vector fields resulting from these registrations, resulting in a reduced set of possible deformations during respiration, specific to the patient from whom 4DCT images were acquired (Li et al. 2011b). The actual displacement vectors corresponding to each time point are derived using an iterative optimization scheme that compares measured BEV images to DRRs derived from estimated 3D images based on the motion model. The 3D images are updated until their corresponding DRRs match the measured BEV images. Using these images, Cai et al. (2015) showed that errors in the delivered dose estimated based on 4DCT were reduced by more than 50%, using their motion-model–based approach. In the most extreme case published in their study, the minimum dose delivered to 95% of the target (D95) was reduced by nearly 70%. However, this degradation was calculated using a plan with 0 mm margin for setup errors. When a more clinically realistic 7 mm margin was used, the reduction in D95 was 10%.

Motion-model–based approaches to delivered dose calculation that are based on 4DCT do not account for changes in anatomy or breathing patterns that occur between 4DCT acquisition and treatment. For this reason, researchers are developing techniques that use motion models derived from 4D cone-beam computed tomography (CBCT) images acquired immediately prior to treatment. Dhou et al. (2015) showed that 3D images derived from 4DCT-based motion models do not perform well in the presence of setup errors or tumor baseline position shifts. Cai et al. (2016) used motion models derived from 4DCBCT to compute delivered dose using methods similar to those described earlier for 4DCT-based motion models. The 4DCBCT approach is promising in that it can account for interfractional changes in anatomy or breathing patterns caused by atelectasis, pleural effusion, weight gain, or other factors. However, 4DCBCT suffers from inferior image quality, unreliable electron density measurements, and smaller fields of view compared to 4DCT. Cai et al. (2016) addressed these issues by deforming previous 4DCT images to 4DCBT images acquired immediately prior to treatment, using the 4DCT images to correct the electron density measurements, reduce imaging artifacts, and extend the field of view of 4DCBCT. Their preliminary work was tested on digital phantoms with irregular and regular breathing patterns, with errors of less than 1.5% in tumor coverage (D95) and less than 3% in the 3D dose distribution.

10.4 SUMMARY AND EXISTING CHALLENGES

In this chapter, we reviewed several methods of calculating delivered dose based on BEV imaging. The basic elements of a BEV-based delivered dose calculation system were presented, and existing studies in this area were summarized. At the time of writing this work, a commercial solution for BEV-based delivered dose verification is not available from any major vendor, but academic research is active.

Every delivered dose calculation method described in this chapter relies on the ability to track anatomical motion in the EPID image provided by the exit radiation of the treatment field. This can be particularly challenging for IMRT or VMAT treatments, where the field is often partially blocked by MLC leaves. Most research in the area has only been applied to 3D conformal fields, where anatomical structures or implanted fiducial markers are more likely to be visible. As stated in Section 10.3, Azcona et al. found that even with three fiducial markers implanted in each prostate cancer patient, no fiducial markers were visible 43.8% of the time on average. How applicable the existing BEV-based delivered dose estimation methods are to IMRT and VMAT treatments remains an open question. Some researchers have attempted to develop optimization algorithms that incorporate the requirement for at least one fiducial marker to remain visible at all times (Ma et al. 2009).

In addition to the studies described here, a large number of papers have been published both on methods of tracking targets in BEV images (described in Chapter 9), and on methods of incorporating motion into dose calculations. Improvements to EPID imaging, motion model development, and 4DCBCT image reconstruction are active areas of research. Continued technological developments in each of these areas create the potential for new, improved methods of delivered dose verification.

REFERENCES

Adamson, J., Q. Wu, and D. Yan. 2011. Dosimetric effect of intrafraction motion and residual setup error for hypofractionated prostate intensity-modulated radiotherapy with online cone beam computed tomography image guidance, *Int J Radiat Oncol Biol Phys*, 80: 453–461.

Aristophanous, M., J. Rottmann, L. E. Court, and R. I. Berbeco. 2011. EPID-guided 3D dose verification of lung SBRT, *Med Phys*, 38: 495–503.

Azcona, J. D., R. Li, E. Mok, S. Hancock, and L. Xing. 2013a. Automatic prostate tracking and motion assessment in volumetric modulated arc therapy with an electronic portal imaging device, *Int J Radiat Oncol Biol Phys*, 86: 762–768.

Azcona, J. D., R. Li, E. Mok, S. Hancock, and L. Xing. 2013b. Development and clinical evaluation of automatic fiducial detection for tumor tracking in cine megavoltage images during volumetric modulated arc therapy, *Med Phys*, 40: 031708.

Azcona, J. D., L. Xing, X. Chen, K. Bush, and R. Li. 2014. Assessing the dosimetric impact of real-time prostate motion during volumetric modulated arc therapy, *Int J Radiat Oncol Biol Phys*, 88: 1167–1174.

Belec, J., and B. G. Clark. 2013. Monte Carlo calculation of VMAT and helical tomotherapy dose distributions for lung stereotactic treatments with intra-fraction motion, *Phys Med Biol*, 58: 2807–2821.

Berbeco, R. I., F. Hacker, C. Zatwarnicki, S. J. Park, D. Ionascu, D. O'Farrell, and H. J. Mamon. 2008. A novel method for estimating SBRT delivered dose with beam's-eye-view images, *Med Phys*, 35: 3225–3231.

Bharat, S., P. Parikh, C. Noel, M. Meltsner, K. Bzdusek, and M. Kaus. 2012. Motion-compensated estimation of delivered dose during external beam radiation therapy: Implementation in Philips' Pinnacle (3) treatment planning system, *Med Phys*, 39: 437–443.

Cai, W., S. Dhou, F. Cifter, M. Myronakis, M. H. Hurwitz, C. L. Williams, R. I. Berbeco, J. Seco, and J. H. Lewis. 2016. 4D cone beam CT-based dose assessment for SBRT lung cancer treatment, *Phys Med Biol*, 61: 554–568.

Cai, W., M. H. Hurwitz, C. L. Williams, S. Dhou, R. I. Berbeco, J. Seco, P. Mishra, and J. H. Lewis. 2015. 3D delivered dose assessment using a 4DCT-based motion model, *Med Phys*, 42: 2897–2907.

Chetty, I. J., M. Rosu, N. Tyagi, L. H. Marsh, D. L. McShan, J. M. Balter, B. A. Fraass, and R. K. T. Haken. 2003. A fluence convolution method to account for respiratory motion in three-dimensional dose calculations of the liver: A Monte Carlo study, *Med Phys*, 30: 1776–1780.

Craig, T., J. Battista, and J. V. Dyk. 2003. Limitations of a convolution method for modeling geometric uncertainties in radiation therapy. II. The effect of a finite number of fractions, *Med Phys*, 30: 2012–2020.

Dhou, S., M. Hurwitz, P. Mishra, W. Cai, J. Rottmann, R. Li, C. Williams, M. Wagar, R. Berbeco, D. Ionascu, and J. H. Lewis. 2015. 3D fluoroscopic image estimation using patient-specific 4DCBCT-based motion models, *Phys Med Biol*, 60: 3807–3824.

Floriano, A., I. Santa-Olalla, and A. Sanchez-Reyes. 2013. Initial evaluation of intrafraction motion using frameless CyberKnife VSI system, *Rep Pract Oncol Radiother*, 18: 173–178.

George, R., P. J. Keall, V. R. Kini, S. S. Vedam, J. V. Siebers, Q. Wu, M. H. Lauterbach, D. W. Arthur, and R. Mohan. 2003. Quantifying the effect of intrafraction motion during breast IMRT planning and dose delivery, *Med Phys*, 30: 552–562.

Glide-Hurst, C. K., M. M. Shah, R. G. Price, C. Liu, J. Kim, M. Mahan, C. Fraser, I. J. Chetty, I. Aref, B. Movsas, and E. M. Walker. 2015. Intrafraction variability and deformation quantification in the breast, *Int J Radiat Oncol Biol Phys*, 91: 604–611.

Hamamoto, Y., H. Inata, N. Sodeoka, S. Nakayama, S. Tsuruoka, H. Takeda, T. Manabe, T. Mochizuki, and M. Umeda. 2015. Observation of intrafraction prostate displacement through the course of conventionally fractionated radiotherapy for prostate cancer, *Jpn J Radiol*, 33: 187–193.

Han, Z., J. C. Bondeson, J. H. Lewis, E. G. Mannarino, S. A. Friesen, M. M. Wagar, T. A. Balboni, B. M. Alexander, N. D. Arvold, D. J. Sher, and F. L. Hacker. 2016. Evaluation of initial setup accuracy and intrafraction motion for spine stereotactic body radiation therapy using stereotactic body frames, *Pract Radiat Oncol*, 6: e17–e24.

Keall, P. J., G. S. Mageras, J. M. Balter, R. S. Emery, K. M. Forster, S. B. Jiang, J. M. Kapatoes et al. 2006. The management of respiratory motion in radiation oncology report of AAPM Task Group 76, *Med Phys*, 33: 3874–3900.

Kilby, W. and C. Savage. 2003. The effect of the Varian amorphous silicon electronic portal imaging device on exit skin dose, *Phys Med Biol*, 48: 3117–3128.

Kron, T., J. Thomas, C. Fox, A. Thompson, R. Owen, A. Herschtal, A. Haworth, K. H. Tai, and F. Foroudi. 2010. Intra-fraction prostate displacement in radiotherapy estimated from pre- and post-treatment imaging of patients with implanted fiducial markers, *Radiother Oncol*, 95: 191–197.

Li, H. S., I. J. Chetty, C. A. Enke, R. D. Foster, T. R. Willoughby, P. A. Kupellian, and T. D. Solberg. 2008. Dosimetric consequences of intrafraction prostate motion, *Int J Radiat Oncol Biol Phys*, 71: 801–812.

Li, R., B. P. Fahimian, and L. Xing. 2011a. A Bayesian approach to real-time 3D tumor localization via monoscopic x-ray imaging during treatment delivery, *Med Phys*, 38: 4205–4214.

Li, R., J. H. Lewis, X. Jia, T. Zhao, W. Liu, S. Wuenschel, J. Lamb, D. Yang, D. A. Low, and S. B. Jiang. 2011b. On a PCA-based lung motion model, *Phys Med Biol*, 56: 6009–6030.

Lovelock, D. M., A. P. Messineo, B. W. Cox, M. A. Kollmeier, and M. J. Zelefsky. 2015. Continuous monitoring and intrafraction target position correction during treatment improves target coverage for patients undergoing SBRT prostate therapy, *Int J Radiat Oncol Biol Phys*, 91: 588–594.

Lujan, A. E., E. W. Larsen, J. M. Balter, and R. K. Ten Haken. 1999. A method for incorporating organ motion due to breathing into 3D dose calculations, *Med Phys*, 26: 715–720.

Ma, Y., L. Lee, O. Keshet, P. Keall, and L. Xing. 2009. Four-dimensional inverse treatment planning with inclusion of implanted fiducials in IMRT segmented fields, *Med Phys*, 36: 2215–2221.

Marchant, T. E., A. M. Amer, and C. J. Moore. 2008. Measurement of inter and intra fraction organ motion in radiotherapy using cone beam CT projection images, *Phys Med Biol*, 53: 1087–1098.

Minn, A. Y., D. Schellenberg, P. Maxim, Y. Suh, S. McKenna, B. Cox, S. Dieterich, L. Xing, E. Graves, K. A. Goodman, D. Chang, and A. C. Koong. 2009. Pancreatic tumor motion on a single planning 4D-CT does not correlate with intrafraction tumor motion during treatment, *Am J Clin Oncol*, 32: 364–368.

Naqvi, S. A., and W. D. D'Souza. 2005. A stochastic convolution/superposition method with isocenter sampling to evaluate intrafraction motion effects in IMRT, *Med Phys*, 32: 1156–1163.

Olsen, J. R., P. J. Parikh, M. Watts, C. E. Noel, K. W. Baker, L. Santanam, and J. M. Michalski. 2012. Comparison of dose decrement from intrafraction motion for prone and supine prostate radiotherapy, *Radiother Oncol*, 104: 199–204.

Poulsen, P. R., M. L. Schmidt, P. Keall, E. S. Worm, W. Fledelius, and L. Hoffmann. 2012. A method of dose reconstruction for moving targets compatible with dynamic treatments, *Med Phys*, 39: 6237–6246.

Quon, H., D. A. Loblaw, P. C. Cheung, L. Holden, C. Tang, G. Pang, G. Morton, A. Mamedov, and A. Deabreu. 2012. Intra-fraction motion during extreme hypofractionated radiotherapy of the prostate using pre- and post-treatment imaging, *Clin Oncol* (*R Coll Radiol*), 24: 640–645.

Ramakrishna, N., F. Rosca, S. Friesen, E. Tezcanli, P. Zygmanszki, and F. Hacker. 2010. A clinical comparison of patient setup and intra-fraction motion using frame-based radiosurgery versus a frameless image-guided radiosurgery system for intracranial lesions, *Radiother Oncol*, 95: 109–115.

Rico, M., E. Martinez, S. Pellejero, B. Bermejo, P. Navarrete, M. Barrado, M. Campo, F. Maneru, E. Villafranca, and J. Aristu. 2016. Influence of different treatment techniques and clinical factors over the intrafraction variation on lung stereotactic body radiotherapy, *Clin Transl Oncol*, 18: 1011–1018.

Rottmann, J., M. Aristophanous, A. Chen, L. Court, and R. Berbeco. 2010. A multi-region algorithm for markerless beam's-eye view lung tumor tracking, *Phys Med Biol*, 55: 5585–5598.

Seco, J., G. C. Sharp, Z. Wu, D. Gierga, F. Buettner, and H. Paganetti. 2008. Dosimetric impact of motion in free-breathing and gated lung radiotherapy: A 4D Monte Carlo study of intrafraction and interfraction effects, *Med Phys*, 35: 356–366.

St James, S., J. Seco, P. Mishra, and J. H. Lewis. 2013. Simulations using patient data to evaluate systematic errors that may occur in 4D treatment planning: A proof of concept study, *Med Phys*, 40: 091706.

Waghorn, B. J., S. L. Meeks, and K. M. Langen. 2011. Analyzing the impact of intrafraction motion: correlation of different dose metrics with changes in target D95%, *Med Phys*, 38: 4505–4511.

Waghorn, B. J., A. P. Shah, W. Ngwa, S. L. Meeks, J. A. Moore, J. V. Siebers, and K. M. Langen. 2010. A computational method for estimating the dosimetric effect of intra-fraction motion on step-and-shoot IMRT and compensator plans, *Phys Med Biol*, 55: 4187–4202.

Yin, W. J., Y. Sun, F. Chi, J. L. Fang, R. Guo, X. L. Yu, Y. P. Mao, Z. Y. Qi, Y. Guo, M. Z. Liu, and J. Ma. 2013. Evaluation of inter-fraction and intra-fraction errors during volumetric modulated arc therapy in nasopharyngeal carcinoma patients, *Radiat Oncol*, 8: 78.

Yorke, E., Y. Xiong, Q. Han, P. Zhang, G. Mageras, M. Lovelock, H. Pham, J. P. Xiong, and K. A. Goodman. 2015. Kilovoltage imaging of implanted fiducials to monitor intrafraction motion with abdominal compression during stereotactic body radiation therapy for gastrointestinal tumors, *Int J Radiat Oncol Biol Phys*, 95: 1042–1049.

Zhuang, T. 2015. On the effect of intrafraction motion in a single fraction step-shoot IMRT, *Med Phys*, 42: 4310–4319.

11

EPID-based *in vivo* transit dosimetry

BEN MIJNHEER

11.1 INTRODUCTION

11.1.1 WHY *IN VIVO* TRANSIT DOSIMETRY

The rationale for *in vivo* dose measurements is to provide an accurate and effective independent dose verification of a radiotherapy treatment procedure starting from treatment planning to treatment delivery. It will enable the identification of potential errors in dose calculation, data transfer, dose delivery, patient setup, and changes in patient anatomy. The clinical use of *in vivo* dosimetry has been addressed in a large number of papers and reports, and has been reviewed in some recent publications (IAEA, 2013; Mijnheer, 2013; Mijnheer et al., 2013). These clinical applications of *in vivo* dosimetry mainly concern entrance dose measurements using a large variety of point detectors. Electronic portal imaging device (EPID) dosimetry has, however, proliferated in the past 10–15 years as given in the review by van Elmpt et al. (2008) and Chapter 6 of this book. Great progress has been made in determining and understanding the dosimetric characteristics of

EPIDs, and in the implementation in the clinic as a dosimeter. Most current applications focus on the use of EPIDs for pretreatment dose verification, whereas their use for *in vivo* dosimetry is still limited. However, a number of groups have shown that EPID-based *in vivo* transit dosimetry can be used to detect a number of errors that could not be detected by other quality assurance (QA) checks, leading to the avoidance of serious mistreatments (Nijsten et al., 2007; Mans et al., 2010; Fidanzio et al., 2015; Mijnheer et al., 2015). Furthermore, in a recent paper it was shown that EPID-based *in vivo* transit dosimetry measurements show the most promise for detection of errors that had been reported clinically (Bojechko et al., 2015). The results of that analysis indicated that the incidents with the highest occurrence in that center were related to patient setup errors and errors in the patient CT dataset, and would not be detected by pretreatment QA when using a phantom.

As discussed in Chapter 6, the quantities that are compared during pretreatment verification are generally not the 3D dose distributions but 2D fluence or dose distributions, using alert criteria that are not based on the actual tumor or organs-at-risk (OAR) geometry in a specific patient. Clinical judgment of the importance of observed dose differences is therefore difficult, if not impossible, if assessed from phantom measurements alone. Recently the work of the Imaging and Radiation Oncology Core (IROC) Houston group (Kry et al., 2014; McKenzie et al., 2014) showed that there is indeed a fundamental problem with patient-specific pretreatment intensity-modulated radiation therapy (IMRT) verification. Just to quote one of the conclusions in the paper of McKenzie et al.: "Patient-specific IMRT QA techniques in general should be thoroughly evaluated for their ability to correctly differentiate acceptable and unacceptable plans"; and from the paper of Kry et al.: "Moreover, the particularly poor agreement between IMRT QA and the IROC Houston phantoms highlights surprising inconsistency in the QA process." It is obvious that the current methods of pretreatment verification need additional tools, such as EPID-based *in vivo* transit dosimetry, to guarantee the correct delivery of a specific treatment plan to a patient under clinical conditions.

EPID-based *in vivo* transit dosimetry is now only clinically used in selected centers during verification of IMRT and volumetric modulated arc therapy (VMAT). With the introduction of commercial solutions and their implementation as part of a fully automated workflow, most likely many more centers will use EPID-based transit dosimetry systems for patient-specific QA. The field is also moving toward adaptive radiotherapy (ART) requiring verification of the adapted plan, maybe even online for some applications. It can therefore be expected that in the future patient-specific QA measurements will shift more and more from pretreatment methods to *in vivo* approaches, with EPID-based transit dosimetry as one of the most promising techniques.

11.1.2 Differences with other approaches

Many 2D and semi-3D devices are used for pretreatment patient-specific QA. These instruments are able to detect errors in the transfer of a plan from the treatment planning system to the accelerator, in the dose calculation (with some systems), and in the delivery of the beams. For instance, the wrong position of a multileaf collimator (MLC) leaf can be detected with most of these devices, although the sensitivity for detecting delivery errors may vary depending on the type of instrument and the characteristics of the patient plan. However, the clinical relevance of a single-leaf error might be limited if the under- or overdose in a certain voxel at a specific angle is compensated by the dose delivered at other angles as may happen, for instance, during VMAT delivery with arcs having a large angle range. In addition, the phantoms used for these pretreatment QA measurements are not related to the actual geometry of a specific patient, further limiting the relevance of the results. Ultimately, patient-specific QA should resemble the clinical situation as closely as possible.

As discussed in Chapter 6, the characteristics that make amorphous silicon (a-Si)-type EPIDs suitable as dosimeters are the linearity of the response with dose and dose rate, good long-term stability, high spatial resolution, no dead time, and real-time readout. However, there are technical challenges to using EPIDs as dosimeters in clinical practice. Some of these issues are related to a specific type of EPID or to image acquisition software from a particular vendor, whereas others are related to the basic properties of (a-Si)-type EPIDs such as their nonwater equivalent energy response. Many research groups have for a long time worked on solutions for those problems, which resulted in a number of different approaches for EPID dosimetry as reviewed by van Elmpt et al. (2008) and discussed in detail in Chapter 6. In this chapter, we will discuss *in vivo* transit dose measurements made with EPIDs during the irradiation of a patient, which are either compared directly with predicted dose distributions at the EPID level, or inside the patient after back-projection.

Currently EPID-based transit dosimetry is used to measure the dose of a single IMRT field, the total dose of all IMRT fields, or the total dose of a VMAT arc during one fraction. It therefore verifies the integrated dose per fraction and is not able to measure the variation of the dose distribution during one fraction, for instance, due to tumor and/or OAR, as discussed in Chapter 10 of this book. Intrafraction EPID-based dosimetry during VMAT delivery has been investigated, as will be discussed briefly in Section 11.4 and is covered more extensively in Chapter 6 (for pretreatment verification).

11.2 *IN VIVO* TRANSIT DOSIMETRY

11.2.1 Techniques for EPID-based transit dosimetry

Transit dose measurements using EPIDs can be analyzed in several ways as discussed by van Elmpt et al. (2008). Briefly, fluence or dose comparisons can be made at the EPID level or in the patient/phantom. In the first approach, measured grayscale distributions are compared directly or after conversion to dose values, with predicted grayscale or dose distributions. Image grayscale distributions, which are often called portal dose images, have been calculated using in-house developed algorithms by several groups (Chytyk-Praznik et al., 2013; Bedford et al., 2014; Berry et al., 2014). Other groups converted the portal dose images to dose distributions at the EPID level, and compared these with dose distributions using the dose calculation algorithm from a commercial treatment planning system (e.g., Reich et al., 2006; Baek et al., 2014).

In the second approach, the dose at the EPID level is correlated with the dose in the patient and then compared at a point, in a plane, or in a volume with the predicted dose in that region of interest in the patient. For that purpose 1D, 2D, and 3D back-projection techniques have been developed as will be discussed in the following paragraphs.

Nijsten et al. (2007) correlated the dose measured with an EPID on the central beam axis with dose values at 5 cm depth using a back-projection model having a (semi-) empirical relationship between these two quantities. François et al. (2011) investigated a simple transit dosimetry method that correlates the response of the central pixels of an EPID with the dose at specified points in a patient. The method developed by this group has been implemented in a commercial system (EPIgray, DOSIsoft, Cachan, France) and recently clinically evaluated for IMRT fields (Ricketts et al., 2016). A similar approach, based on a different formalism, was described by Piermattei et al. (2015), and was recently also implemented in a commercial system (SOFTDISO, Best Medical, Italy). Peca and Brown have extended the isodose check method developed by Piermattei and colleagues to reconstruct two-dimensional dose maps of the entire radiation field at the depth of isocenter in a plane parallel to the EPID (Peca and Brown, 2014; Peca et al., 2015).

EPID-based 3D *transit* dose reconstruction methods have been developed by many groups (3D dose calculations in patients from *in-air* measurements with EPIDs are discussed in Chapter 6). Two approaches of dose reconstruction can be distinguished. In the first method, the photon fluence measured with the EPID is back-projected through the phantom or patient and then used to calculate the 3D dose distribution in the phantom or patient geometry (van Elmpt et al., 2007; Van Uytven et al., 2015). In the second approach, the dose distribution measured with the EPID is directly back-projected in the phantom or patient CT dataset (Wendling et al., 2009). The dose calculation algorithms used in these models vary in sophistication. A pencil-beam type of dose calculation model has been described by Wendling et al. (2009). Van Uytven et al. (2015) combined the incident fluence with the predicted extrafocal fluence to calculate the 3D patient dose distribution via a collapsed-cone convolution method, whereas a Monte Carlo (MC) approach was described by van Elmpt et al. (2007). In principle, MC-based dose calculations are the most accurate but require MC expertise in a department during the development stage, whereas the calculation times are still rather long. These may decrease; however, in the future with further development of multicore CPU and GPU technology. Dose calculation algorithms based on a pencil-beam model are fast and relatively easy to implement, but for some treatment sites the accuracy is inferior to dose calculations performed by most modern treatment planning systems, particularly with respect to tissue heterogeneity corrections. In order to bypass that problem, which is particularly important for verification of lung treatments, a modification of the pencil-beam type of dose calculation has been developed (Wendling et al., 2012). The key feature of the *in aqua vivo* method is that the dose reconstruction in the patient is based on EPID images obtained during the actual treatment, but these images are converted to a situation as if the patient consisted entirely of water. The method has the same sensitivity for detecting errors as *in vivo* dosimetry of sites without large tissue heterogeneities.

In addition to these in-house developed methods, commercial software has been released for 3D dose reconstruction, which includes a verification of the dose at the isocenter (Dosimetry Check [DC], Math Resolutions, Columbia, MD, USA). The commissioning of, and initial experience with, the DC software has been reported for phantom irradiations with various treatment techniques including VMAT (Gimeno et al., 2014) and helical tomotherapy (Mezzenga et al., 2014). With the increasing clinical use of VMAT and other advanced treatment techniques, it can be expected that in the future more commercial EPID-based 3D *in vivo* dose verification systems will become available.

It should be noted that all approaches of *in vivo* transit dosimetry are sensitive to changes in internal anatomy, which have to be taken into account when analyzing transit dose measurements. Although this complicates the interpretation of observed differences between measured and predicted 3D dose distributions, it is at the same time one of the main advantages of transit dosimetry compared to other patient-specific QA methods, because it indicates that actions may be required to improve the actual dose delivery to a patient.

11.2.2 Dose comparison methods, action levels, and follow up actions

As discussed in Chapter 6, differences between measured and predicted dose distributions are generally quantified using gamma evaluation, whereas other methods such as dose-volume histogram (DVH) examination are increasingly used. A variety of metrics are applied, and criteria for accepting deviations between measured and planned dose distributions, based on pretreatment dose verification analysis, have been formulated by several groups (NCS, 2013, 2015; upcoming AAPM Task Group report). As EPID-based *in vivo* dosimetry is based on reconstruction of dose distributions in planning CT data, there are in practice often variations in patient anatomy and patient setup that cause a larger spread in the *in vivo* dose determinations compared to phantom measurements.

At several hospitals the action level for isocenter dose difference is 5%, and an investigation is conducted to trace the origin of the difference (François et al., 2011; Hanson et al., 2014; Fidanzio et al., 2015; Mijnheer et al., 2015; Fidanzio et al., 2015; Piermattei et al., 2015; Ricketts et al., 2016). Gamma evaluation criteria using global 3%/3-mm γ metrics, or other metrics applied for pretreatment verification, are often chosen (e.g., Bedford et al., 2014; Berry et al., 2014; Mijnheer et al., 2015). The alert criteria are often different for different treatment sites, and are subject to modification with growing clinical experience.

When an EPID-based dose measurement raises an alert, one of the first follow-up actions is to check if an image acquisition error has occurred and a second *in vivo* dose measurement should be carried out to confirm the deviating result. If there is no obvious explanation for the observed difference, then the next step is to collect information about possible variations in patient setup and anatomy when the *in vivo* dose measurement was performed. For this purpose, visual inspection of the EPID or cone-beam computed tomography (CBCT) images made during that day should be performed. Investigations of situations where the tolerance level has been exceeded should also include an inspection of the treatment plan, as well as additional measurements with a phantom. When the deviation is large or unexplained, a decision must be made as to what further action is necessary. Such an action has to be taken by the physician of record, together with the medical physicist in charge of the *in vivo* dosimetry measurement. This may result in making a new treatment plan for the patient based on a new CT scan, or an adaptation of an existing CT scan using CBCT information in combination with a deformable image registration procedure.

11.3 CURRENT CLINICAL EXPERIENCE

11.3.1 Overall results of EPID-based *in vivo* dosimetry

In this section, we will discuss the current clinical experience with EPID-based *in vivo* transit dose verification and give some examples of different types of errors discovered in some centers. When performing QA measurements, the word *error* generally has the meaning that the difference with a reference value is outside the tolerance level. Criteria for acceptance of a particular patient-specific *in vivo* QA measurement should be related to the uncertainty in that type of *in vivo* dose determination. Figure 11.1 shows the distribution of the ratios between the measured and prescribed doses at the isocenter in two centers determined with EPID-based *in vivo* dosimetry. The upper part shows the results of the dose verification of patients treated at Institut Curie, Paris, France, with 3D conformal radiotherapy (3D CRT) using 4, 6, 10, and 20 MV beams at various treatment sites (François et al., 2011). The lower part of Figure 11.1 shows the observed deviations for patients treated with 3D CRT and IMRT using 6, 10, and 15 MV beams between 2008 and 2013 at the Royal Marsden Hospital NHS Trust (RMH), London, UK (Hanson et al., 2014). This cohort of patients includes almost all treatment sites, with the exception of CNS and TBI patients. The EPID-based dose reconstruction software used at Insitut Curie was a prototype of the EPIgray system, whereas the software used at the RMH was developed at The Netherlands Cancer Institute (NKI), Amsterdam, the Netherlands.

Both sets of data presented in Figure 11.1 show that on average the results are excellent and no systematic error has been detected. The standard deviation in these ratios is about 3.5%, which is typical for these types of *in vivo* measurements. The overall results of EPID-based *in vivo* dosimetry

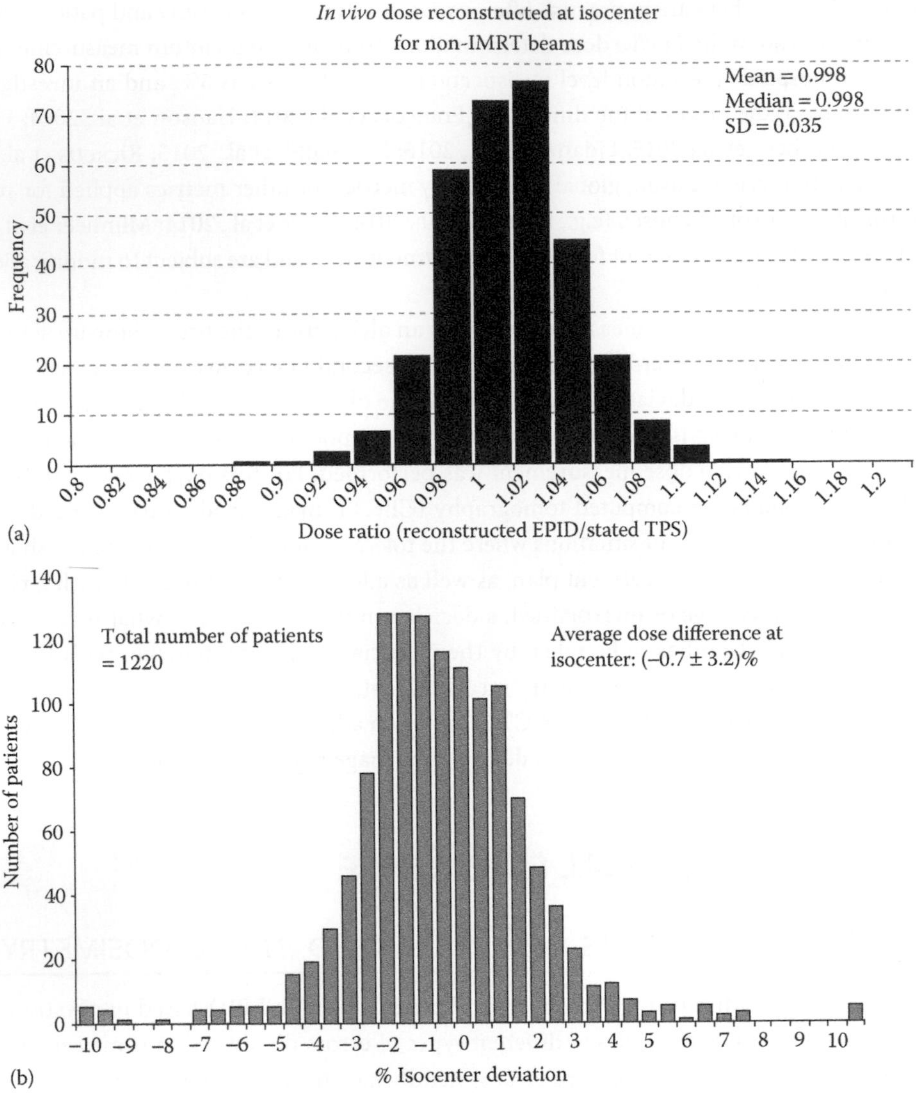

Figure 11.1 Distribution of the ratio of the dose at the isocenter reconstructed from EPID-based *in vivo* transit dose measurements and the prescribed dose. (a) Shows the results for patients treated with 3D CRT techniques at Institut Curie. (From François, P., Boissard, P., Berger, L., and Mazal, A., *Phys. Med.*, 27, 1–10, 2011.) (b) Gives similar results for 3D CRT and IMRT treatments performed at RMH. (From Hanson, I.M., Hansen, V.N., Olaciregui-Ruiz, I., and van Herk, M., *Phys. Med. Biol.*, 59, N171–N179, 2014.)

in some other centers showed standard deviations varying between about 2.5% for specific patient groups (Fidanzio et al., 2015), and values between 2.9% and 5.2% for various patient groups verified at NKI (Mijnheer et al., 2015). Obviously when choosing an action level of 5%, as is often done for *in vivo* isocenter dose verification, many alerts can be expected, depending on the site and treatment technique, which then need further inspection.

Most alerts originated because thresholds were exceeded. The majority of the alerts observed in these centers concern situations that cause errors of at least 5% at the isocenter or reference point, a large drop in gamma pass rate, or a change in another metric when dose distributions are analyzed in 2D or 3D. Often dose differences up to 10% or more in part of the planning target volume (PTV) or an OAR can arise, which frequently happens when there are changes in anatomy in the time period between acquisition of the planning CT and the treatment of the patient. For instance, an underdose of the order of 10% in part of the PTV may occur during treatment in the case that large gas pockets are present in the rectum in the planning CT, whereas an increase in dose of about 10% in part of the PTV may happen in the case of severe weight loss. (Dis)appearance of atelectasis during lung cancer treatment may result in changes in mean lung dose up to 10%. Other errors observed by many groups are related to human errors made in the clinic, such as the use of incorrect CT numbers or not taking the patient couch or fixation devices into account during the planning process. The discussion in the department of these errors should result in improved procedures, thus avoiding these types of errors in the future.

11.3.2 Examples of errors detected by EPID-based *in vivo* dosimetry

Many groups have shown that during a course of lung cancer treatments, very often anatomical changes in both target volume and OARs occur, leading to considerable dose differences compared to the planned situation (Piermattei et al., 2009; Mans et al., 2010; Wendling et al., 2012; Persoon et al., 2013; Berry et al., 2014; Kwint et al., 2014; Fidanzio et al., 2015; Mijnheer et al., 2015). Figure 11.2 shows an example of such a dose difference as detected with EPID-based *in vivo* dosimetry at NKI during the treatment of a lung cancer patient. The data presented in the upper part of the figure show that the results of the 3D gamma evaluation of the total dose distribution of a seven-field IMRT treatment were outside the action level. Inspection of the CBCT scan made that day, shown at the bottom part of the figure, indicated that a large amount of fluid was present in the lung resulting in an underdosage of part of the target volume. As a result of these observations, a new CT scan was made and a new plan was generated. It should be noted that the dose deviation at the isocenter is very small and would not have triggered an alert. Obviously 2D or 3D EPID-based dosimetry is needed to observe these dose differences due to anatomical changes during the treatment course of lung cancer patients.

Several other types of patient-related errors cannot be detected by pretreatment verification. Figure 11.3 shows as an example the result of two measurements of the dose at the isocenter at the Universita Cattolicà del Sacre Cuore, Rome, Italy, using the EPID-based SOFTDISO software

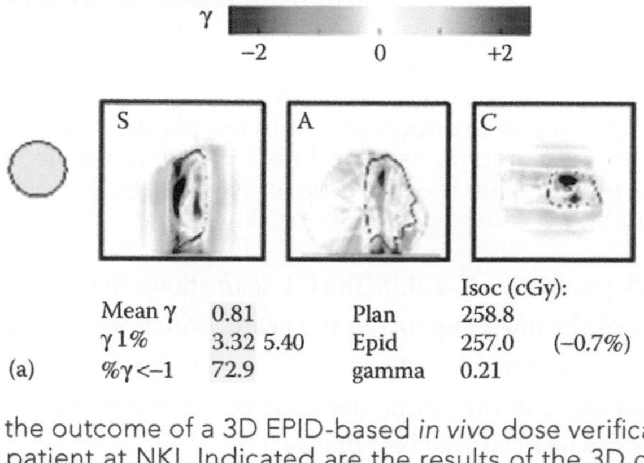

Figure 11.2 (a) Shows the outcome of a 3D EPID-based *in vivo* dose verification of an IMRT treatment of a lung cancer patient at NKI. Indicated are the results of the 3D gamma evaluation in a sagittal, axial, and coronal plane through the isocenter. A signed gamma display is used: the yellow and red color indicate regions where the EPID dose is higher than the planned dose, whereas the green and blue color indicates regions where the EPID dose is equal to or lower than the planned dose. The 50% isodose line is shown in black. The yellow dot means that at least one of the alert criteria is outside the action level. *(Continued)*

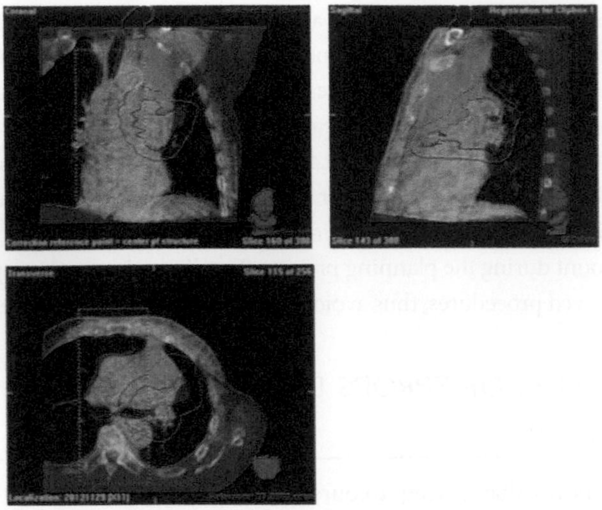

(b)

Figure 11.2 (Continued) (b) Shows a CBCT scan (green) made that day compared with the planning CT scan (purple) in the three orthogonal planes, showing an increase in fluid in the lung.

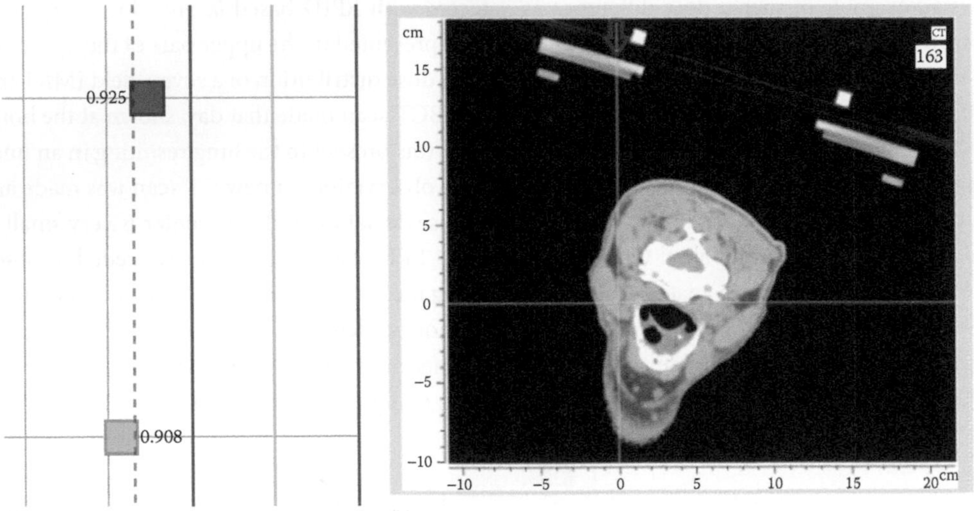

(a) (b)

Figure 11.3 (a) The ratio of the *in vivo* reconstructed and planned dose at the isocenter is given for two fractions of a patient treated for head-and-neck cancer using the SOFTDISO software and (b) shows the CT slice at the isocenter level. The red arrow indicates the direction of the central axis of a 3D CRT beam. (Courtesy of Fidanzio and Piermattei.)

(Fidanzio et al., 2015; Piermattei et al., 2015). The CT scan shows that the underdosage of about 8% was due to the presence of the mask support in the beam, which was not taken into account in the TPS dose calculation. As a result of these observations, a new plan was generated using other gantry angles that avoided attenuators in the beam. The effect on the patient dose due to obstruction from immobilization devices can be measured best *in vivo*.

Results of 3D EPID-based *in vivo* dose verification measurements can be presented in several ways. In Figure 11.2 gamma evaluation data in three orthogonal planes through the isocenter are shown, but DVHs can also be used to illustrate differences between measured and predicted 3D dose distributions. Figure 11.4 shows the result of an EPID-based 3D *in vivo* dose verification of a helical

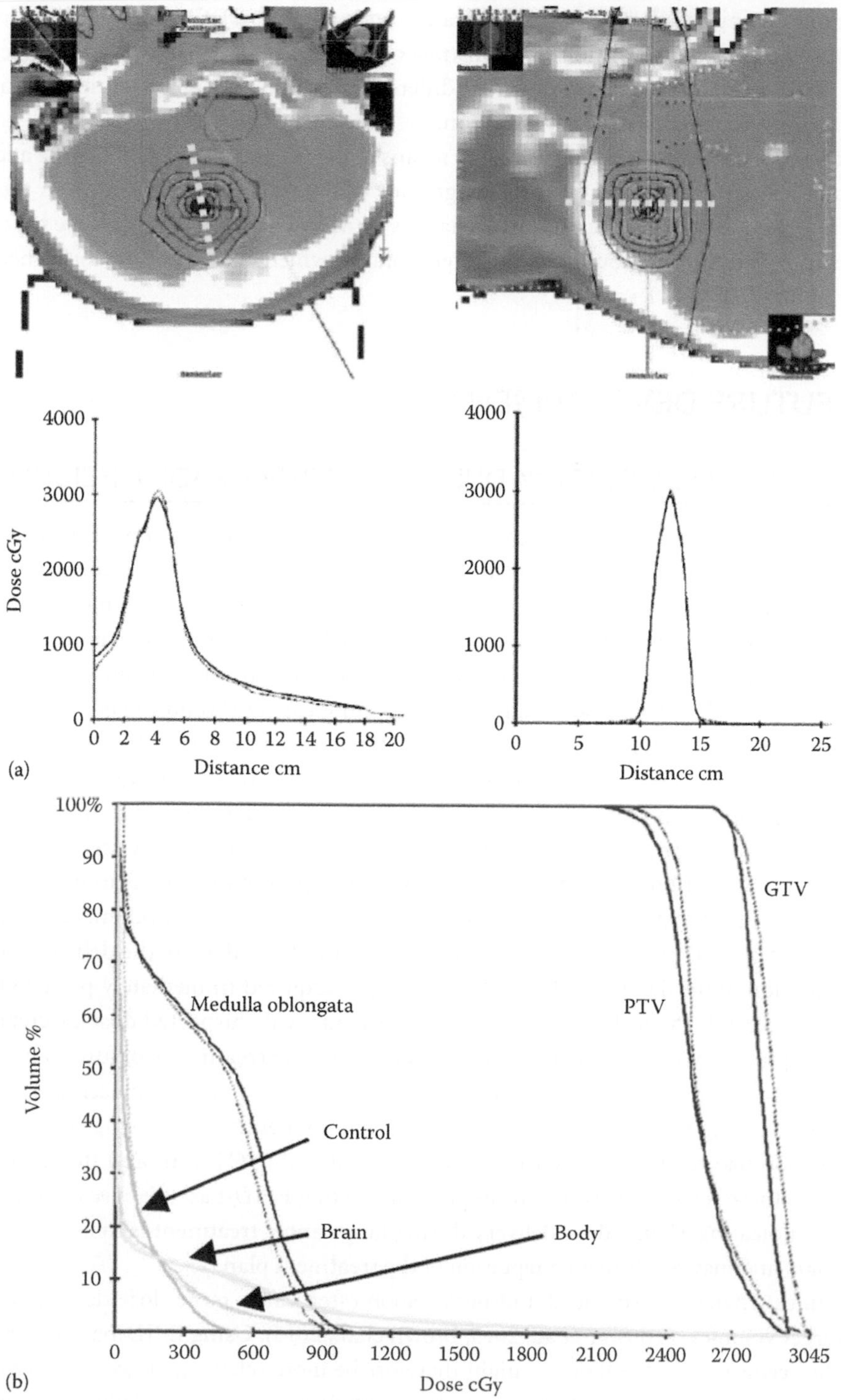

Figure 11.4 EPID-based 3D *in vivo* dose verification results obtained with the Dosimetry Check (DC) software of a tomotherapy brain treatment. (a) Shows isodose lines superimposed on a patient CT slice, obtained with DC (green) and TPS (magenta). Below each CT slice, representing an axial (left) and sagittal (right) plane, dose profiles along the yellow dotted lines, are shown. The corresponding target and organ-at-risk DVHs are presented in (b) (dotted lines–TPS, solid lines–DC). (From Mezzenga, E., Cagni, E., Botti, A., Orlandi, M., and Iori, M., *J. Inst.*, 9, C04039, 2014.)

tomotherapy treatment of a patient with a brain tumor using the DC software (Mezzenga et al., 2014). The reconstructed isodose lines demonstrate overall good agreement with those calculated by the TPS. However, the DVHs show systematic differences, both in the target volume and in some of the OARs, which required further investigation.

Anatomy changes can also be observed by means of portal imaging or CBCT. The added value of *in vivo* dosimetry is that it assesses the magnitude of the resulting deviation from the planned dose distribution. Furthermore, anatomical changes far away from the target volume may also have an influence on the dose distribution in the target volume, which may not be obvious when inspecting a portal image or CBCT scan.

11.4 FUTURE DEVELOPMENTS

11.4.1 *IN VIVO* DOSIMETRY DURING ADAPTIVE RADIOTHERAPY

Due to variation in anatomy during the course of radiotherapy, the dose actually delivered to a patient may deviate considerably from the planned dose distribution. Especially underdosage of the target volume or an overdosage in an OAR might require an adaptation of the treatment plan based on imaging performed during a patient's treatment course (ART), to account for anatomical changes.

EPID-based *in vivo* dosimetry is used in several centers to detect the dosimetric effects of anatomical changes during fractionated radiotherapy of lung cancer (Wendling et al., 2012; Persoon et al., 2013; Berry et al., 2014). At NKI EPID-based 3D *in vivo* dose verification is fully automated (Olaciregui-Ruiz et al., 2013), and performed for all lung cancer treatments using the *in aqua vivo* EPID dosimetry approach (Wendling et al., 2012). The method raises alerts resulting from deviations in the dose delivery due to changes in the anatomy (Figure 11.2). It is used routinely for all lung treatments and for breast treatments if a considerable part of the lung is influencing the dose distribution. At the MAASTRO clinic in Maastricht, the Netherlands, portal images are automatically processed by a 3D portal dose reconstruction algorithm to calculate the delivered 3D patient dose distribution of the day from the kV-CBCT images acquired immediately prior to treatment (Persoon et al., 2013). The technique is able to flag patients with suspected dose discrepancies for potential adaptation of the treatment plan, and is used during treatments of lung cancer patients with atelectasis apparent in the pretreatment images. Berry et al. (2014) presented a case study in which their 2D transit dosimetry algorithm was able to identify that a lung patient's bilateral pleural effusion had resolved in the time between the planning CT scan and the treatment. The experience in these centers illustrates the important role that EPID-based *in vivo* transit dosimetry can play in indicating when a dose delivery during lung cancer treatment is inconsistent with the original plan, and may result in an adaptation of the treatment plan.

Anatomical changes in the head-and-neck region often cause more dose deviations in OARs than in target volumes (Brouwer et al., 2015; Brown et al., 2015). Using EPID-based *in vivo* dosimetry for the verification of OAR doses might therefore be more relevant. However, OARs are often located in regions with a large dose gradient, thus requiring accurate knowledge of the position of an OAR with respect to the beam geometry. Verification of the OAR position is therefore a prerequisite for assessing the actual dose in an OAR. If the position of an OAR is known, for example, from CBCT information, then a first approximation of the *true* 3D dose distribution in the OAR can be obtained from the TPS by using the actual position of the OAR. The result of that calculation will indicate if *in vivo* dose verification is really required to get proper OAR dose estimation.

For instance, if the dose is well below the optimization constraints for that OAR, then further action is generally not necessary. Furthermore, accurate knowledge of the complete 3D dose distribution in a serially organized OAR is often not needed and only information about the maximum dose is of importance for the prediction of a biological effect in that OAR. For some small OARs, only point dose information is required. For all of these reasons, only limited information about the use of EPID-based *in vivo* dose verification of OAR doses is available.

ART is resource intensive because it requires a replanning process necessitating additional use of planning equipment and staff time. Furthermore, any adaptation of a plan also requires an additional patient-specific verification, either with a phantom measurement before the next fraction is delivered or *in vivo* during that fraction, thus further increasing the workload.

Currently a number of ART clinical trials are underway to assess the advantage of adding extra dose to specific parts of the tumor while sparing sensitive areas and normal tissues during the course of a treatment. To identify these areas in the tumor, multiparametric magnetic resonance imaging (MRI) is used in some centers for delivering a microboost in prostate cancer (e.g., Lips et al., 2011). PET imaging is applied to detect areas with a high uptake of fluorodeoxyglucose (FDG), showing resistance to treatment, for lung cancer and head-and-neck cancer patients (e.g., Duprez et al., 2011). EPID-based *in vivo* dosimetry can be used in these trials for delivered dose verification. However, attention should be paid to the accuracy for small boost fields.

A general challenge for ART is the accounting for changing target and OAR volumes. In the case of treatment adaptation, the radiation oncologist often has to define a new clinical target volume (CTV) on a CT or CBCT scan, which is time consuming. Furthermore, when using kV-CBCT instead of CT, tumor redelineation is often difficult because of the difference in image quality. This difference might be reduced in the future by optimizing the reconstruction and acquisition methods of kV-CBCT. Another approach is to redelineate the CTV on CBCT images and use deformable image registration (DIR) to adapt the planning CT contours. However, the accuracy, reproducibility, and computational performance of the DIR algorithms that are commercially available vary. The results of a multi-institution DIR accuracy study showed large discrepancies in reported shifts, although the majority of deformable registration algorithms performed at an accuracy equivalent to the voxel size (Brock, 2010). In addition, DIR for tumor registration is less precise than for normal tissues due to limited contrast and complex nonelastic tumor response (Mencarelli et al., 2014). Consequently, this method will introduce an additional uncertainty depending on the location and size of the deformation, and caution should therefore be exercised when using DIR to evaluate accumulated tumor doses. In addition, it is still not clear how to incorporate disappearing voxels, for instance, resulting from a shrinking tumor, in deriving the total dose delivered to a changing CTV.

11.4.2 Real-time dosimetry

Two groups have developed EPID-based real-time dose verification systems to detect gross treatment delivery errors. In the method of the Australian group, *WatchDog*, a reference dataset of predicted EPID images using parameters exported from the treatment planning system is compared with a cumulative EPID image acquired during treatment. The comparison is performed within the time of acquisition of an image (~0.1 s) and allows for both geometric and dosimetric verification of dynamic IMRT as a function of control point and gantry angle (Fuangrod et al., 2013). Figure 11.5 shows the system rapidly detecting an MLC error; the MLC leaves at fraction 26 were retracted and remained there until the end of the treatment. In this case study, the system detected an error at frame 49 (~0.1 s delay) using the individual frame comparison, and

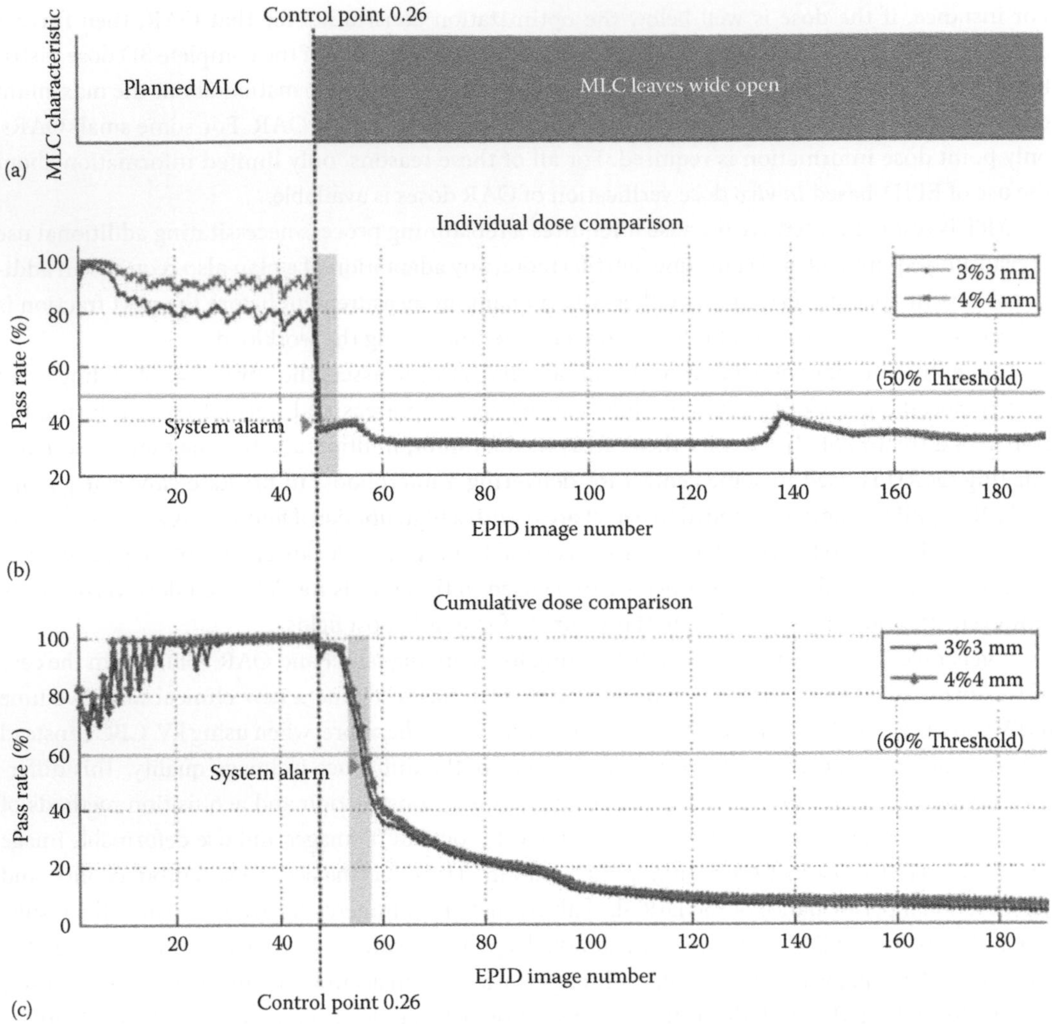

Figure 11.5 Result of automated error detection for an introduced MLC leaf position error during treatment: (a) MLC characteristics indicating retracted positions starting at control point 0.26, (b) individual dose gamma comparison pass rates, and (c) cumulative dose comparison pass rates. (From Fuangrod, T. et al., *Med. Phys.*, 40, 091907, 2013.)

frame 57 (~1.3 s delay) using the cumulative frame comparison method. The authors reported on the first clinical demonstration of their real-time dose delivery verification system for 28 patients undergoing IMRT and VMAT treatments using the cumulative frame comparison approach (Woodruff et al., 2015). The main aim of that study was to implement and gain initial experience with the system and to detect only major deviations from the treatment course. Therefore, 4%/4 mm global criteria were chosen and a treatment field was flagged as failed if more than four consecutive frames fell below a pass rate of 40%. No errors were detected using these criteria, and the average pass rate was 91.1% ±11.5% (1SD).

At NKI, a software package for real-time 3D EPID-based dose verification was derived from their clinical offline 3D dose verification system (Spreeuw et al., 2016). Portal images are processed faster than the frame rate of the portal imager by precomputing all input for 3D EPID-based dose reconstruction, and speeding up the reconstruction algorithm via a new, multithreaded implementation. After a portal image is acquired, the dose distribution is reconstructed in 3D in the

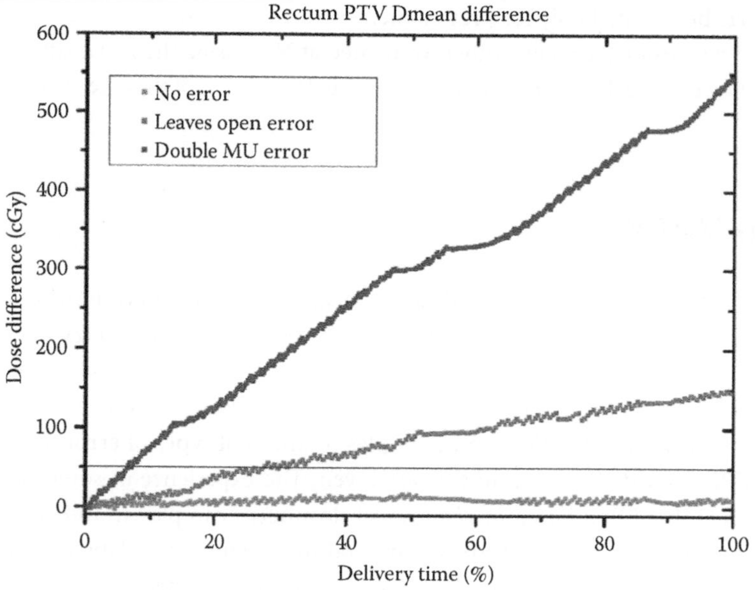

Figure 11.6 Difference between reconstructed and planned mean dose in the PTV for three scenarios (no error, leaves open error, and double MU error), as a function of delivery time of a single-arc 10 MV rectum plan. The horizontal line indicates the detection threshold. (From Spreeuw, H. et al., *Med. Phys.*, 43, 3969–3974, 2016.)

patient planning CT data, and compared with the 3D dose distribution predicted by the treatment planning system using dose-volume histogram parameters. Whenever dose differences outside tolerance levels are detected, an alert is generated, which could be used as a trigger to stop the linear accelerator (LINAC) automatically. The software was tested by irradiating an Alderson phantom with VMAT arcs after introducing some serious delivery errors. The result of one of these tests is shown in Figure 11.6, demonstrating that the LINAC could indeed be halted quickly and automatically when cumulative reconstructed and total planned 3D dose distributions diverged beyond a certain limit.

Defining criteria to detect serious errors and not small errors due to, for instance, transient anatomical changes such as gas in the rectum as observed by Woodruff et al. (2015) will be a topic of further study by both groups. One challenge is that the importance of a detected dose difference depends on the accumulated total dose already received by the patient during fractionated radiotherapy. Testing real-time dose verification methods may therefore be started, and is probably most relevant, for hypofractionated treatments consisting of only a few high dose fractions.

An interesting new application of EPID-based verification is the possibility to perform time-dependent measurements during VMAT to verify the field shape and intensity of sub-arcs (Liu et al., 2013; Woodruff et al., 2013; Podesta et al., 2014). These models were used to verify pretreatment delivered MLC fluences, and the approach of Woodruff et al. (2015) has been incorporated in their real-time *in vivo* approach. The model developed by Podesta et al. (2014) was able to verify time-dependent absolute dose distributions. Their study showed that EPID-based dose verification measurements can be performed on a control point basis for VMAT plans delivered with both flattened and flattening filter free beams. By measuring in a time dependent manner, dose deviations from predicted time-dependent 2D dose distributions can be observed that might be hidden in integrated dose images, allowing the detection of systematic dose errors related to the treatment delivery. So far, their method has been tested only for pretreatment verification and not yet *in vivo*. Time-resolved DVH-based *in vivo* dosimetry is currently investigated at NKI using their real-time 3D EPID-based software (Spreeuw et al., 2016). A more comprehensive description of these techniques can be found in Chapter 6.

11.5 SUMMARY

Several 2D and 3D EPID-based *in vivo* dose verification methods have been developed during the past decade. After discussing the relationship between *in vivo* dosimetry and other types of patient-specific QA, the clinical implementation of EPID-based *in vivo* dose verification has been elucidated. Several issues such as the definition of tolerance/action levels and follow-up actions in case an alert is raised are clarified, and examples of different types of error detected by various groups using EPID-based *in vivo* dosimetry are given. The experience in some centers illustrates the important role that EPID-based *in vivo* transit dosimetry can play in indicating when a dose delivery error occurs during a patient treatment, and may require an adaptation of the treatment plan. Most of these errors are patient related and cannot be detected by means of pretreatment dose verification using phantom measurements. The correlation of 3D dose verification with 3D in-room imaging will therefore become more important in the future with the increasing use of ART. Finally, it can be expected that real-time EPID-based *in vivo* dosimetry approaches will be further developed in order to ensure safe radiation therapy delivery.

REFERENCES

Baek TS, Chung EJ, Son J and Yoon M. (2014). Feasibility study on the verification of actual beam delivery in a treatment room using EPID transit dosimetry. *Radiat Oncol* 9: 273.

Bedford JL, Hanson IM and Nordmark HV. (2014). Portal dosimetry for VMAT using integrated images obtained during treatment. *Med Phys* 41: 021725.

Berry SL, Polvorosa C, Cheng S, Deutsch I, Chao KSC and Wuu C-S. (2014). Initial clinical experience performing patient treatment verification with an electronic portal imaging device transit dosimeter. *Int J Radiat Oncol Biol Phys* 88: 289–309.

Bojechko C, Phillips M, Kalet A and Ford EC. (2015). A quantification of the effectiveness of EPID dosimetry and software-based plan verification systems in detecting incidents in radiotherapy. *Med Phys* 42: 5363–5369.

Brock KK. (2010). Results of a multi-institution deformable registration accuracy study (MIDRAS). *Int J Radiat Oncol Biol Phys* 76: 583–596.

Brouwer CL, Steenbakkers RJHM, Langendijk JA and Sijtsema NM. (2015). Identifying patients who may benefit from adaptive radiotherapy: Does the literature on anatomic and dosimetric changes in head and neck organs at risk during radiotherapy provide information to help? *Radiother Oncol* 115: 285–294.

Brown E et al. (2015). Predicting the need for adaptive radiotherapy in head and neck cancer. *Radiother Oncol* 116: 57–63.

Chytyk-Praznik K, VanUytven E, vanBeek TA, Greer PB and McCurdy BM. (2013). Model-based prediction of portal dose images during patient treatment. *Med Phys* 40: 031713.

Duprez F, De Neve W, De Gersem W, Coghe M and Madani I. (2011). Adaptive dose painting by numbers for head-and-neck cancer. *Int J Radiat Oncol Biol Phys* 80: 1045–1055.

Fidanzio A, Azario L, Greco F, Cilla S and Piermattei A. (2015). Routine EPID *in-vivo* dosimetry in a reference point for conformal radiotherapy treatments. *Phys Med Biol* 60: N141–N150.

François P, Boissard P, Berger L and Mazal A. (2011). *In vivo* dose verification from backprojection of a transit dose measurement on the central axis of photon beams. *Phys Med* 27: 1–10.

Fuangrod T et al. (2013). A system for EPID-based real-time treatment delivery verification during dynamic IMRT treatment. *Med Phys* 40: 091907.

Gimeno J et al. (2014). Commissioning and initial experience with a commercial software for *in vivo* volumetric dosimetry. *Phys Med* 30: 954–959.

Hanson IM, Hansen VN, Olaciregui-Ruiz I and van Herk M. (2014). Clinical implementation and rapid commissioning of an EPID based in-vivo dosimetry system. *Phys Med Biol* 59: N171–N179.

IAEA (2013). Development of procedures for *in vivo* dosimetry in radiotherapy. IAEA Human Health Report No. 8. International Atomic Energy Agency, Vienna, Austria.

Kry SF et al. (2014). Institutional patient-specific IMRT QA does not predict unacceptable plan delivery. *Int J Radiat Oncol* 90: 1195–1201.

Kwint M et al. (2014). Intra thoracic anatomical changes in lung cancer patients during the course of radiotherapy. *Radiother Oncol* 113: 392–397.

Lips IM et al. (2011). Single blind randomized phase III trial to investigate the benefit of a focal lesion ablative microboost in prostate cancer (FLAME-trial): Study protocol for a randomized controlled trial. *Trials* 12: 255.

Liu B, Adamson J, Rodrigues A, Zhou F, Yin F and Wu Q. (2013). A novel technique for VMAT QA with EPID in cine mode on a Varian TrueBeam linac. *Phys Med Biol* 58: 6683–6700.

Mans A et al. (2010). Catching errors with *in vivo* dosimetry. *Med Phys* 37: 2638–2644.

McKenzie EM, Balter PA, Stingo FC, Jones J, Followill DS and Kry S. (2014). Toward optimizing patient-specific IMRT QA techniques in the accurate detection of dosimetrically acceptable and unacceptable patient plans. *Med Phys* 41: 121702.

Mencarelli A et al. (2014). Deformable image registration for adaptive radiation therapy of head and neck cancer: accuracy and precision in the presence of tumor changes. *Int J Radiat Oncol Biol Phys* 90: 680–687.

Mezzenga E, Cagni E, Botti A, Orlandi M and Iori M. (2014). Pre-treatment and in-vivo dosimetry of helical tomotherapy treatment plans using the dosimetry check system. *J Inst* 9: C04039.

Mijnheer B. (2013). *In vivo* dosimetry. In: *The Modern Technology of Radiation Oncology*, Vol. 3, Van D. J. (Ed.), pp. 301–336. Medical Physics Publishing, Madison, WI, USA.

Mijnheer B, Beddar S, Izewska J and Reft C. (2013). *In vivo* dosimetry in external beam radiotherapy. *Med Phys* 40: 070903.

Mijnheer BJ, González P, Olaciregui-Ruiz I, Rozendaal RA, van Herk M and Mans A. (2015). Overview of three year experience with large scale EPID-based 3D transit dosimetry. *Pract Radiat Oncol* 5: e679–e687.

NCS (2013). Code of practice for the quality assurance and control for intensity modulated radiotherapy. NCS Report 22. Netherlands Commission on Radiation Dosimetry (http://www.radiationdosimetry.org).

NCS (2015). Code of practice for the quality assurance and control for volumetric modulated arc therapy. NCS Report 24. Netherlands Commission on Radiation Dosimetry (http://www.radiationdosimetry.org).

Nijsten SMJJG, Mijnheer BJ, Dekker ALAJ, Lambin P and Minken AWH. (2007). Routine individualized patient dosimetry using electronic portal imaging devices. *Radiother Oncol* 83: 65–75.

Olaciregui-Ruiz I, Rozendaal R, Mijnheer B, van Herk M and Mans A. (2013). Automatic *in vivo* portal dosimetry of all treatments. *Phys Med Biol* 58: 8253–8264.

Peca S and Brown DW. (2014). Two-dimensional *in vivo* dose verification using portal imaging and correlation ratios. *J Appl Clin Med Phys* 15: 117–128.

Peca S, Brown D and Smith WL. (2015). *In vivo* EPID dosimetry detects interfraction errors in 3D-CRT of rectal cancer. In: *IFMBE Proceedings*, Vol. 51, Jaffray, D.A. (Ed.), pp. 531–534. Springer International Publishing, Switzerland.

Persoon LCGG, Egelmeer AGTM, Ollers MC, Nijsten, SMJJG, Troost EGC and Verhaegen F. (2013). First clinical results of adaptive radiotherapy based on 3D portal dosimetry for lung cancer patients with atelectasis treated with volumetric-modulated arc therapy (VMAT). *Acta Oncol* 52: 1484–1489.

Piermattei A et al. (2009). Integration between *in vivo* dosimetry and image guided radiotherapy for lung tumors. *Med Phys* 36: 2206–2214.

Piermattei A et al. (2015). aSi EPIDs for the in-vivo dosimetry of static and dynamic beams. *Nucl Instrum Meth Phys Res A* 796: 93–95.

Podesta M, Nijsten SMJJG, Persoon LCGG, Scheib SG, Baltes C and Verhaegen F. (2014). Time dependent pre-treatment EPID dosimetry for standard and FFF VMAT. *Phys Med Biol* 59: 4749–4768.

Reich P, Bezak E, Mohammadi M and Fog L. (2006). The prediction of transmitted dose distributions using a 3D treatment planning system. *Austral Phys Eng Sci Med* 29: 18–29.

Ricketts K et al. (2016). Clinical experience and evaluation of patient treatment verification with a transit dosimeter. *Int J Radiat Oncol Biol Phys* 95: 1513–1519.

Spreeuw H et al. (2016). Online 3D EPID-based dose verification: Proof of concept. *Med Phys* 43: 3969–3974.

van Elmpt W, McDermott L, Nijsten S, Wendling M, Lambin P and Mijnheer B. (2008). A literature review of electronic portal imaging for radiotherapy dosimetry. *Radiother Oncol* 88: 289–309.

van Elmpt WJ, Nijsten SM, Dekker AL, Mijnheer BJ and Lambin P. (2007). Treatment verification in the presence of inhomogeneities using EPID-based three-dimensional dose reconstruction. *Med Phys* 34: 2816–2826.

Van Uytven E, Van Beek T, McCowan PM, Chytyk-Praznik K, Greer PB and McCurdy BM. (2015). Validation of a method for *in vivo* 3D dose reconstruction for IMRT and VMAT treatments using on-treatment EPID images and a model-based forward-calculation algorithm. *Med Phys* 42: 6945–6954.

Wendling M et al. (2012). In aqua vivo EPID dosimetry. *Med Phys* 39: 367–77.

Wendling M, McDermott LN, Mans A, Sonke J-J, van Herk M and Mijnheer B. (2009). A simple back-projection algorithm for 3D EPID dosimetry of IMRT treatments. *Med Phys* 36: 3310–3321.

Woodruff HC et al. (2015). First experience with real-time EPID-based delivery verification during IMRT and VMAT treatments. *Int J Radiat Oncol Biol Phys* 93: 516–522.

Woodruff HC, Fuangrod T, Rowshanfarzad P, McCurdy BMC and Greer PB. (2013). Gantry-angle resolved VMAT pretreatment verification using EPID image prediction. *Med Phys* 40: 081715.

12

Advanced technologies for beam's eye view imaging

JOSH STAR-LACK

12.1 INTRODUCTION

Over the past decade, image-guided radiation therapy (IGRT) has become an essential component of contemporary clinical practice (Sonke et al. 2005, Dawson and Jaffray 2007, Li and Xing 2007, Boda-Heggemann et al. 2011, Mahmood et al. 2015). On the majority of systems, imaging capabilities are provided either by a kilovoltage (kV) source-detector pair, such as the X-ray volumetric imager (XVI, Elekta Oncology Systems) or on-board imager (OBI, Varian Medical Systems) that is orthogonally positioned relative to the direction of the treatment beam (Yoo et al. 2006, Lehmann et al. 2007, Gardner et al. 2014), or by stationary stereoscopic imagers such as are used in the Exactrac (Brainlab AG) or by the Cyberknife (Accuray Inc). With the maturation of these technologies and the recognition of some of their inherent limitations, there is renewed interest in high quality in-line or (treatment) beam's eye view (BEV) imaging to serve as an alternative or compliment. Specific applications and needs include (1) improved image-based intrafraction motion management (Mao et al. 2008b, Maltz et al. 2009, Yip et al. 2015, Hunt et al. 2016, Zhang et al. 2016), (2) rapid (single breath hold) cone-beam computed tomography (CBCT) acquisition enabled by combining projections from a BEV unit and an orthogonal unit (Wertz et al. 2010, Hunt et al. 2016), (3) rapid patient setup using simultaneously acquired orthogonal radiographs, (4) improved metal artifact reduction and electron density measurement accuracy enabled by MV imaging (Ruchala et al. 1999, Groh et al. 2002, Meeks et al. 2005, Sawant et al. 2006, Chang et al. 2007, Zhang and Yin 2007, Wu et al. 2014), and (5) a less costly IGRT alternative, in general.

High-quality BEV imaging can, in principle, be achieved using either the megavoltage (MV) treatment beam or a separate in-line kV X-ray system. Although a kV system has the advantage of offering good soft tissue contrast resolution at lower doses, there are considerable technical difficulties associated with developing such a system. On the other hand, use of the MV treatment beam for soft tissue imaging, while being simpler conceptually and requiring less new hardware, necessitates greatly increasing the sensitivity of electronic portal imaging devices (EPIDs), which is also a challenging task. This chapter will review progress on both the kV and MV approaches for achieving high-caliber BEV imaging capabilities while speculating on future research directions.

12.2 MV IMAGING

Previous studies have shown that a zero-frequency detective quantum efficiency [DQE(0)] of at least 20% at 6 MV is desired to obtain acceptable image quality at acceptably low doses, and that spatial resolutions of 1 mm or better at isocenter are required (Groh et al. 2002, Rathee et al. 2006, Sillanpaa et al. 2006, Sawant et al. 2006, Monajemi et al. 2006). Significant progress has been made over the past two decades in developing such high DQE MV detectors as described in Section 12.2.2. Note, that while the 1D ion chamber-based detector used in the tomotherapy system (Accuray Inc) for single-slice CT meets the 20% DQE(0) threshold and has been important for showing the viability of soft tissue imaging at MV energies (Ruchala et al. 1999, Meeks et al. 2005), it will not be further examined because it does not provide for the 2D planar imaging capabilities necessary for radiography, fluoroscopy, CBCT, and dosimetry.

12.2.1 Current EPID technology

Nearly all commercial EPIDs use an indirect detection mechanism (Boyer 1992, Antonuk et al. 2002, Van Elmpt and McDermott 2008). Typically, a 1 mm-thick copper build-up plate is coupled to a thin terbium-doped gadolinium oxysulfide scintillator screen (Cu-GOS) that, in turn, is mounted onto an active matrix flat-panel imager (AMFPI) utilizing amorphous silicon(aSi) thin film photodiodes and switching transistors. The Cu-GOS screen converts the incoming X-ray photons into optical photons that are then detected by the AMFPI. On account of the high amounts of optical scattering present in the GOS layer, it is preferred that the GOS area density be kept below 400 mg/cm^2 thus limiting the imager's DQE(0) to a range of 1%–2% (Bissonnette et al. 1997, Munro and Bouius 1998, Kausch et al. 1999, El-Mohri et al. 2001). Consequently, an alternative to a single-layer Cu-GOS imager is required to meet the sensitivity goals stated earlier.

12.2.2 Prospective high-DQE technologies

A promising means of increasing EPID sensitivity is to replace the Cu-GOS assembly with an array of high quantum efficiency (QE) thick pixelated scintillators that can be coupled to the AMFPI. The arrays are commonly fabricated using the slicing and gluing technique described by Uribe et al. (2003) and illustrated in Figure 12.1. Briefly, a block of scintillator is cut into slabs and both sides of each slab are polished. A reflector is then inserted between the slabs that are glued back together to reform the block. The same cutting, polishing, and gluing processes are then repeated in the orthogonal direction. Typically, larger area arrays comprise multiple such blocks that are then glued

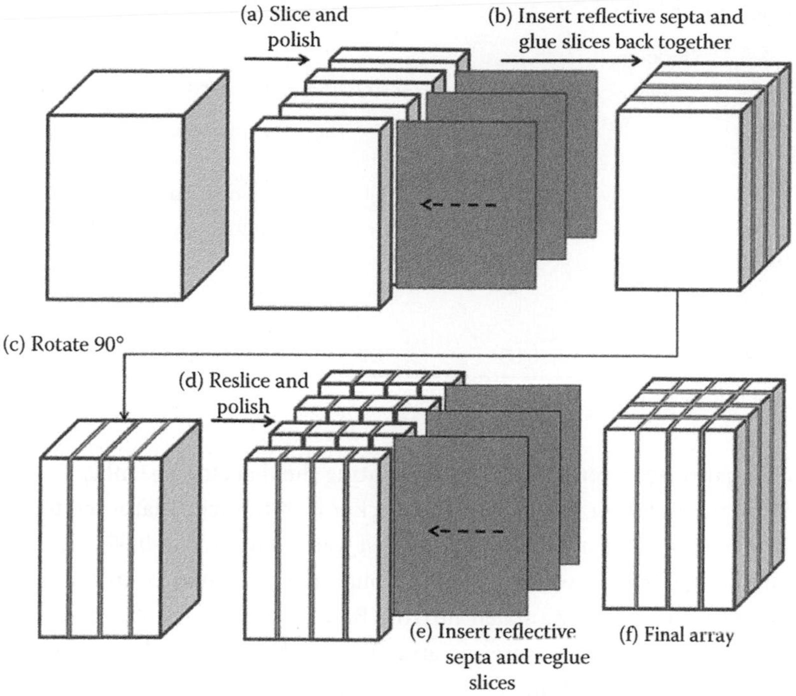
(a) Slice and polish
(b) Insert reflective septa and glue slices back together
(c) Rotate 90°
(d) Reslice and polish
(e) Insert reflective septa and reglue slices
(f) Final array

Figure 12.1 Scintillator array fabrication technique employing slicing and regulating operations. 2(N-1) cuts and 4(N-1) polishing and gluing steps are required to construct an array comprising N^2 pixels. (Reprinted with permission from Star-Lack J. et al. *Med. Phys.*, 42, 5084–5099, 2015.)

to each other. Table 12.1 shows key properties of several scintillators that have been studied and Table 12.2 shows properties of different reflectors that have been used. These are further examined in Section 12.2.3.

In early proof-of-concept studies, Morton et al. (1991) built a 1D array consisting of 128 zinc tungstate ($ZnWO_4$) crystals, each measuring $5 \times 5 \times 25$ mm^3 in size, which were coupled to discrete

Table 12.1 Properties of scintillators used for high DQE portal imaging

Scintillator	Density (g/cm^3)	Index of refraction	Light yield (photons/MeV)	Output stability and radiation hardness	Cleave plane	Optical transparency	Hygroscopic	Cost
$Bi_4Ge_3O_{12}$ (BGO)	7.2	2.2	8000	Low	No	High	No	$$$
$CdWO_4$ (CWO)	7.9	2.3	15000	High	Yes	High	No	$$$
CsI:Tl (CsI)	4.5	1.8	65000	Moderate	No	High	Yes	$$
Gd_2O_2S:Tb (Ceramic GOS)	7.3	2.3	60000	High	No	Low	No	$$
LKH-5 Scintillating Glass	3.8	1.6	4000	High	No	High	No	$

Adapted from Van Eijk, CWE. *Phys. Med. Biol.* 47, R85–R106, 2002; Knoll, G.F., *Radiation Detection and Measurement*, New York, Wiley, 2000; Mao, R. et al., *IEEE Trans. Nucl. Sci.*, 55, 2425–2431, 2008a.

Table 12.2 Properties of reflectors used to help optically separate the pixels

Reflector	Thickness (mm)	Reflectivity	Transmission	Specular/ lambertian	References
TiO_2 in epoxy	0.3	–	–	Lambertian	(Mosleh-Shirazi et al. 1998)
Polystyrene	0.075	>.975	–	Lambertian	(Breitbach et al. 2011)
Gelcoat polyester with epoxy resin	0.04	>0.95	–	Lambertian	(Rathee et al. 2006)
Reflective Polymer	0.1	>0.94	–	–	(Sawant et al. 2006, Wang et al. 2009)
Aluminized Mylar	0.012	0.88	0%	Specular	(Star-Lack et al. 2015)

Note: Parameters that were not specified were labeled with a "– ".

photodiodes. Planar images were created by translating the detector assembly in the longitudinal direction. Although resolution was coarse, the thickness of the scintillator led to a very high QE and results showed that 2.5 mm holes of 1% contrast were visualized at 0.55 cGy. Some of the first snapshot 2D high DQE images were created by coupling a pixelated cesium iodide (CsI) screen to a video camera using a 45° angulated mirror (Bissonnette et al. 1992, Mosleh-Shirazi et al. 1998). Although these early results were encouraging and demonstrated the potential of snapshot high-DQE MV imaging, the bulkiness of the systems and their low-optical coupling efficiencies reduced their appeal especially with the advent of radiation hard AMFPI technology.

The first large area, AMFPI-based high-DQE portal imager was built by Seppi et al. (2003). An 8 mm-thick thallium-doped CsI scintillator was pixelated with a 0.388 mm pitch and coupled to Varian flat-panel imager (Varian Medical Systems, Palo Alto CA) covering a 40 cm × 30 cm area. DQE(0) was estimated to be 10% at 6 MV but never directly measured. Results showed that image quality was significantly improved relative to that obtained with a conventional Cu-GOS screen, and CBCT scans taken with the imager were shown to be useful for monitoring interfractional lung tumor changes (Sillanpaa et al. 2006, Chang et al. 2007) (Figure 12.2). Despite the small pixel size of 0.396 mm, the spatial resolution at the isocenter (detector magnification = 1.5X) was only on the order of 1 mm indicating that there was significant light sharing between the scintillator pixels.

Sawant et al. (2006) built a prototype 16 × 16 cm^2 pixelated CsI imager that was 40 mm thick and had a pitch of 1 mm. DQE(0) was measured to be 22% at 6 MV. Compared to a Cu-GOS screen, large improvements in contrast detail were observed. However, as with the Seppi imager, spatial resolution was degraded by optical crosstalk. By comparing the performance of 120 × 60 pixel 1 mm-pitch bismuth germanate (BGO) and CsI imagers, Wang et al. (2009) and El-Mohri et al. (2011) showed that BGO may be a preferred scintillation material due to its higher density and higher index of refraction. For a given quantum efficiency, the increase in density allows the pixel height to be reduced, thus ameliorating beam divergence effects, and the higher index of refraction increases the probability that optical photons undergo total internal reflection at the pixel-glue boundaries, thus reducing crosstalk. However, preirradiation of the BGO scintillator with 2000 cGy was required to achieve acceptable output stability. Figure 12.3 shows two of the detectors built and several associated CBCT images.

An alternative scintillator is sintered GOS which is commonly used in diagnostic CT detectors (van Eijk 2002). Breitback et al. (2011) studied ceramic GOS for imaging at 6 MV using a pixelated array of area 40 cm × 10 cm and thickness of 1.8 mm. A modest 2.5× sensitivity increase was achieved when compared to Cu-GOS performance. Although suitable for lower energy systems, the

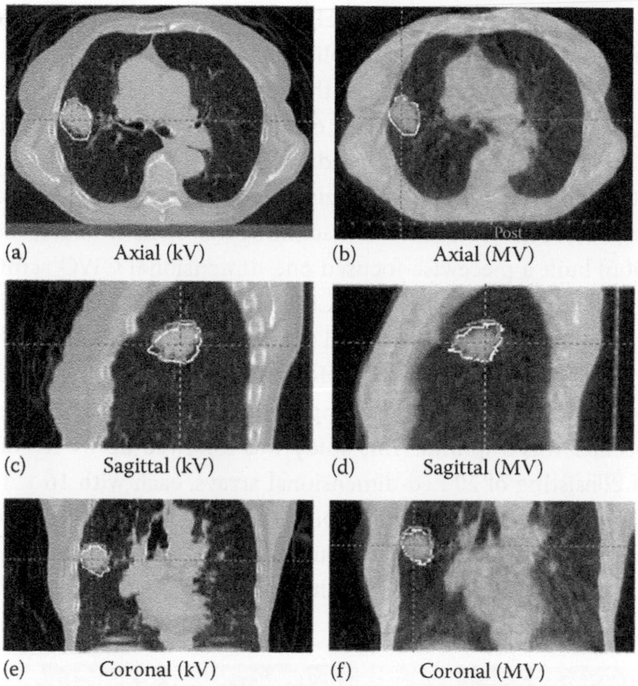

Figure 12.2 Image fusion of the simulation computed tomography (CT) scan and megavoltage cone-beam computed tomography (MV CBCT) axial, sagittal, and coronal views obtained using the prototype detector developed by Seppi et al. (2003). The gross tumor volume defined by the simulation CT scan (CT GTV) is outlined in green and the gross tumor volume defined by the MV CBCT scan (MV GTV) is in white. (Reprinted from Chang J. et al., *Int. J. Radiat. Oncol. Biol. Phys.*, 67, 1548–1558, 2007. With permission.)

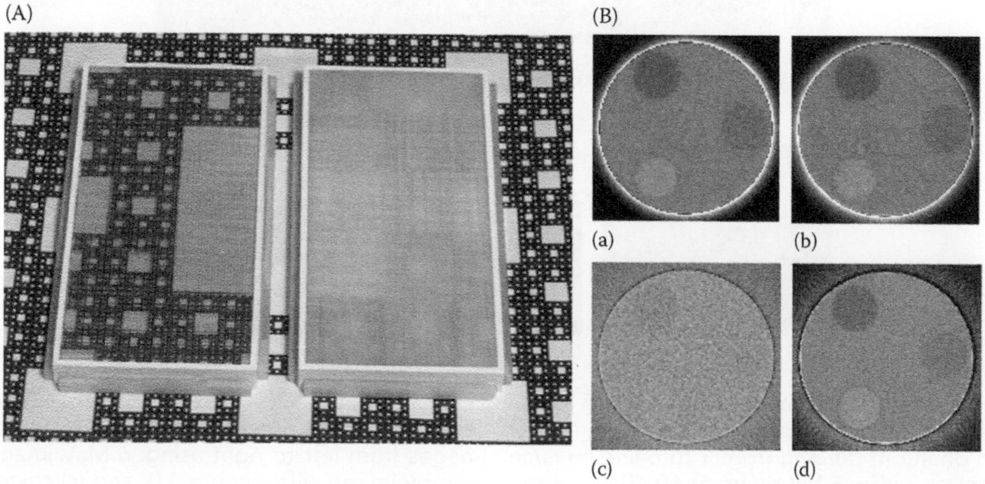

Figure 12.3 (A) Photograph showing the BGO (clear) and CsI:Tl segmented scintillator prototypes from Wang et al. (2009) and El-Mohri et al. (2011). (B) Reconstructed images of a contrast phantom embedded with tissue-equivalent objects corresponding to relative electron densities of (clockwise from top): 0.954, 0.988 and 1.049. The images were obtained at a scan dose of ~4 cGy with (a) BGO, (b) CsI, (c) Cu-GOS, and (d) Cu-GOS 160 cGy (average of 40 scans of c). (Reprinted from Wang et al., *Med. Phys.*, 36, 3227–3238, 2009; El-Mohri et al., *Phys. Med. Biol.*, 56, 1509–1527, 2011. With permission.)

low-optical transmissivity of the ceramic material may restrict the construction of pixels with high aspect ratios as are required for high-energy applications.

Cadmium tungstate (CWO) shares many of the advantages of BGO and ceramic GOS including a high density (7.9 gm/cm^3) and high index of refraction (n = 2.3), but with some additional benefits. Compared to BGO, CWO has a high radiation resistivity (10^7 rad) and has twice the light output. Compared to sintered GOS, CWO is highly transparent. The main challenges of working with CWO are its high cost and difficulties in machining due to its crystal structure.

Rathee et al. (2006) built a piecewise-focused one-dimensional CWO scintillator array with a 2.75 mm transaxial pitch and 10 mm thickness. Custom electronics arrayed in a piecewise-focused arc were developed for readout and digitization. DQE(0) was measured to be 19% when tested with a bremsstrahlung X-ray beam produced by 6 MeV electrons impinging on the scatter foils and hardened through 4 cm of water (6 MeV imaging beam). The spatial resolution of the CT reconstruction at the isocenter was 5 lp/mm. This study was followed up by Kirvan et al. (2010) who built an arc detector consisting of 20 two-dimensional arrays, each with 16 × 16 pixels with 1 mm pitch. The total imaging area comprised 5100 pixels and measured 16 mm × 320 mm. Custom photodiodes and electronics were used for readout and CBCT scans were acquired by rotating the phantom. Figure 12.4 shows the system and resulting images.

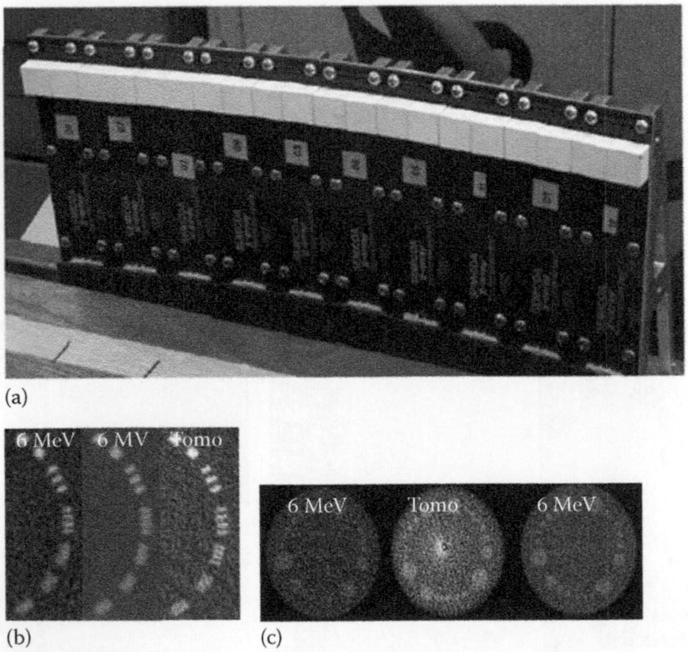

Figure 12.4 (a) Prototype 320 × 16 mm^2 imager from Kirvan et al. (2010) comprising 16 2D CWO arrays coupled to discrete photodiodes, (b) reconstruction of a portion of a CATPHAN500 resolution phantom ranging from 1 to 6 line pairs/cm. Images from left to right using: 6 MeV imaging beam at 2 cGy, 6 MV beam at 60 cGy, and tomotherapy image with pitch = 1.0, and (c) custom-designed low contrast phantom of CATPHAN500 showing plugs of 3.0%, 2.5%, and 1.5% contrast (clockwise from left). For each contrast level there are cylinders of 20, 4, 5, 6, 7, 8, and 15 mm diameter. Images from left to right: 6 MeV beam at 2 cGy and 8 mm slice thickness, Tomo beam with pitch = 1.0 and 5 mm slice thickness, and 6 MeV imaging beam at 4 cGy and 8 mm slice thickness. (Reprinted from Kirvan et al., *Med. Phys.*, 37, 249–257, 2010. With permission.)

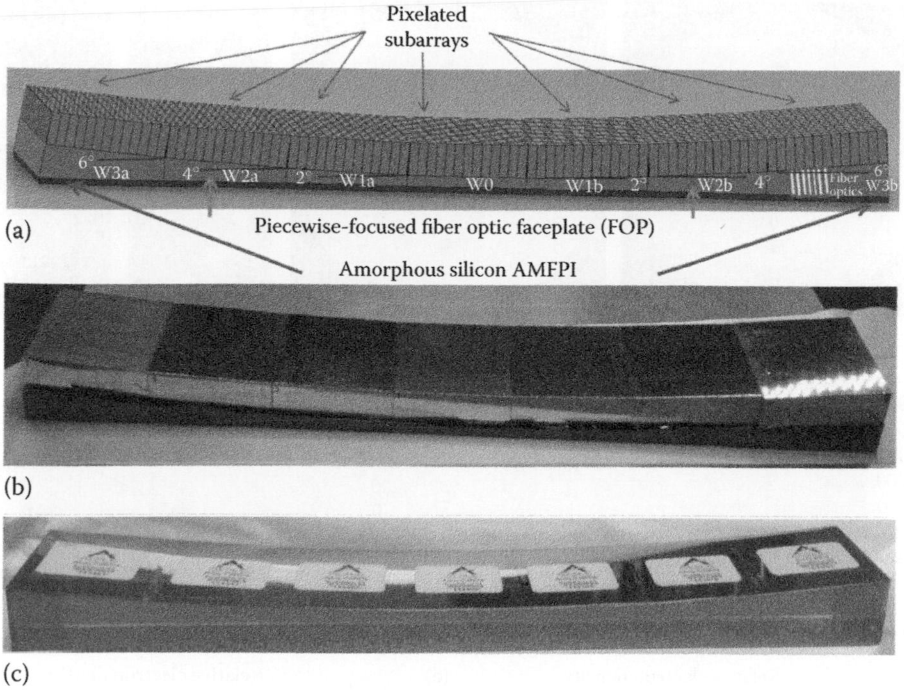

Figure 12.5 Piecewise-focused strip assembly developed by Star-Lack et al. (2015). The total area covered is 361 mm × 52 mm and comprises 28,072 scintillator pixels of pitch 0.784 mm: (a) Schematic showing the pixelated subarrays mounted to a custom FOP comprising seven sections (W0, W1a, b, W2a, b, W3a, b) which, in turn, are coupled to the AMFPI, (b) photograph of the scintillator–FOP structure after partial assembly. The BGO subarray is in position W3b with its top covered by a reflector. The CWO subarrays have not yet been covered, and (c) the scintillator–FOP structure after final assembly. (Reprinted from Star-Lack et al., *Med. Phys.*, 42, 5084–5099, 2015. With permission.)

Recently, Star-Lack et al. (2015) developed a prototype focused high resolution strip imager that coupled pixelated arrays through a piecewise linear arc-shaped fiber optic plate (FOP) to a Varian AS1000 AMFPI (Figure 12.5). The 361 mm × 52 mm scintillator assembly contained a total of 28,072 pixels and comprised seven subarrays, each 15 mm thick. Six of the subarrays were fabricated from CWO with a pixel pitch of 0.784 mm, whereas one array was constructed from BGO for comparison. Measured CWO DQE(0) at 6 MV was 22% and the average ratio of CWO DQE(f) to Cu-GOS DQE(f) measured across the frequency range of 0.0–0.62 mm^{-1} was 23. As shown in Figure 12.6, the CWO CBCT images demonstrated a spatial resolution of 7 lp/cm with no deleterious effects from beam divergence. The improved contrast-to-noise ratio (CNR) of the CWO CBCT images compared to the Cu-GOS CBCT images reflected a 22× sensitivity improvement. Excellent linearity was achieved with the CWO scintillator material showing significantly higher stability and light yield than the BGO material.

Despite the excellent performance of crystalline-based high aspect ratio scintillators, they may be difficult to commercialize due to high raw material and manufacturing costs. This has spurred investigation into amorphous scintillating glasses such as LKH-5 (Collimated Holes Inc, Campbell, CA), which are relatively easy to cut and polish and are significantly less expensive to produce. Although the LKH-5 density and light output are reduced compared to those of crystalline scintillators (Table 12.1), performance may still be adequate for the clinical tasks at hand. To test this hypothesis,

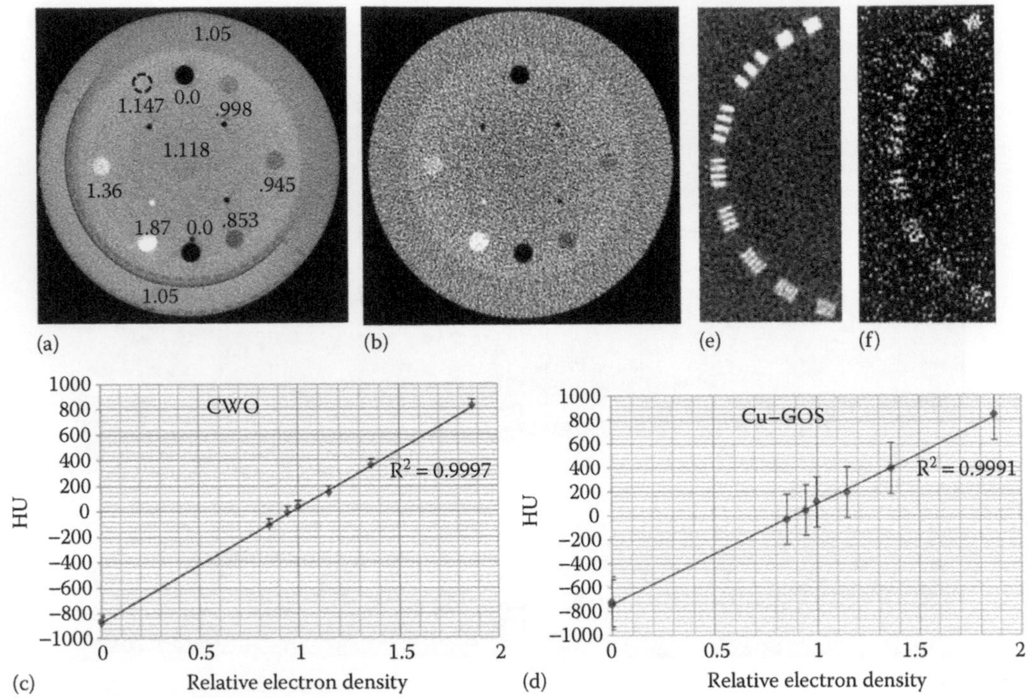

Figure 12.6 The CWO piecewise focused strip array (Star-Lack et al. 2015) provides a 22x sensitivity improvement over Cu-GOS in CBCT reconstructions (transaxial pixel size = 0.5 mm, slice thickness = 2 mm): (a, b) 6MV CWO and Cu-GOS images of the CTP404 CatPhan module (window width = 1000). The relative electron densities of the inserts are labeled (c, d) excellent linearity is achieved. The area in (a) outlined with the dashed arc was subtended by the BGO array in a 210° scan and exhibits a 4% reconstruction error as evidenced by the brightness increase. (e, f) Reconstructions of the high-contrast spatial resolution CatPhan module (window width = 600 HU). CWO spatial resolution is 7 lp/cm, whereas Cu-GOS spatial resolution is difficult to assess due to low SNR. Data were acquired with 5.4 MU producing a dose of 4 cGy at the phantom center. (Reprinted from Star-Lack et al., *Med. Phys.*, 42, 5084–5099, 2015. With permission.)

Shedlock et al. (2016) constructed and evaluated a prototype large area LKH-5-based imager. The 12 mm thick scintillator assembly contained a total of 78,400 pixels with 1.51 mm pitch covering an area 42.4 × 42.4 cm^2. The scintillator was coupled to the AMFPI that is used in the Varian aS1200 EPID. Using the 6MV beam, the LKH-5 DQE(0) was measured to be 13% reflecting a 10× improvement over Cu-GOS DQE(0) with both imagers possessing similar effective spatial resolutions at isocenter of 5 lp/cm (Figure 12.7). Despite LKH-5's lower light output, it does not appear that electronic noise floor issues were encountered even at an extremely low dose of 0.0056 MU/CBCT projection.

Another approach seeks to develop an EPID for simultaneous imaging and dose verification using thick plastic scintillators that can exhibit both high DQE and a water-equivalent response (Beddar et al. 1992). Initial work has been done with square BCF-99-06A scintillator fibers with polymethyl methacrylate (PMMA) cladding (Saint-Gobain Crystals) (Blake et al. 2014, 2013a).

Finally, the maturation of AMFPI technology and increased availability of radiation hard-low noise electronics has led to the novel approach of increasing DQE by stacking together multiple conventional EPID layers (Rottmann et al. 2016). Figure 12.8 shows a prototype multilayer imager (MLI) that was constructed comprising four layers with each layer consisting of a Cu-GOS scintillator screen coupled to AS1200 AMPFI. As predicted, DQE nearly quadrupled compared to that of the (conventional) single-layer device and there was no substantial degradation of

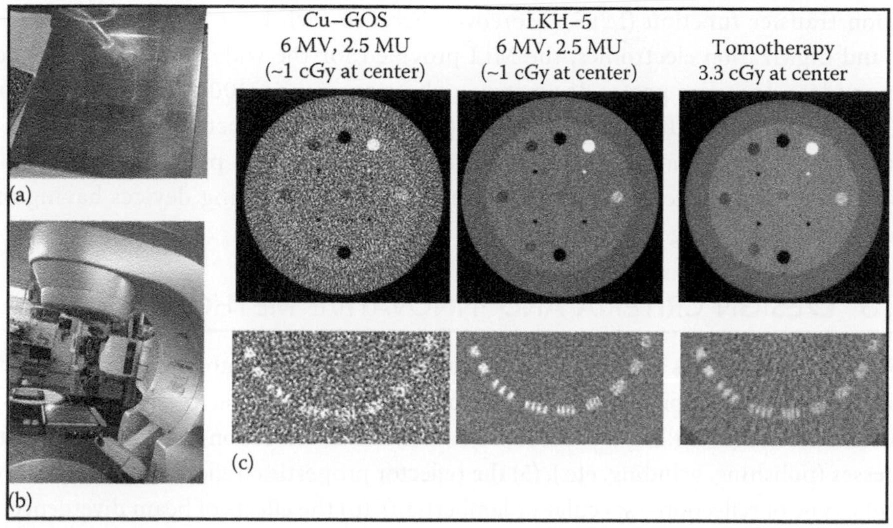

Figure 12.7 (a) Large area 42.4 × 42.4 cm² LKH-5 scintillator array (Shedlock et al. 2016) composed of a matrix of 49 38 × 38 pixel subarrays abutted to each other. The 12 mm-thick scintillator assembly contains a total of 78,400 pixels with 1.51 mm pitch. (b) Prototype imager mounted to a Truebeam (Varian Medical Systems) for testing. (c) Reconstructed Cu-GOS and LKH-5 CBCT images (1 mm slice thickness) of the CatPhan along with a CatPhan image from a tomotherapy system (2 mm slice thickness). The LKH-5 and tomotherapy images possess comparable CNRs and spatial resolutions. The tomotherapy imaging dose was approximately three times the LKH-5 imaging dose. Window width of top row images = 1000. Window width of bottom row images = 1500. (Reprinted from Shedlock et al., *EPI2k16 Symposium, Electronic Patient Imaging*, Saint Louis, MO, 2016. With permission.)

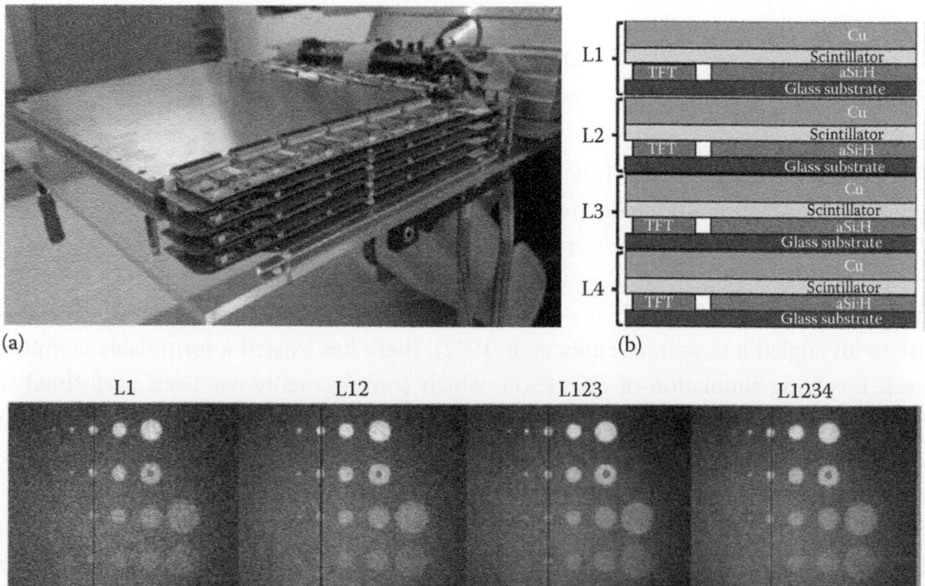

Figure 12.8 (a) The four-layer EPID multilayer imager (MLI) prototype from Rottman et al. (2016) before enclosure into its housing. (b) Schematic drawing. (c) Las Vegas contrast phantom imaged with 1, 2, 3, and 4 layers combined. The images are offset and gain corrected but not pixel defect corrected. Noise is reduced from left to right with CNR(L1234)=1.9xCNR(L1). (Reprinted from Rottmann et al., *Phys. Med. Biol.*, 61, 6297–6306, 2016. With permission.)

modulation transfer function (MTF). Moreover, because each layer had its own set of amplification and digitization electronics, the MLI provided for the wide dynamic ranges required to accommodate dose-rate portal dosimetry applications (up to 4000 MU/min) and low dose CBCT (0.02 MU/frame). Although the prototype MLI was constructed using Cu-GOS scintillator screens, it should be noted the architecture is compatible with pixelated scintillator arrays thus potentially leading the construction of very high DQE imaging devices having minimal liabilities from beam divergence.

12.2.3 Design criteria and innovative methods

The main design parameters affecting performance are (1) the scintillator material's density (ρ) and atomic number (Z), (2) the pixel geometry (size, aspect ratio, and fill factor), (3) the optical properties of the scintillator (scintillation yield, clarity, and index of refraction), (4) the scintillator finishing processes (polishing, grinding, etc.), (5) the reflector properties (reflectivity and transmissivity coefficients, type of reflection—specular or lambertian), (6) the effects of beam divergence, and (7) the number of imaging layers.

To streamline design and optimization processes, the impact of these parameters can be assessed *in silico* through Monte Carlo (MC) simulations of X-ray and optical photon transport (Kausch et al. 1999, Keller et al. 2002, Monajemi et al. 2004, Kirkby and Sloboda 2005, Liaparinos et al. 2006, Sawant et al. 2006, Wang et al. 2010, Blake et al. 2013b). The most accepted figure-of-merit for imager performance is its frequency-dependent detective quantum efficiency [DQE(f)], which is the spectral representation in Fourier domain of the signal-to-noise characteristics of a given detector

$$\mathrm{DQE(f)} = \frac{\mathrm{MTF(f)}^2}{q \cdot \mathrm{NNPS(f)}} \tag{12.1}$$

where:

MTF is the modulation transfer function
q is the X-ray photon (gamma) fluence in units of gammas/mm^2
NNPS is the normalized noise power spectrum

Although MC modeling of the MTF is straightforward and can be readily achieved by analyzing the response to an angled line source (Fujita et al. 1992), there has existed a formidable computational bottleneck involving simulation of qNNPS(f), which conventionally has been performed using a flood-field ensemble requiring a high amount of incident photons to be launched (>10^8) to mimic experimental conditions (Dobbins et al. 2006). Recently, it was shown that the fluence level can be dramatically reduced for a simulation without loss of accuracy to the limit that each *flood field* becomes a single-gamma event (Star-Lack et al. 2014). The Fujita–Lubberts–Swank (FLS) method that emerged (Figure 12.9) combines the principles described by Fujita et al. 1992, Lubberts 1968, and Swank 1973 to compute MTF, NPS and DQE simultaneously. The technique has been shown to dramatically reduce computation times so that DQE can be simulated on a laptop computer in minutes.

In the FLS simulation, *N* gamma photons are directed toward the detector along an angled line and the point spread function (PSF) for each event is recorded. The shape of the 1D NPS produced by each detected gamma photon *p* along (in this case) the *x*-axis is computed by summing the PSF in the *y*-direction and taking the square magnitude of its Fourier transform:

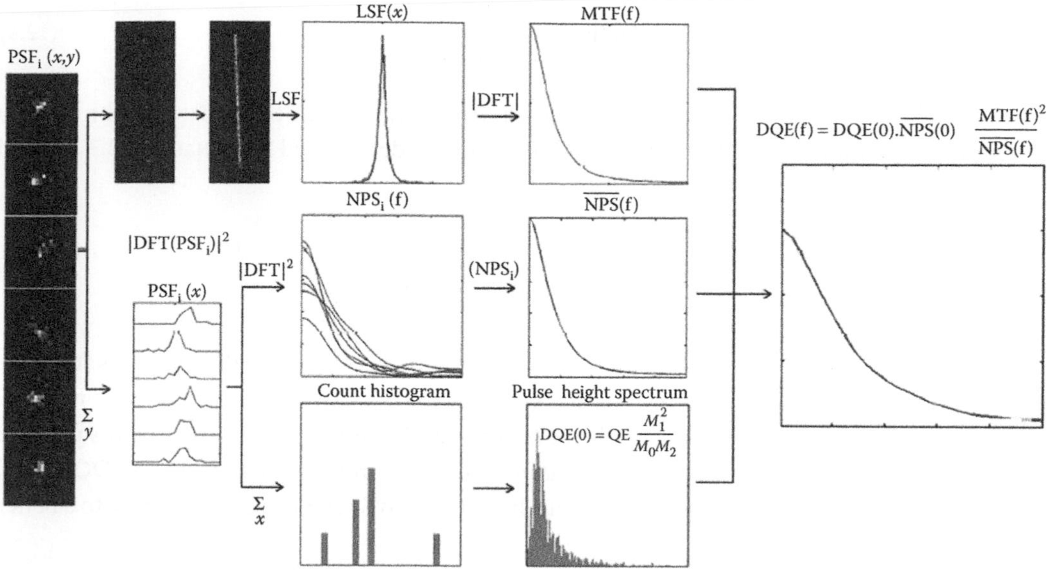

Figure 12.9 Schematic flow chart for the Fujita–Lubberts–Swank (FLS) simulation method (Star-Lack et al. 2014). Each gamma photon that interacts with the imager produces a typically unique 2D point-spread function. To compute the NPS, each PSF is summed along one dimension to yield a PSF projection that is Fourier transformed and squared to generate an NPS. The NPS' from all events are summed and normalized to compute the NPS shape (middle row). Each PSF is also individually summed to give the total received counts for that event, which is then tallied into a pulse height spectrum from which the DQE(0) is calculated using the Swank formalism (bottom row). By directing the gammas along an angled line, the MTF can be determined following the Fujita procedure (top row). The results of these calculations are then combined to yield the final DQE(f) (right). (Reprinted from Star-Lack et al., *Med. Phys.*, 41, 031916, 2014. With permission.)

$$\mathrm{NPS}_p(f) = \left| \mathrm{DFT}_i \left(\sum_j \mathrm{PSF}_p(i,j) \right) \right|^2 \tag{12.2}$$

Here, *i*, *j* label detector pixels along the *x*, *y* directions respectively, and the DFT operator refers to the 1D discrete Fourier transform taken, in this case, in the *i*-direction. According to the Lubberts relation (1968), the shape of the resulting NPS curve, up to a scale factor, is obtained by averaging $NPS_p(f)$ over all detected events N_p:

$$\overline{NPS(f)} = \frac{1}{N_p} \sum_{p=1}^{N_p} \mathrm{NPS}_p(f) \tag{12.3}$$

To generate the NPS scaling factor, it is noted that the product of *q* and the zero-frequency NNPS [NNPS(0)] is the reciprocal of the zero-frequency detective quantum efficiency [DQE(0)], which can be determined from the measured quantum efficiency ($QE = N_p/N_\gamma$) and the pulse height spectrum using the Swank formula (Swank 1973) to account for the effects of the X-ray spectrum and the energy-specific detector response on SNR:

$$\mathrm{DQE}(0)=\mathrm{QE}\frac{M_1^2}{M_0 M_2} \tag{12.4}$$

where M_k label kth moment of the pulse height spectrum. Hence the denominator of the DQE(f) expression (Equation 12.1) is the normalized NPS shape as determined by Equation 12.3 divided by DQE(0) from Equation 12.4:

$$q\cdot\mathrm{NNPS(f)}=\frac{\overline{\mathrm{NPS(f)}}}{\mathrm{DQE}(0)\cdot\overline{\mathrm{NPS}(0)}} \tag{12.5}$$

Figure 12.10 shows results from a simulation analyzing the impact of a single design parameter—the reflector type—on MTF and DQE. The scintillator material was LKH-5 with a height of 12 mm and pitch of 1.568 mm as reported in work by Shedlock et al. (2016). Although the two reflectors produce similar MTFs, it is seen that specular reflection is preferred as it provides for an increased DQE(f) of, in this case, 30%. The improvement largely results from reduced Swank noise thus leading to a better

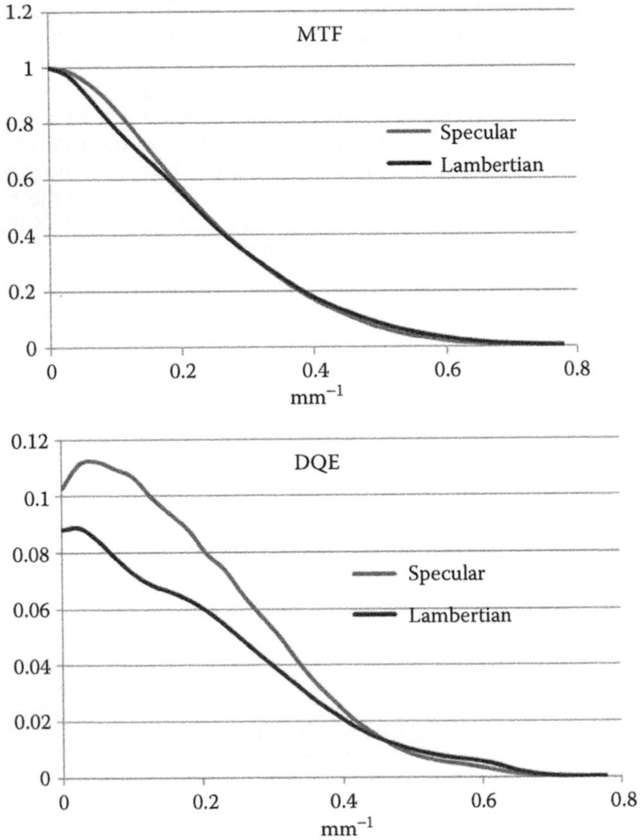

Figure 12.10 Simulations of the effects of septa having specular versus Lambertian reflection profiles on MTF and DQE for a LKH-5 pixelated scintillator array (height = 12 mm, pixel pitch = 1.5 mm). The MTFs are similar because both reflectors have zero transmission and hence there is no optical crosstalk. However, Lambertian reflection increases Swank noise that can be seen to have a deleterious effect on DQE.

quality NPS confirming previous results obtained using flood fields for NPS simulations (Constantin et al. 2012). These previous simulations required seven orders of magnitude longer computation times than the then FLS method whose computation time was only three CPU minutes. The above analysis of reflector type is a simple example of how MC techniques can be used to optimize design without incurring the costs of building multiple systems. Comprehensive analysis of scintillator type, pixel geometry, reflector type, scintillator finishing properties, and number of layers are currently underway to optimize EPID subject to cost and image quality constraints with the goal of developing a high quality MV imager that is both affordable and effective.

12.2.4 Future methods for MV EPID performance assessment

Unlike for diagnostic imagers operating in the kV energy range, there exists no standard technique for measuring MTF and DQE for MV EPIDs. Creating such a method should be a priority going forward especially because EPID development is starting to rapidly advance and it becomes more necessary to objectively compare the performance of different designs.

At kV energies, the DQE measurement protocol has been standardized by the International Electrotechnical Commission (IEC) (2003), the beam is first hardened by a specified amount of aluminum to represent an *average* patient, the MTF is then measured using an edge-based technique (Samei et al. 1998) and the NPS is measured using the flood-field approach (Dobbins et al. 2006) at prescribed doses. The DQE is then computed using Equation 12.1.

The flood field-based NPS measurement method is relatively ubiquitous and straightforward to implement, independent of beam energy. However, there is yet no consensus on an approach for measuring MTF at MV energies. Most previous MTF measurements have been made by irradiating an angulated slit phantom to generate an (oversampled) line-spread function (LSF) input to the detector (Munro and Bouius 1998, Sawant et al. 2006, 2007). The phantom, which is created from a pair of narrowly separated, highly attenuating blocks of material, typically has a large mass (>25 kg), is expensive to construct and cumbersome to use, and can be subject to inaccuracies from radiation leakage. It is not clear why there has been reluctance to use the more convenient edge-based technique, but it may be due to its lack of compete attenuation, which is the expectation at kV energies. However, recently it was shown that, despite the modest amounts of attenuation provided by a thin edge phantom, this method may indeed be adapted to MV energies (Star-Lack et al. 2015).

Figure 12.11 shows measured MTFs, obtained with both a slit and an edge phantom, of the Varian AS1000 portal imager employing a Cu-GOS screen. Measurements were made at 6 MV with the flattening filter (FF) in place. The slit phantom comprised 150 mm-thick tungsten *jaws* spaced 200 μm apart as described by Munro and Bouius (1998) and the edge phantom was a 0.5 mm thick tantalum sheet. Excellent agreement is achieved between the two techniques so long as forward scatter from the edge phantom is taken into account.

Although these initial results are promising, more comprehensive studies should be undertaken to optimize the edge-based method and validate the technique for MV imaging. This includes investigating different edge phantom materials and thicknesses as well as optimizing the distance from the edge to the imager. The signal processing methodology used to create an oversampled edge-spread function (ESF) and, from that, the LSF used to reconstruct the MTF should also be standardized. In addition, similar to the RQA standards that have been created for kV energies (IEC 2003), a set of beam quality standards needs to be created.

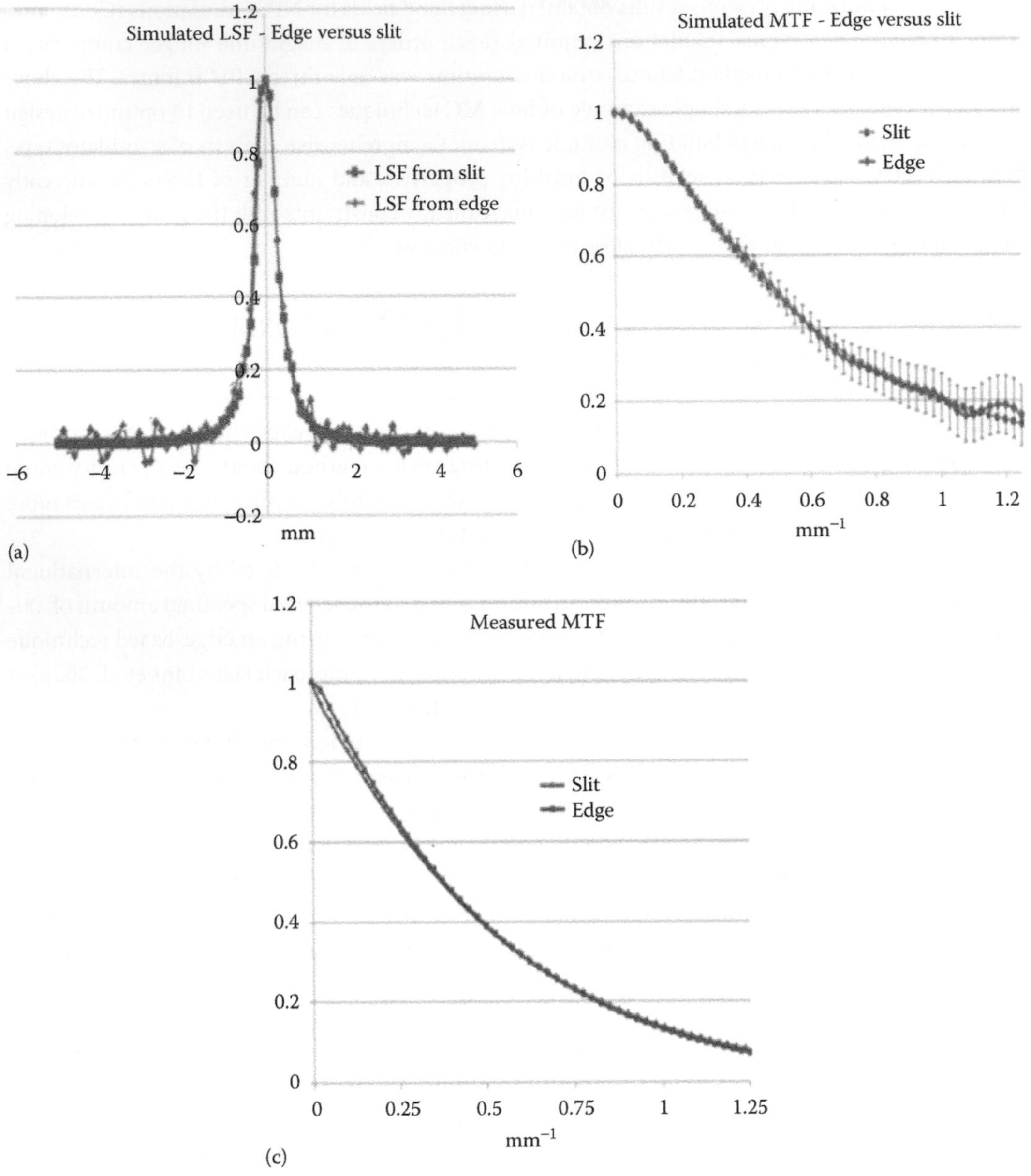

Figure 12.11 At MV energies, the MTF's measured with the slit and edge methods are equivalent (Star-Lack et al 2015). (a) 4x oversampled LSF's derived from either the simulated AS1000 response to a 0.5 mm-thick Ta edge or to an ideal slit (FLS simulation). A good match is achieved, although high-frequency noise, mostly beyond the Nyquist frequency of 1.25 mm^{-1}, is seen in the edge-generated LSF. Note that more than 500,000 CPU-hours of simulation time, which was prohibitively expensive, would have been required to match experimental fluence levels. (b) MTF's computed from the simulated LSF's match within the quantum noise-induced error of the edge simulation. (c) An excellent match is obtained in the MTFs of the Varian AS1000 imager measured at 6 MV using the slit (17 cGy exposure) and edge (100 cGy exposure) methods. (Reprinted from Star-Lack et al., *Med. Phys.*, 42, 5084–5099, 2015. With permission.)

12.3 kV IMAGING

Several kV-based BEV imaging approaches have been proposed (Figure 12.12). A prototype ARTISTE system (Siemens) was developed comprising an in-line source-detector pair (Oelfke et al. 2006). A retractable aSi flat-panel detector situated underneath the treatment head was irradiated by a kV source situated at position nominally 180° away underneath the retractable (MV) EPID. Using phantom studies, it was demonstrated that the imaging framework was capable of automatically detecting fiducial marker positions during a simulated treatment session during which the kV imager was heavily irradiated from its backside by the treatment beam (Fast et al. 2012a). In another approach designed for the ARTISTE, Maltz et al. (2009) proposed using a

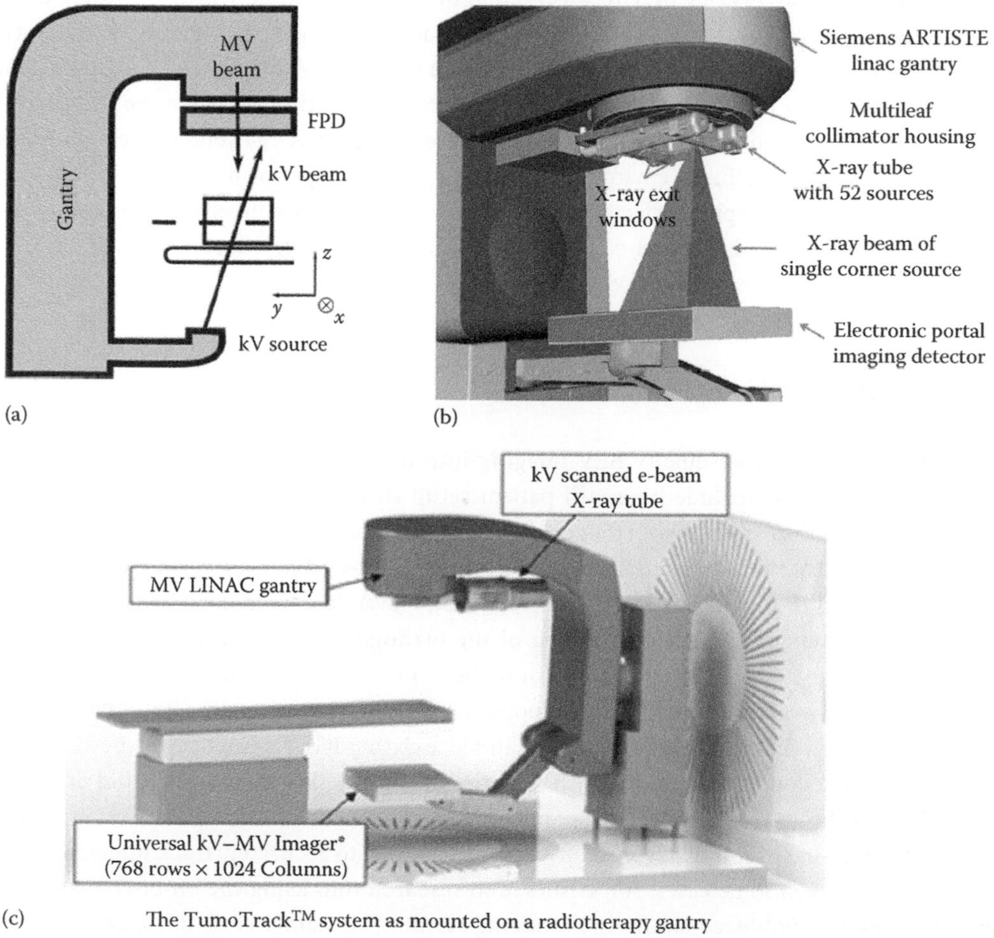

Figure 12.12 (a) Modified ARTISTE (Siemens) system with flat-panel detector (FPD) situated under the treatment head and a kV source located nominally 180° (Oelfke et al. 2006, Fast et al. 2012b). (b) Perspective view of the proposed nanotube multisource X-ray tube for digital tomosynthesis (Maltz et al. 2009). The X-ray beam from a single corner source is shown incident on the flat-panel electronic portal imaging detector. (c) Depiction of the proposed TumoTrak (Telecurity Sciences) system for tomosynthesis imaging employing a scanning beam multisource X-ray tube (Partain et al. 2015). (Reprinted from Oelfke et al., *Med. Dosim.*, 31, 62–70, 2006; Fast et al., *Med. Phys.*, 39, 109, 2012; Maltz et al., *Med. Phys.*, 36, 1624, 2009; Partain et al., 42, 3276–3276, 2015. With permission.)

compact multiple source X-ray tube surrounding the treatment head to produce a tomosynthesis image. The electron sources within the tube are realized using cold cathode carbon nanotube technology and the X-ray images are captured by the same EPID employed for portal imaging. A related approach (Tumotrak, Telesecurity Inc) has been proposed by Partain et al. (2015) where, instead of using fixed nanotube X-ray source technology, a scanning electron beam is utilized for X-ray generation.

These three approaches can be compared from the perspectives of ease of implementation and prospective image quality/IGRT utility. The retractable aSi flat-panel/kV source configuration described by Oelfke et al. (2006) has the advantage of providing for a direct BEV imaging capability without requiring tomosynthetic reconstruction. However, an extra kV flat panel is required, which can be subject to intense amounts of MV radiation when it is not retracted. This intense radiation can create lag and other charge-trapping artifacts that can deleteriously affect image quality (Siewerdsen and Jaffray 1999). CBCT reconstruction in particular will be sensitive to these nonlinearities (Starman et al. 2011). The tomosynthetic approaches proposed by Maltz et al. and Partain et al. have the advantage of using the same EPID for both kV and MV imaging. Although the images are more complex to reconstruct, visualization and target identification may be improved by resolving depth information. However, significantly more complex X-ray sources are required that are as yet unproven in a clinical environment. Also, as with the approach of Oelfke (Oelfke et al. 2006), undesired charge trapping effects such as detector lag may be amplified due to the wide range of input fluxes encountered. Another consideration is whether these geometries will support CBCT imaging due to their limited fields-of-view.

12.4 DISCUSSION

The ability to perform high-quality BEV imaging introduces several intriguing possibilities for enhancing IGRT. These include improved patient setup through breath hold CBCT enabled by combining projections from two orthogonal imaging devices, robust metal artifact reduction enabled by the use of the MV beam for imaging, and/or improved intrafraction motion management facilitated by in-line imaging. Alternatively, a BEV system may make it possible to achieve significant cost savings through elimination of the orthogonal onboard kV source-image pair (plus associated robotic arms), thereby providing the opportunity for state-of-the-art IGRT capabilities to become available to underserved populations. However, the viability of each of these potential use cases still must be proven. For example, existing IGRT systems have been shown to be largely adequate for patient setup and there is the possibility that their costs could decrease as flat panel and robotic arm technologies continue to mature. For intrafraction motion management, existing kV-based stationary systems have been shown to accurately track fiducial markers, bony anatomy, and in some cases, soft tissues, and research into using the on-board systems for target tracking has produced similarly promising results (Mostafavi et al. 2010, De Los Santos et al. 2013, Keall et al. 2015, Hazelaar et al. 2016). Thus, the principal challenge is to show a realistic potential for BEV imaging to provide IGRT capabilities that cannot be provided by existing technologies. Fortunately, there now exist several prototype high-quality BEV devices that can be evaluated clinically to help determine whether they can facilitate the development of sufficiently compelling IGRT capabilities.

If important clinical applications are identified, much work lies ahead in charting a technical path forward. For future kV-based BEV systems, focus must be on practical and cost-effective approaches. For MV-based systems, of paramount importance is determining (1) the EPID MTF and DQE required to generate acceptable image quality at acceptably low doses, and (2) the optimal size of the active high-DQE imaging area (i.e., the entire EPID, a central region, or a strip). Other technical considerations such as readout speed, linearity, and perhaps energy range may be of high importance. It is also critical that the portal imager design does not compromise performance in other areas related to dosimetry or machine QA. Finally, as new MV imagers are developed and tested, it is essential that the DQE measurement method be standardized.

Up until now, perhaps the major limitation to the commercialization of high-quality MV imagers has been the cost. Crystalline scintillators are expensive to produce and the arrays are expensive to manufacture. The recent work with LKH-5 scintillating glass may offer a path forward, particularly if the technology can be combined in a cost-effective way with the multilayer EPID approach to provide for sufficient sensitivity while mitigating beam divergence effects and maintaining the required dynamic range.

ACKNOWLEDGMENTS

The author thanks Daniel Shedlock, Daniel Morf, Adam Wang, Paul Baturin, Peter Munro, and Hassan Mostafavi for their contributions. This work was supported in part by NIH Grant R01-CA188446 Academic-Industrial Partnership.

REFERENCES

Antonuk LE. et al. 2002. Electronic portal imaging devices: a review and historical perspective of contemporary technologies and research. *Phys. Med. Biol.* **47**: 302.

Beddar AS, Mackie TR, and Attix FH. 1992. Water-equivalent plastic scintillation detectors for high-energy beam dosimetry: I. Physical characteristics and theoretical considerations *Phys. Med. Biol.* **37**: 1883–1900. Online: http://stacks.iop.org/0031-9155/37/i=10/a=006?key=crossref.5613f934867964b717f68f943ab55376.

Bissonnette JP, Cunningham IA, and Munro P. 1997. Optimal phosphor thickness for portal imaging. *Med. Phys.* **24**: 803–114.

Bissonnette J-P, Jaffray DA, Fenster A, and Munro P. 1992. Physical characterization and optimal magnification of a portal imaging system. Ed. R Shaw (International Society for Optics and Photonics). pp. 182–188. Online: http://proceedings.spiedigitallibrary.org/proceeding.aspx?articleid=985781.

Blake SJ, McNamara AL, Deshpande S, Holloway L, Greer PB, Kuncic Z, and Vial P. 2013a. Characterization of a novel EPID designed for simultaneous imaging and dose verification in radiotherapy. *Med. Phys.* **40**: 091902. Online: http://scitation.aip.org/content/aapm/journal/medphys/40/9/10.1118/1.4816657.

Blake SJ, Vial P, Holloway L, Greer PB, McNamara AL and Kuncic Z. 2013b. Characterization of optical transport effects on EPID dosimetry using Geant4. *Med. Phys.* **40**: 041708. Online: http://scitation.aip.org/content/aapm/journal/medphys/40/4/10.1118/1.4794479.

Blake S, Vial P, Holloway L, and Kuncic Z. 2014. WE-E-18A-08: Towards a next-generation electronic portal device for simultaneous imaging and dose verification in radiotherapy. *Med. Phys.* **41**: 511–511. Online: http://scitation.aip.org/content/aapm/journal/medphys/41/6/10.1118/1.4889460.

Boda-Heggemann J, Lohr F, Wenz F, Flentje M, and Guckenberger M. 2011. kV Cone-Beam CT-Based IGRT. *Strahlentherapie und Onkol.* **187**: 284–291. Online: http://link.springer.com/10.1007/s00066-011-2236-4.

Boyer AL. 1992. A review of electronic portal imaging devices (EPIDs). *Med. Phys.* **19**: 1. Online: http://scitation.aip.org/content/aapm/journal/medphys/19/1/10.1118/1.596878.

Breitbach EK, Maltz JS, Gangadharan B, Bani-Hashemi A, Anderson CM, Bhatia SK, Stiles J, Edwards DS and Flynn RT. 2011. Image quality improvement in megavoltage cone beam CT using an imaging beam line and a sintered pixelated array system. *Med. Phys.* **38**: 5969.

Chang J, Mageras GS, Yorke E, De Arruda F, Sillanpaa J, Rosenzweig KE, Hertanto A, Pham H, Seppi E, Pevsner A, Ling CC and Amols H. 2007. Observation of interfractional variations in lung tumor position using respiratory gated and ungated megavoltage cone-beam computed tomography. *Int. J. Radiat. Oncol. Biol. Phys.* **67**: 1548–1558.

Constantin D, Sun M, Abel E, Star-Lack J, and Fahrig R. 2012. WE-C-217BCD-11: Coupled radiative and optical Geant4 simulation of MV EPIDs based on thick pixelated scintillating crystals. *Med. Phys.* **39**: 3951. Online: http://scitation.aip.org/content/aapm/journal/medphys/39/6/10.1118/1.4736127.

Dawson LA. and Jaffray DA. 2007. Advances in image-guided radiation therapy. *J. Clin. Oncol.* **25**: 938–946. Online: http://jco.ascopubs.org/content/25/8/938.short.

Dobbins JT, Samei E, Ranger NT, and Chen Y. 2006. Intercomparison of methods for image quality characterization. II. Noise power spectrum *Med. Phys.* **33**: 1466. Online: http://scitation.aip.org/content/aapm/journal/medphys/33/5/10.1118/1.2188819.

El-Mohri Y, Antonuk LE, Zhao Q, Choroszucha RB, Jiang H, and Liu L. 2011. Low-dose megavoltage cone-beam CT imaging using thick, segmented scintillators. *Phys. Med. Biol.* **56**: 1509–1527.

El-Mohri Y, Jee KW, Antonuk LE, Maolinbay M, and Zhao Q. 2001. Determination of the detective quantum efficiency of a prototype, megavoltage indirect detection, active matrix flat-panel imager. *Med. Phys.* **28**: 2538–2550.

Fast MF, Koenig T, Oelfke U, Nill S et al. 2012a. Performance characteristics of a novel megavoltage cone-beam-computed tomography device. *Phys. Med. Biol.* **57**: N15–N24. Online: http://stacks.iop.org/0031-9155/57/i=3/a=N15?key=crossref.f2518469b6fbf24533045df0fac35e5b.

Fast MF, Krauss A, Oelfke U, and Nill S. 2012b. Position detection accuracy of a novel linac-mounted intrafractional x-ray imaging system. *Med. Phys.* **39**: 109. Online: http://scitation.aip.org/content/aapm/journal/medphys/39/1/10.1118/1.3665712.

Fujita H, Tsai DY, Itoh T, Doi K, Morishita J, Ueda K and Ohtsuka A. 1992. A simple method for determining the modulation transfer function in digital radiography. *IEEE Trans. Med. Imaging.* **11**: 34–39. Online: http://ieeexplore.ieee.org/articleDetails.jsp?arnumber=126908.

Gardner SJ, Studenski MT, Giaddui T, Cui Y, Galvin J, Yu Y, and Xiao Y. 2014. Investigation into image quality and dose for different patient geometries with multiple cone-beam CT systems. *Med. Phys.* **41**: 031908. Online: http://www.ncbi.nlm.nih.gov/pubmed/24593726.

Groh BA, Siewerdsen JH, Drake DG, Wong JW, and Jaffray DA. 2002. A performance comparison of flat-panel imager-based MV and kV cone-beam CT. *Med. Phys.* **29**: 967–975.

Hazelaar C, Dahele M, Mostafavi H, van der Weide L, Slotman BJ, and Verbakel WFAR. 2016. Subsecond and submillimeter resolution positional verification for stereotactic irradiation of spinal lesions. *Int. J. Radiat. Oncol. Biol. Phys.* **94**: 1154–1162. Online: http://www.sciencedirect.com/science/article/pii/S0360301616000110.

Hunt MA, Sonnick M, Pham H, Regmi R, Xiong J-P, Morf D, Mageras GS, Zelefsky M, and Zhang P. 2016. Simultaneous MV-kV imaging for intrafractional motion management during volumetric-modulated arc therapy delivery. *J. Appl. Clin. Med. Phys.* **17**: 5836. Online: http://www.ncbi.nlm.nih.gov/pubmed/27074467.

International Electrotechnical Commission (IEC) 2003. *Medical Electrical Equipment–Characteristics of Digital X-Ray Imaging Devices–Part 1: Determination of the Detective Quantum Efficiency*. Geneva, Switzerland.

Kausch C, Schreiber B, Kreuder F, Schmidt R, and Dössel O. 1999. Monte Carlo simulations of the imaging performance of metal plate/phosphor screens used in radiotherapy. *Med. Phys.* **26**: 2113–2124.

Keall P, Ng JA, Caillet V, Huang CY, Colvill E, Simpson E, Poulsen PR, Kneebone A, Eade T and Booth J. 2015. Sub-mm accuracy results measured from the first prospective clinical trial of a novel real-time IGRT system, Kilovoltage Intrafraction Monitoring (KIM). *Int. J. Radiat. Oncol.* **93**: S192–S193. Online: http://linkinghub.elsevier.com/retrieve/pii/S036030161501192X.

Keller H, Glass M, Hinderer R, Ruchala K, Jeraj R, Olivera G, and Mackie TR. 2002. Monte Carlo study of a highly efficient gas ionization detector for megavoltage imaging and image-guided radiotherapy. *Med. Phys.* **29**: 165–175.

Kirkby C. and Sloboda R. 2005. Comprehensive Monte Carlo calculation of the point spread function for a commercial a-Si EPID. *Med. Phys.* **32**: 1115–1127.

Kirvan PF, Monajemi TT, Fallone BG, and Rathee S. 2010. Performance characterization of a MVCT scanner using multislice thick, segmented cadmium tungstate-photodiode detectors. *Med. Phys.* **37**: 249–257.

Knoll GF. 2000. *Radiation Detection and Measurement*. New York: Wiley. Online: http://books.google.com/books?id=HKBVAAAAMAAJ&pgis=1.

Lehmann J, Perks J, Semon S, Harse R and Purdy JA. 2007. Commissioning experience with cone beam CT for image guided radiation therapy. *J. Appl. Clin. Med. Phys.* **8**(3): 21–36. Online: http://www.jacmp.org/index.php/jacmp/article/view/2354.

Li T. and Xing L. 2007. Optimizing 4D cone-beam CT acquisition protocol for external beam radiotherapy. *Int. J. Radiat. Oncol. Biol. Phys.* **67**: 1211–1219. Online: http://www.sciencedirect.com/science/article/pii/S0360301606033384.

Liaparinos PF, Kandarakis IS, Cavouras DA, Delis HB and Panayiotakis GS. 2006. Modeling granular phosphor screens by Monte Carlo methods. *Med. Phys.* **33**: 4502. Online: http://scitation.aip.org/content/aapm/journal/medphys/33/12/10.1118/1.2372217.

De Los Santos J, Popple R, Agazaryan N, Bayouth JE, Bissonnette J-P, Bucci MK, Dieterich S. et al. 2013. Image Guided Radiation Therapy (IGRT) Technologies for radiation therapy localization and delivery. *Int. J. Radiat. Oncol.* **87**: 33–45. Online: http://linkinghub.elsevier.com/retrieve/pii/S0360301613002137.

Lubberts G. 1968. Random noise produced by X-ray fluorescent screens. *J. Opt. Soc. Am.* **58**: 1475.

Mahmood U, Huo J, Koshy M, Pugh TJ, McGuire SE, Choi S, Frank SJ, Lee A, Kuban DA, Giordano S, Buchholz TA, Smith BD and Hoffman KE. 2015. Image guidance is associated with decreased gastrointestinal toxicity in patients receiving definitive external beam radiation therapy for prostate cancer. *Int. J. Radiat. Oncol.* **93**: E242. Online: http://linkinghub.elsevier.com/retrieve/pii/S0360301615018878.

Maltz JS, Sprenger F, Fuerst J, Paidi A, Fadler F and Bani-Hashemi AR. 2009. Fixed gantry tomosynthesis system for radiation therapy image guidance based on a multiple source x-ray tube with carbon nanotube cathodes. *Med. Phys.* **36**: 1624. Online: http://scitation.aip.org/content/aapm/journal/medphys/36/5/10.1118/1.3110067.

Mao R, Zhang L and Zhu R.Y. 2008a. Optical and scintillation properties of inorganic scintillators in high energy physics. *IEEE Trans. Nucl. Sci.* **55**: 2425–2431.

Mao W, Wiersma RD. and Xing L. 2008b. Fast internal marker tracking algorithm for onboard MV and kV imaging systems. *Med. Phys.* **35**: 1942. Online: http://scitation.aip.org/content/aapm/journal/medphys/35/5/10.1118/1.2905225.

Meeks SL, Harmon JF, Langen KM, Willoughby TR, Wagner TH, and Kupelian PA. 2005. Performance characterization of megavoltage computed tomography imaging on a helical tomotherapy unit. *Med. Phys.* **32**: 2673–2681.

Monajemi TT, Steciw S, Fallone BG, and Rathee S. 2004. Modeling scintillator-photodiodes as detectors for megavoltage CT. *Med. Phys.* **31**: 1225–1234.

Monajemi TT, Tu D, Fallone BG, and Rathee S. 2006. A bench-top megavoltage fan-beam CT using CdWO4-photodiode detectors. II. Image performance evaluation. *Med. Phys.* **33**: 1090–1100.

Morton EJ, Swindell W, Lewis DG, and Evans PM, 1991. A linear array, scintillation crystal–photodiode detector for megavoltage imaging *Med. Phys.* **18**: 681. Online: http://scitation.aip.org/content/aapm/journal/medphys/18/ 4/10.1118 / 1.596661.

Mosleh-Shirazi MA, Evans PM, Swindell W, Symonds-Tayler JRN, Webb S, and Partridge M, 1998. Rapid portal imaging with a high-efficiency, large field-of-view detector *Med. Phys.* **25**: 2333. Online: http://scitation.aip.org/content/aapm/journal/medphys/25/12/10.1118/1.598443.

Mostafavi H, Sloutsky A, and Jeung A. 2010. WE-D-204B-08: Tracking 3D trajectory of internal markers using radiographic sequential stereo imaging: Estimation of breathing motion. *Med. Phys.* **37**: 3430–3430. Online: http://scitation.aip.org/content/aapm/journal/medphys/37/6/10.1118/1.3469405.

Munro P. and Bouius DC. 1998. X-ray quantum limited portal imaging using amorphous silicon flat-panel arrays. *Med. Phys.* **25**: 689–702. Online: http://scitation.aip.org/content/aapm/journal/medphys/25/5/10.1118/1.598252.

Oelfke U, Tücking T, Nill S, Seeber A, Hesse B, Huber P, and Thilmann C. 2006. Linac-integrated kV-cone beam CT: Technical features and first applications *Med. Dosim.* **31**: 62–70.

Partain L, Kwon J, Rottmann J, Zentai G, Berbeco R, and Boyd D. 2015. SU-E-J-56: Static gantry digital tomosynthesis from the Beam's-eye-view. *Med. Phys.* **42**: 3276–3276. Online: http://scitation.aip.org/content/aapm/journal/medphys/42/6/10.1118/1.4924143.

Rathee S, Tu D, Monajemi TT, Rickey DW, and Fallone BG. 2006. A bench-top megavoltage fan-beam CT using CdWO[sub 4]-photodiode detectors. I. System description and detector characterization. *Med. Phys.* **33**: 1078.

Rottmann J, Morf D, Fueglistaller R, Zentai G, Star-Lack J, and Berbeco R. 2016. A novel EPID design for enhanced contrast and detective quantum efficiency. *Phys. Med. Biol.* **61**: 6297–6306. Online: http://iopscience.iop.org/article/10.1088/0031-9155/61/17/6297.

Ruchala KJ, Olivera GH, Schloesser EA and Mackie TR. 1999. Megavoltage CT on a tomotherapy system. *Phys. Med. Biol.* **44**: 2597–2621. Online: http://stacks.iop.org/0031-9155/44/i=10/a=316.

Samei E, Flynn MJ and Reimann DA. 1998. A method for measuring the presampled MTF of digital radiographic systems using an edge test device. *Med. Phys.* **25**: 102–113. Online: http://www.ncbi.nlm.nih.gov/pubmed/9472832.

Sawant A, Antonuk L E, El-Mohri Y, Zhao Q, Wang Y, Li Y, Du H, and Perna L. 2006. Segmented crystalline scintillators: Empirical and theoretical investigation of a high quantum efficiency EPID based on an initial engineering prototype CsI(TI) detector. *Med. Phys.* **33**: 1053–1066.

Sawant A, Antonuk L, and El-Mohri Y. 2007. Slit design for efficient and accurate MTF measurement at megavoltage x-ray energies *Med. Phys.* **34**: 1535. Online: http://scitation.aip.org/content/aapm/journal/medphys/34/5/10.1118/1.2717405.

Seppi EJ, Munro P, Johnsen SW, Shapiro EG, Tognina C, Jones D, Pavkovich JM, Webb C, Mollov I, Partain LD, and Colbeth RE. 2003. Megavoltage cone-beam computed tomography using a high-efficiency image receptor. *Int. J. Radiat. Oncol. Biol. Phys.* **55**: 793–803. Online: http://www.ncbi.nlm.nih.gov/pubmed/12573767.

Shedlock D, Wang AS, Humber D, Morf D, Fueglistaller R, Vaigner K, Star-Lack J, Meade R, Lee M, and Star-Lack J. 2016. A high DQE pixelated EPID constructed from LKH-5 scintillating glass. *EPI2k16 Symposium, Electronic Patient Imaging.*

Siewerdsen JH. and Jaffray DA. 1999. A ghost story: Spatio-temporal response characteristics of an indirect-detection flat-panel imager. *Med. Phys.* **26**: 1624–1641. Online: http://scitation.aip.org/content/aapm/journal/medphys/26/8/10.1118/1.598657.

Sillanpaa J, Chang J, Mageras G, Yorke E, De Arruda F, Rosenzweig KE, Munro P, Seppi E, Pavkovich J, and Amols H. 2006. Low-dose megavoltage cone-beam computed tomography for lung tumors using a high-efficiency image receptor. *Med. Phys.* **33**: 3489–3497.

Sonke J-J, Zijp L, Remeijer P, and van Herk M. 2005. Respiratory correlated cone beam CT *Med. Phys.* **32**: 1176. Online: http://scitation.aip.org/content/aapm/journal/medphys/32/4/10.1118/1.1869074.

Star-Lack J, Shedlock D, Swahn D, Humber D, Wang A, Hirsh H, Zentai G, Sawkey D, Kruger I, Sun M, Abel E, Virshup G, Shin M, and Fahrig R. 2015. A piecewise-focused high DQE detector for MV imaging. *Med. Phys.* **42**: 5084–5099. Online: http://scitation.aip.org/content/aapm/journal/medphys/42/9/10.1118/1.4927786.

Star-Lack J, Sun M, Meyer A, Morf D, Constantin D, Fahrig R, and Abel E. 2014. Rapid Monte Carlo simulation of detector DQE(f). *Med. Phys.* **41**: 031916. Online: http://www.ncbi.nlm.nih.gov/pubmed/24593734.

Starman J, Star-Lack J, Virshup G, Shapiro E, and Fahrig R, 2011. Investigation into the optimal linear time-invariant lag correction for radar artifact removal. *Med. Phys.* **38**: 2398–2411.

Swank RK. 1973. Absorption and noise in x-ray phosphors *J. Appl. Phys.* **44**: 4199. Online: http://scitation.aip.org/content/aip/journal/jap/44/9/10.1063/1.1662918.

Uribe J, Baghaei H, Farrell R, Aykac M, and Bilgen D. 2003. An efficient detector production method for position-sensitive scintillation detector arrays with 98% detector packing fraction *IEEE Trans. Nucl. Sci.* **50**: 1469–1476. Online: http://ieeexplore.ieee.org/articleDetails.jsp?arnumber=1236951.

van Eijk CWE. 2002 Inorganic scintillators in medical imaging. *Phys. Med. Biol.* **47**: R85–R106. Online: http://www.ncbi.nlm.nih.gov/pubmed/12030568.

Van Elmpt W. and McDermott L. 2008. A literature review of electronic portal imaging for radiotherapy dosimetry *Radiother Oncol.* **88**: 289–309. Online: http://www.sciencedirect.com/science/article/pii/S0167814008003721.

Wang Y, Antonuk L E, El-Mohri Y, and Zhao Q. 2009. A Monte Carlo investigation of Swank noise for thick, segmented, crystalline scintillators for radiotherapy imaging. *Med. Phys.* **36**: 3227–3238.

Wang Y, El-Mohri Y, Antonuk LE, and Zhao Q. 2010. Monte Carlo investigations of the effect of beam divergence on thick, segmented crystalline scintillators for radiotherapy imaging. *Phys. Med. Biol.* **55**: 3659–3673.

Wertz H, Stsepankou D, Blessing M, Rossi M, Knox C, Brown K, Gros U, Boda-Heggemann J, Walter C, Hesser J, Lohr F, and Wenz F. 2010. Fast kilovoltage/megavoltage (kVMV) breath-hold cone-beam CT for image-guided radiotherapy of lung cancer. *Phys. Med. Biol.* **55**: 4203–4217.

Wu M, Keil A, Constantin D, Star-Lack J, Zhu L, and Fahrig R. 2014. Metal artifact correction for x-ray computed tomography using kV and selective MV imaging. *Med. Phys.* **41**: 121910. Online: http://www.ncbi.nlm.nih.gov/pubmed/25471970.

Yip S, Rottmann J, Chen H, Morf D, Fueglistaller R, Star-Lack J, Zentai G, and Berbeco R. 2015. TH-EF-BRB-02: Combination of Multiple EPID Imager Layers Improves Image Quality and Tracking Performance of Low Contrast Objects *Med. Phys.* **42**: 3742–3742. Online: http://scitation.aip.org/content/aapm/journal/medphys/42/6/10.1118/1.4926300.

Yoo S, Kim G-Y, Hammoud R, Elder E, Pawlicki T, Guan H, Fox T, Luxton G, Yin F-F, and Munro P. 2006. A quality assurance program for the on-board imager[sup®]. *Med. Phys.* **33**: 4431. Online: http://scitation.aip.org/content/aapm/journal/medphys/33/11/10.1118/1.2362872.

Zhang J and Yin F-F. 2007. Minimizing image noise in on-board CT reconstruction using both kilovoltage and megavoltage beam projections. *Med. Phys.* **34**: 3665–3673.

Zhang P, Happersett L, Ravindranath B, Zelefsky M, Mageras G, and Hunt M. 2016. Optimizing fiducial visibility on periodically acquired megavoltage and kilovoltage image pairs during prostate volumetric modulated arc therapy. *Med. Phys.* **43**: 2024–2029. Online: http://scitation.aip.org/content/aapm/journal/medphys/43/5/10.1118/1.4944737.

Index

Note: Page numbers followed by f and t refer to figures and tables, respectively.

F

S

T

Printed and bound by CPI Group (UK) Ltd, Croydon, CR0 4YY
01/11/2024
01782604-0007